Excel

YEAR 12

Mathematics Standard 1 Revision & Exam Workbook

ESSENTIAL skills

Get the Results You Want!

Allyn Jones & AS Kalra

ISBN 978 1 74125 690 1

Pascal Press
PO Box 250
Glebe NSW 2037
(02) 8585 4044
www.pascalpress.com.au

Publisher: Vivienne Joannou
Project editor: Rosemary Peers
Edited by Rosemary Peers
Answers checked by Peter Little
Additional material written by Jim Stamell and Chelsea Smithies
Page design and typesetting by Replika Press, Precision Typesetting (Barbara Nilsson), DiZign Pty Ltd and lj Design (Julianne Billington)
Printed by Vivar Printing/Green Giant Press

Students
All care has been taken in the preparation of this study guide, but please check with your teacher or the NSW Education Standards Authority about the exact requirements of the course you are studying as these can change from year to year.

All efforts have been made to gain permission for the copyright material reproduced in this book. In the event of any oversight, the publisher welcomes any information that will enable rectification of any reference or credit in subsequent editions.

Contents

Introduction

This workbook has been written for the Year 12 Mathematics Standard 1 course.

For students to successfully complete this course they must undertake assessment tasks in each year of Stage 6 and sit for the HSC Examination—this workbook will support them in the revision for the course.

This workbook, combined with the ***Excel*** *Year 11 Mathematics Standard Revision & Exam Workbook*, will provide an excellent resource for students to use to maximise their marks in the HSC Examination.

Features of this book include:

- **graded exercises** that follow the topics of the syllabus—students can practise the skills they need to complete the course
- **space for students to write their answers** in a workbook format—there should be sufficient space to answer each question, setting out clearly and working down each page
- **answers at the back of the book** to check students' work
- **timed topic tests at the end of each chapter** that have been designed to completely cover the content of each topic and to test the understanding of all skills needed, with HSC-style questions and marks allocated for every question
- **three HSC-style Sample Examinations**. The HSC Mathematics Standard 1 Examination consists of topics studied throughout Stage 6. This means the majority of questions will be based on Year 12 topics but some questions will be on topics students have studied in Year 11.

Remember

The best way to study Mathematics is by working through examples.

Using this workbook for revision will allow you to have all the questions, working and answers together in one book.

You can write notes in the margins and have a complete personalised review book.

You can also use this book as a diagnostic tool to quickly assess areas of concern and determine any weaknesses in your revision.

Any student who has worked through all these questions and understands the content should feel confident of doing well in the Year 12 Mathematics Standard 1 course.

Some useful hints for the HSC Examination

Know what to expect

- Find out from your teacher exactly what topics will be assessed in the examination.
- Determine what format is being used for the exam. Are there multiple-choice questions?

Show all working

- Read the instructions on the examination paper—marks may be allocated for working.
- Even when marks are not allocated, working is important and it is rewarded, even though your answer may be incorrect.
- Never use correction fluid—just put a line through any incorrect working. Let the examiner see all your work.

Allocate your time

- Ensure you are working through the paper efficiently and not spending too much time on each question, only to find you have run out of time at the end of the exam.

Understand mark allocations

- The questions worth the most marks are often the most complex and difficult. Showing working is crucial for these types of questions.

Re-read and check

- Once you have completed the question, rather than moving to the next, re-read that question to make sure you have in fact answered the correct question. This will only take a second or two.
- If you finish the examination with time still remaining, check your answers—often mistakes are found and can be corrected.

Use quality diagrams

- Diagrams should be of good quality, large and drawn with a lead pencil. Use an eraser for deleting mistakes, not correction fluid.

Be ready

- Finally, you need to be prepared for the examination.
- Always study Mathematics actively. Active study means using pen and paper to make notes, writing down difficult questions and their solutions, and recording rules to learn.

NSW GOVERNMENT

NSW Education Standards Authority

2020 HIGHER SCHOOL CERTIFICATE EXAMINATION

Mathematics Standard 1
Mathematics Standard 2

REFERENCE SHEET

Measurement

Limits of accuracy

Absolute error $= \frac{1}{2} \times$ precision

Upper bound = measurement + absolute error

Lower bound = measurement – absolute error

Length

$$l = \frac{\theta}{360} \times 2\pi r$$

Area

$$A = \frac{\theta}{360} \times \pi r^2$$

$$A = \frac{h}{2}(a + b)$$

$$A \approx \frac{h}{2}(d_f + d_l)$$

Surface area

$$A = 2\pi r^2 + 2\pi rh$$

$$A = 4\pi r^2$$

Volume

$$V = \frac{1}{3}Ah$$

$$V = \frac{4}{3}\pi r^3$$

Trigonometry

$$\sin A = \frac{\text{opp}}{\text{hyp}}, \quad \cos A = \frac{\text{adj}}{\text{hyp}}, \quad \tan A = \frac{\text{opp}}{\text{adj}}$$

$$A = \frac{1}{2}ab\sin C$$

$$\frac{a}{\sin A} = \frac{b}{\sin B} = \frac{c}{\sin C}$$

$$c^2 = a^2 + b^2 - 2ab\cos C$$

$$\cos C = \frac{a^2 + b^2 - c^2}{2ab}$$

Financial Mathematics

$$FV = PV(1 + r)^n$$

Straight-line method of depreciation

$$S = V_0 - Dn$$

Declining-balance method of depreciation

$$S = V_0(1 - r)^n$$

Statistical Analysis

An outlier is a score

less than $Q_1 - 1.5 \times IQR$

or

more than $Q_3 + 1.5 \times IQR$

$$z = \frac{x - \mu}{\sigma}$$

Normal distribution

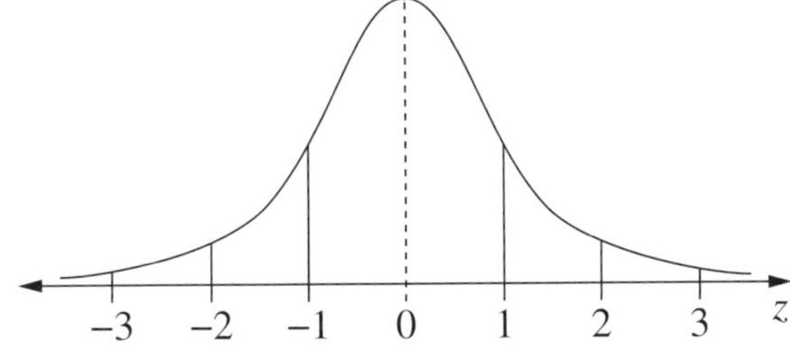

- approximately 68% of scores have z-scores between –1 and 1
- approximately 95% of scores have z-scores between –2 and 2
- approximately 99.7% of scores have z-scores between –3 and 3

CHAPTER 1
Financial mathematics: Investments

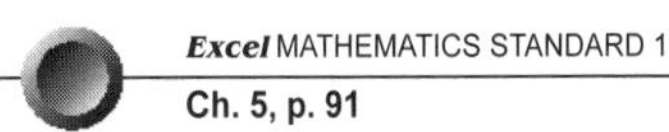

Interest rates

QUESTION **1** An interest rate of 6% p.a. is what rate:

a monthly?

b quarterly?

c six-monthly?

d four-monthly?

QUESTION **2** Find the monthly interest rate if the annual rate is:

a 9%

b 7.5%

QUESTION **3** Find the quarterly interest rate if the annual rate is:

a 8%

b 5%

QUESTION **4** Find the number of:

a months in 5 years

b quarters in 3 years

c six-monthly periods in 8 years

d four-monthly periods in 2 years

QUESTION **5** Interest on an investment is to be paid quarterly. If the principal is invested for 4 years and the annual interest rate is 9% find:

a the number of quarters

b the quarterly interest rate

QUESTION **6** Find the annual interest rate when it is:

a 2.5% per quarter

b 0.9% per month

c 6.5% per six-monthly period

d 0.046% per day

Simple interest 1

QUESTION 1 Find the simple interest for each of the following:

a \$4500 at 8% p.a. for 2 years

b \$8000 at 7% p.a. for 6 years

c \$20 000 at 9% p.a. for 8 years

d \$5900 at 12% p.a. for 6 months

e \$20 500 at $7\frac{1}{2}$% p.a. for 3 months

f \$36 000 at 10.25% p.a. for 4 years

g \$65 000 for 5 years at 6.5% p.a.

h \$82 000 for 2 years at 8.25% p.a.

QUESTION 2 \$3000 is invested at 5% p.a. simple interest for 4 years. Find:

a the total amount of interest earned

b the total value of the investment at the end of the 4 years

QUESTION 3 Find the length of time for:

a \$432 to be the interest on \$1800 at 6% p.a.

b \$960 to be the interest on \$2400 at 8% p.a.

QUESTION 4 Find the rate per cent per annum for:

a \$1134 to be the interest on \$5400 for 5 years

b \$189 to be the interest on \$2700 for 2 years

Simple interest 2

QUESTION **1** Find the principal required for the simple interest to be:

a \$900 on an amount invested for 2 years at 10% p.a.

b \$306 on an amount invested for 1 year at 9% p.a.

QUESTION **2**

a Complete the table to show the amount of simple interest earned (I) if \$500 is invested for n years at each of the given rates.

	n	0	1	2	3	4	5	6	7	8	9	10
4% p.a.	I											
7% p.a.	I											
9% p.a.	I											

b Draw the graph of I against n for each of the above interest rates.

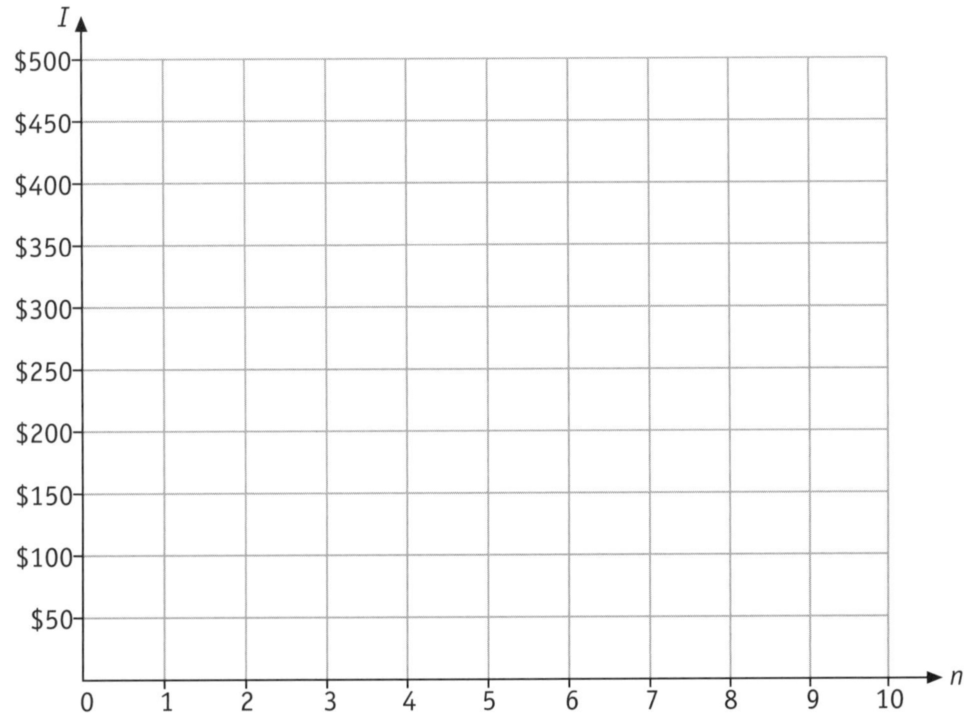

Compound interest

QUESTION 1 $8000 is invested for 5 years at 6.5% p.a. interest, compounded annually. Find:

a the future value

b the compound interest earned

QUESTION 2 Find the amount of compound interest earned from the following investments:

a $6000 at 9% p.a. for 4 years, compounded annually

b $18 000 at 14% p.a. for 2 years, compounded six-monthly

c $48 000 at 10% p.a., compounded quarterly for 5 years

d $32 000 for 3 years at 7.2% p.a., compounded monthly

e $120 000 for 25 years at 3.6% p.a. interest, compounded monthly

f $3650 for 5 years at 6.5% p.a., compounded quarterly

Financial mathematics: Investments

Future value 1

QUESTION **1** Find FV when:

a $PV = \$4000$, $r = 6\%$, $n = 3$

b $PV = \$9500$, $r = 2\%$, $n = 24$

QUESTION **2** Find the future value if the following amounts are invested for the given time at the given interest rate, compounded annually:

a \$5000 at 8% p.a. for 2 years

b \$8500 at $9\frac{1}{2}\%$ p.a. for 5 years

c \$15 000 at 10% p.a. for 3 years

d \$6000 at 8% p.a. for 12 years

QUESTION **3** Find the final balance if the given amount is invested for the given number of years at the given interest rate, compounded monthly:

a \$4000 at 12% p.a. for 3 years

b \$18 000 at 9% p.a. for 6 years

Future value 2

QUESTION **1** Find the future value if:

a \$6000 is invested for 4 years at 8% p.a., compounded quarterly

b \$2500 is invested for 3 years at 10% p.a., compounded six-monthly

c \$20 000 is invested for 5 years at 7.5% p.a., compounded monthly

d \$32 000 is invested for 7 years at 9% p.a., compounded quarterly

QUESTION **2** Mia's grandparents deposited \$10 000 into a savings account when she was born. The interest rate was fixed at 6% p.a., compounded monthly. Calculate the future value of the investment if Mia withdraws all the money on her:

a 21st birthday

b 32nd birthday, as a deposit for her home

Financial mathematics: Investments

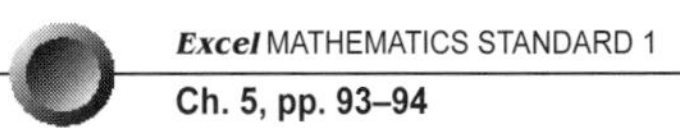

Present value

QUESTION 1 Find the present value to the nearest dollar if:

a future value is \$3345.56 with an interest rate of 6% p.a., compounded annually for 5 years

b future value is \$8531.14 with an interest rate of 4.8% p.a., compounded monthly for 6 years

c future value is \$18 223.59 with an interest rate of 4.2% p.a., compounded quarterly for 10 years

QUESTION 2 What sum of money would need to be invested to be worth \$5000 at the end of 7 years at the given interest rate?

a 6% p.a., compounded annually

b 8% p.a., compounded quarterly

QUESTION 3 What will be their initial deposits if Bill and Ben each want to withdraw \$20 000 from their own savings account in 8 years if:

a Bill's account pays interest of 5.4% p.a., compounded quarterly?

b Ben's account pays compound interest of 5.4% p.a., compounded monthly?

Financial mathematics: Investments

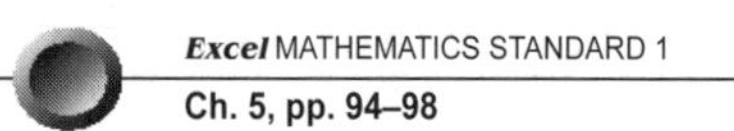

Tables of future values

QUESTION 1 The table shows the future value of \$1 if invested at the given interest rate for the given number of periods, interest compounded per period.

Interest rate per period

Period	1%	2%	2.5%	4%	5%	6%	10%	12%
1	1.0100	1.0200	1.0250	1.0400	1.0500	1.0600	1.1000	1.1200
2	1.0201	1.0404	1.0506	1.0816	1.1025	1.1236	1.2100	1.2544
3	1.0303	1.0612	1.0769	1.1249	1.1576	1.1910	1.3310	1.4049
4	1.0406	1.0824	1.1038	1.1699	1.2155	1.2625	1.4641	1.5735
5	1.0510	1.1041	1.1314	1.2167	1.2763	1.3382	1.6105	1.7623
6	1.0615	1.1262	1.1597	1.2653	1.3401	1.4185	1.7716	1.9738
7	1.0721	1.1487	1.1887	1.3159	1.4071	1.5036	1.9487	2.2107
8	1.0829	1.1717	1.2184	1.3686	1.4775	1.5938	2.1436	2.4760

Use the table to find the future value of:

a \$2000 invested for 7 years at 5% p.a., compounded annually

b \$5500 invested for 2 years at 10% p.a., compounded quarterly

c \$14 400 invested for 3 years at 12% p.a., compounded six-monthly

d \$9750 invested for 5 months at 12% p.a., compounded monthly

QUESTION 2 Use the above table to find the amount of money which could be invested now to give:

a \$10 000 at the end of 8 years, at 4% p.a. interest, compounded annually

b \$15 000 at the end of 18 months, at 8% p.a. interest, compounded quarterly

Graphs of future values

Question 1 The graph shows the future value of $1000 if invested at 18% p.a., compounded monthly. Use this graph to answer the following questions.

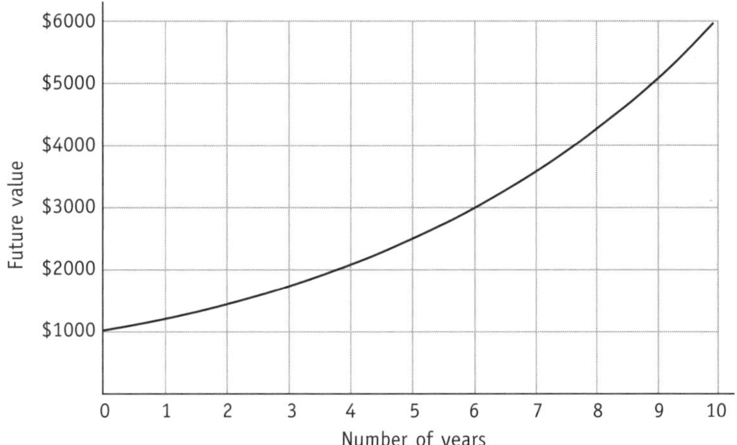

a What is the approximate value of the investment after 3 years? ______

b After approximately how many years is the value $5000? ______

c If $600 was invested at 18% p.a., compounded monthly, what would be its approximate value after 4 years?

______

______

d Give a brief description of what will happen to the future value over the next few years.

______

______

______

Question 2 $1000 is invested at 18% p.a., compounded six-monthly.

a Briefly explain why the future value, $*A*, after *n* six-monthly periods, will be given by $A = 1000(1.09)^n$.

______

______

b Complete the table of values giving *A* to the nearest whole number.

n	2	4	6	8	10	12	14	16	18	20
A										

c Draw another graph on the axes in Question 1 to show the future value of this investment.

d Briefly comment on the expected difference between the two investments over the next few years.

______

______

______

Financial mathematics: Investments

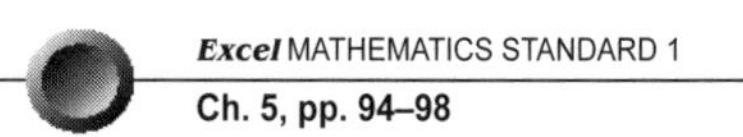
Excel MATHEMATICS STANDARD 1
Ch. 5, pp. 94–98

Modelling future value and present value

QUESTION 1 Carmen uses a spreadsheet to predict the future value of investments at different interest rates and number of years and compounded at different intervals.

Calculation of *FV* given various interest rates/periods/compounding intervals					
	Present value (*PV*):	$10 000.00	$20 000.00	$20 000.00	$20 000.00
	Interest rate p.a.:	3.00	3.00	6.00	6.00
	No. of years:	8	8	8	16
Future value (*FV*) compounded	annually:	$12 667.70	$25 335.40	$31 876.96	$50 807.03
	quarterly:	*A*	$25 402.22	$32 206.49	$51 862.89
	monthly:	$12 708.68	$25 417.37	$32 282.85	$52 109.13
	fortnightly:	$12 710.73	$25 421.47	$32 303.62	$52 176.19
	weekly:	$12 711.61	$25 423.22	*B*	$52 205.03

a What is the difference between investing $20 000 for 8 years at 3% p.a., compounded quarterly and compounded monthly?

b What is the value in the following cells?

i *A*? **ii** *B*?

c What is the amount of interest earned on an investment of $20 000 invested for:

i 8 years at 3% p.a., compounded quarterly? **ii** 16 years at 6% p.a., compounded fortnightly?

QUESTION 2 The spreadsheet is used to show the amounts to be invested (present value) that provide various amounts (future value) of investments at different interest rates and number of years and compounded at different intervals.

Calculation of *PV* given various interest rates/periods/compounding intervals					
	Future value (*FV*):	$50 000.00	$100 000.00	$250 000.00	$300 000.00
	Interest rate p.a.:	6.00	8.00	10.00	12.00
	No. of years:	8	15	20	35
Present value (*PV*) compounded	annually:	*A*	$31 524.17	$37 160.91	$5 681.86
	quarterly:	$31 049.65	*B*	$34 676.14	$4 785.31
	monthly:	$30 976.20	$30 239.61	*C*	$4 593.51

a What amount will need to be invested by Joseph in a savings account today offering 10% p.a. interest, compounded quarterly, if he wants $250 000 available to his grandson in 20 years?

b What is the value in the following cells?

i *A* **ii** *B* **iii** *C*

Appreciation and inflation

Question 1 For the following, calculate the cost of the item after one year:

a a lawnmower costing \$750 with an inflation rate of 2.5% p.a.

b a bottle of milk costing \$2.40 with inflation at 6% p.a.

Question 2 A house appreciates at 4.5% p.a. If it costs \$350 000 now, what will it be worth in 3 years time?

Question 3 The price of a car now is \$38 000. If the inflation rate is 2.75% p.a., what would you expect to pay for the car in 2 years time?

Question 4 The current price of a table is \$650. Calculate its price 5 years ago, if the inflation rate during this time was 2.25% p.a.

Question 5 A block of land increased in value this year from \$460 000 to \$520 880. What is the rate of appreciation?

Question 6 What was the price of an apartment 12 years ago, if its current value is \$980 000 and it has appreciated at 5% p.a.?

Question 7 The cost of a television is \$2100. If the average inflation rate is 3%, what will be the price of the television in 3 years?

Financial mathematics: Investments

TOPIC TEST

SECTION I

Instructions
- This section consists of 5 multiple-choice questions.
- Each question is worth 1 mark.
- Fill in only ONE CIRCLE for each question.

Time allowed: 7 minutes **Total marks: 5**

1 Tyrone invests \$3200 for a term of 15 months. Simple interest is paid at a flat rate of 4.25% p.a. How much will Tyrone's investment be worth at the end of the term?

Ⓐ \$3370 Ⓑ \$3376.80

Ⓒ \$3404 Ⓓ \$4170

2 Florence invests \$16 000 at an interest rate of 3.6% p.a., compounded monthly. What is the future value of the investment after 2 years?

Ⓐ \$16 115.41 Ⓑ \$17 048.69

Ⓒ \$17 172.74 Ⓓ \$17 192.63

3 What amount should Logan invest now at 4% p.a., compounded quarterly, so that in 3 years it will have grown to \$20 000?

Ⓐ \$1650 Ⓑ \$6410

Ⓒ \$17 749 Ⓓ \$17 780

4 Over a 5-year period the annual inflation rate for a basket of groceries was an average of 2.6%. If the price of the groceries was initially \$230, how much has the price increased at the end of the 5 years?

Ⓐ \$30.45 Ⓑ \$31.20

Ⓒ \$31.50 Ⓓ \$32.70

5 Neil has \$10 000 to invest with each of two different financial institutions for a period of 5 years. One institution offers 4.8% p.a., while another 5.4% p.a., both compounded monthly. What is the difference between the two future values of the investments?

Ⓐ \$300 Ⓑ \$348.20

Ⓒ \$385.31 Ⓓ \$1867.92

TOPIC TEST

SECTION II

Instructions
- This section consists of 6 questions.
- Show all working.

Time allowed: 53 minutes **Total marks: 35**

6

Compounded values of $1						
	Interest rate per period					
Period	**0.50%**	**1.00%**	**2.00%**	**5.00%**	**10.00%**	**12.00%**
1	1.005	1.010	1.020	1.050	1.100	1.120
2	1.010	1.020	1.040	1.103	1.210	1.254
3	1.015	1.030	1.061	1.158	1.331	1.405
4	1.020	1.041	1.082	1.216	1.464	1.574
5	1.025	1.051	1.104	1.276	1.611	1.762
6	1.030	1.062	1.126	1.340	1.772	1.974
7	1.036	1.072	1.149	1.407	1.949	2.211
8	1.041	1.083	1.172	1.477	2.144	2.476
9	1.046	1.094	1.195	1.551	2.358	2.773
10	1.051	1.105	1.219	1.629	2.594	3.106
11	1.056	1.116	1.243	1.710	2.853	3.479
12	1.062	1.127	1.268	1.796	3.138	3.896

a Use the table of compounded values of $1 to find the value of an investment of:

i $5000, compounded annually at 5% per annum over 7 years **1 mark**

ii $3400, compounded monthly at 0.5% per month over 1 year **1 mark**

b Use the table to find the amount of interest earned on an investment of:

i $12 000, compounded annually at 10% per annum over 8 years **2 marks**

ii $9600, compounded monthly at 6% per annum over 10 months **2 marks**

7 Find the future value, to the nearest cent, of an investment of:

a \$16 000 over 7 years at 4% p.a., compounded annually **2 marks**

b \$24 000 over 5 years at 6% p.a., compounded monthly **2 marks**

c \$8000 over 4 years at 8% p.a., compounded quarterly **2 marks**

d \$11 400 over 3 years at 9% p.a., compounded monthly **2 marks**

8 Find the present value, to the nearest cent, of an investment worth:

a \$6000 over 12 years at 5% p.a., compounded annually **2 marks**

b \$15 000 over 20 years at 4% p.a., compounded quarterly **2 marks**

c \$9500 over 10 years at 6% p.a., compounded monthly **2 marks**

d \$32 000 over 25 years at 4.8% p.a., compounded monthly **2 marks**

9 Theo's father invested \$5000 in a savings account when his son was born. The interest rate was fixed at 4% p.a., compounded annually. Calculate the future value, to the nearest dollar, of the investment if Theo withdraws his money on his:

a 18th birthday to buy a car **2 marks**

b 65th birthday to help with his retirement **2 marks**

10 A financial institution guarantees 6% p.a. interest, compounded monthly, on any investment for any number of years. What sum of money, to the nearest dollar, will have to be invested today to accumulate an amount of:

a $120 000 after 15 years? **2 marks**

b $500 000 after 30 years? **2 marks**

11 Ella wants to invest $28 000 for a total of 3 years. **5 marks**

She has three investment options:

- option A—simple interest is paid at the rate of 5% p.a.
- option B—compound interest is paid at a rate of 4.8% p.a., compounded annually
- option C—compound interest is paid at a rate of 4.4% p.a., compounded quarterly.

Determine Ella's best investment option.

Support your answer with calculations.

Excel MATHEMATICS STANDARD 1
Ch. 6, pp. 106–110

Declining-balance depreciation 1

QUESTION 1 Convert the following percentages to decimals:

a 22% = ______ b 16.5% = ______ c $18\frac{3}{4}\%$ = ______

QUESTION 2 Convert the following decimals to percentages:

a 0.15 = ______ b 0.235 = ______ c 0.098 = ______

QUESTION 3 Use the declining-balance method of depreciation to calculate the following.

a Benedict buys a car for $28 500. It depreciates at the rate of 20% p.a. What is the value of the car after:

i 2 years? ______

ii 5 years? ______

b Majeda purchased her car 6 years ago and it is now worth $12 600. Given it is depreciating at 12% p.a., what was the value of the car when she bought it? ______

c Geoff bought a van for $37 800, 4 years ago. It is now worth $19 500. What annual rate of depreciation did he use? ______

d Ming's vehicle is depreciating at 15% p.a. It was bought for $45 000 and is now worth $14 400. How long ago did she buy it? ______

QUESTION 4 Complete the following table showing the declining-balance depreciation of a vehicle purchased for $30 000 and depreciating at an annual rate of 12% p.a. (Round all depreciated values to the nearest dollar.)

Year	Net book value ($)	Depreciation ($)	Final value ($)
1	30 000	0.12 × 30 000 = 3600	30 000 − 3600 = 26 400
2	26 400		
3			
4			
5			
6			
7			
8			

a What do you notice about the depreciation amount from year to year? ______

b How is this different from the depreciation amount using the straight-line method? ______

Financial mathematics: Depreciation and loans

Excel MATHEMATICS STANDARD 1
Ch. 6, pp. 106–110

Declining-balance depreciation 2

Question 1 The graph shows the declining-balance depreciation for a vehicle over a given period.

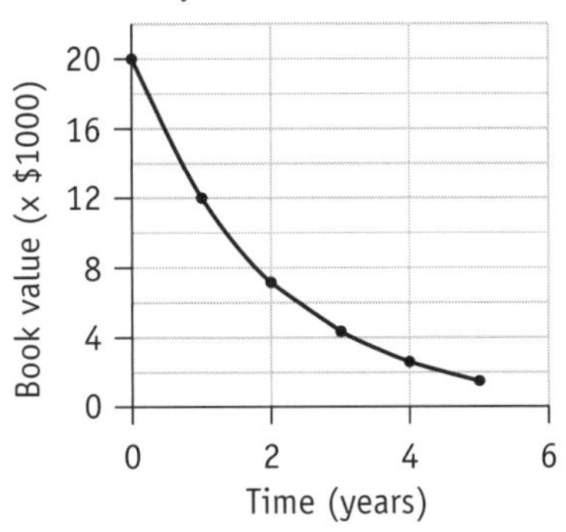

a What was the initial cost of the vehicle? ________

b By how much had the vehicle depreciated in the first year? ________

c Using the graph, and the declining-balance formula, calculate the annual rate of depreciation, *r*, as a percentage.

____

____

d Use the formula to calculate the salvage value of the vehicle after 6 years.

____

____

e Compare your calculated value in part **d** to the one obtained by extrapolating (extending) the graph.

____

____

Question 2 Bella and her brother Ryan each buy a car in the same year. The car Bella buys is depreciating at a rate of 23% p.a. while the one Ryan buys is depreciating at a rate of 18% p.a. Use the graph showing the value of their respective cars over the next 5 years to answer the following questions.

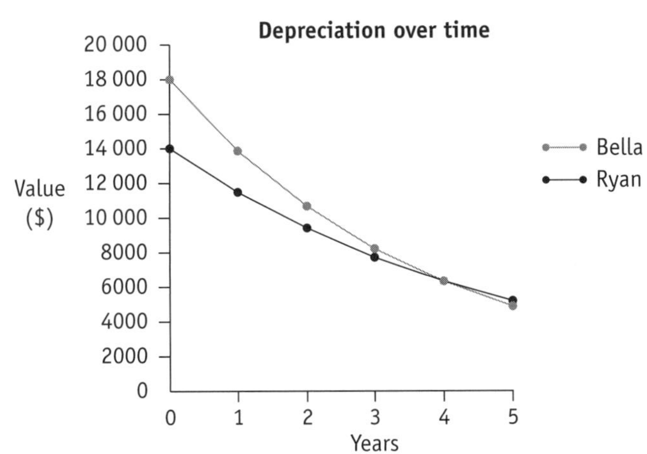

a How much did Bella pay for her car? ______________________________

b How much did Ryan pay for his car? ______________________________

c After how many years is the value of each of their cars approximately equal? ________________

____

d In the long run, Ryan's car was the better purchase because it is worth more than Bella's car. Is this statement correct? Justify your answer.

____

____

____

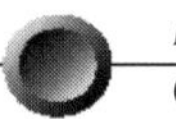

Loans involving flat-rate interest 1

QUESTION 1 James wanted to buy a car and approached a bank for a personal loan of \$20 000. The manager approved the loan at an 8% p.a. flat rate of interest. He has to repay the loan in 5 years.

a What interest will James pay during that period?

b What will be his monthly repayment?

QUESTION 2 John bought a TV for \$3000. He paid 20% deposit and the balance over 3 years, with interest charged at a flat rate of 15% on the balance.

a Find the deposit paid.

b Calculate the balance owing.

c Calculate the interest paid.

d Find the total amount that was repaid on the loan.

e What was the monthly repayment?

QUESTION 3 Chris wants to buy a boat and takes out a loan of \$10 000 on which the interest rate charged is 9.5% flat. There is also a loan protection fee of 30 cents for each \$100 borrowed. The loan is to be repaid over 5 years in equal monthly instalments.

a Calculate the total amount which will be repaid.

b Find the amount of each monthly repayment.

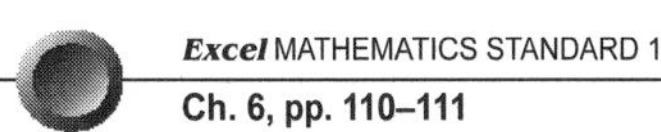

Loans involving flat-rate interest 2

Question 1 Lachlan borrowed $12 900 at 15% p.a. simple interest for 5 years so that he could furnish his apartment. At the end of 5 years both the interest and the principal had been repaid.

a Calculate the amount of interest charged. ______

b How much did he pay back altogether? ______

c If Lachlan repaid the loan in equal monthly instalments, how much was each instalment?

Question 2 Nana borrows $150 000 at a flat rate of 10.5% p.a. The rate is fixed for 5 years. She is to pay back the interest only during this period.

a How much interest is to be paid in the 5 years? ______

b After paying the fixed rate of interest for 4 years, Nana finds that the bank will drop her interest rate to 6.25% if she pays a penalty of $1600. How much will she save by changing to the lower interest rate for the last year?

Question 3 Alex takes a loan of $10 000 over 48 months. The repayment rate is $262.50 per month.

a How much will Alex pay back altogether? ______

b What is the flat interest rate for the loan? ______

Loans involving reducible interest 1

QUESTION 1 Kate's bank agrees to lend her $300 000 towards the purchase of an apartment. The interest rate for the loan is 6.5% p.a. monthly reducible and Kate will repay the loan in monthly instalments of $1860 over 30 years. A table shows the amount ($P + I - R$) Kate still owes on the loan after repayments.

Month	Principal (P)	Interest (I)	$P + I$	$P + I - R$
1	$300 000.00	$1625.00	$301 625.00	$299 765.00
2	$299 765.00	$1623.73	$301 388.73	$299 528.73
3	$299 528.73	$1622.45	$301 151.17	$299 291.17
4	$299 291.17	$1621.16	$300 912.33	$299 052.33
5	***A***	***B***	***C***	***D***
6	***D***	$1618.57	$300 430.77	$298 570.77

a How much does Kate still owe on the loan after 6 months? __________

b What are the values of:

i A? __________ **ii** B? __________

iii C? __________ **iv** D? __________

c After 6 months:

i how much has Kate paid in repayments? __________

ii by what amount has the principal been reduced? __________

QUESTION 2 Mia takes out a loan of $20 000 and is charged a reducible interest rate of 3.2% p.a. At the end of each month she is charged interest and then she makes a repayment of $1500. Mia uses a spreadsheet to calculate the balance owing after each monthly repayment.

Loan table				
		Amount:	**$20 000**	
		Annual interest rate:	**3.20%**	
		Monthly repayment:	**$1500**	
n	Principal (P)	Interest (I)	$P + I$	$P + I - R$
1	$20 000.00	$53.33	$20 053.33	$18 553.33
2	$18 553.33	$49.48	$18 602.81	***E***
3	$17 102.81	$45.61	$17 148.42	$15 648.42
4	$15 648.42	$41.73	$15 690.15	$14 190.15
5	$14 190.15	$37.84	$14 227.99	$12 727.99
6	$12 727.99	$33.94	$12 761.93	$11 261.93
7	$11 261.93	$30.03	$11 291.96	$9791.96
8	***F***	$26.11	$9818.07	$8318.07
9	$8318.07	$22.18	$8340.25	$6840.25
10	$6840.25	$18.24	$6858.49	$5358.49
11	$5358.49	$14.29	***G***	$3872.78
12	$3872.78	$10.33	$3883.11	$2383.11
13	$2383.11	$6.35	$2389.46	$889.46
14	$889.46	$2.37	$891.83	–$608.17
15	–$608.17	–$1.62	–$609.79	–$2109.79

a What is the balance owing after the first repayment? __________

b Calculate the values in the following cells:

i E __________

ii F __________

iii G __________

c What is the total interest charged by the financial institution in the first 3 months?

____

d What is the meaning of the negative numbers in the last two rows of the spreadsheet?

____

e How much does Mia actually pay as her final repayment? __________

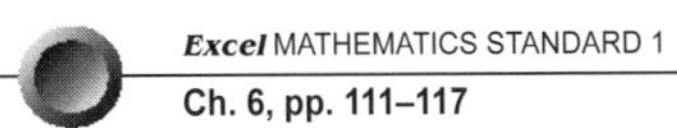

Loans involving reducible interest 2

Question 1 The table below shows the monthly repayments per \$1000 on a bank loan for various annual interest rates.

Term	5.5%	6.0%	6.5%	7.0%	7.5%	8.0%
20 years	\$6.8684	\$7.1643	\$7.4581	\$7.7506	\$8.0560	\$8.3669
30 years	\$5.6754	\$5.9955	\$6.3233	\$6.6503	\$6.9921	\$7.3404

a Calculate the monthly repayment on the following loans, to the nearest cent:

i \$260 000 at 6.5% p.a. for 20 years ______

ii \$312 000 at 8.0% p.a. for 20 years ______

iii \$435 000 at 5.5% p.a. for 30 years ______

b Andrew is borrowing \$327 000 for 30 years at 7.5% p.a. Find:

i his monthly repayment ______

ii the total of his repayments ______

iii the amount of interest he pays ______

c To the nearest dollar, what is the total amount of interest paid over the period of the loan of:

i \$385 000 for 20 years at 6.0% p.a.?

ii \$482 500 for 30 years at 7.5% p.a.?

d Last year Ava was given a loan of \$245 000 to be paid off over 20 years at a fixed interest rate of 6.5% p.a. Now her brother Simon wants to borrow the same amount over 20 years but the interest rate has been increased by 1.5% p.a. How much more will Simon's monthly repayment be than his sister's repayment?

Question 2 Alessandro intends to borrow \$460 000 over 30 years at a reducible interest rate of 6.5%. Use the table in question 1 for the following calculations.

a What will be his monthly repayment? ______

b Find the amount of interest Alessandro will have to pay, to the nearest dollar? ______

Question 3 Gracie borrows \$670 000 over 30 years for her new home. If she is paying an interest rate of 5.5% pa, what amount of interest will she pay during the life of the loan? (Use the table in Question 1.)

Loans involving reducible interest 3

QUESTION 1 Todd borrows $400 000 from a bank at 6% p.a. reducible interest for 30 years. He makes monthly repayments of $2398. The graph represented as **Scenario 1** shows the balance owing throughout the life of the loan.

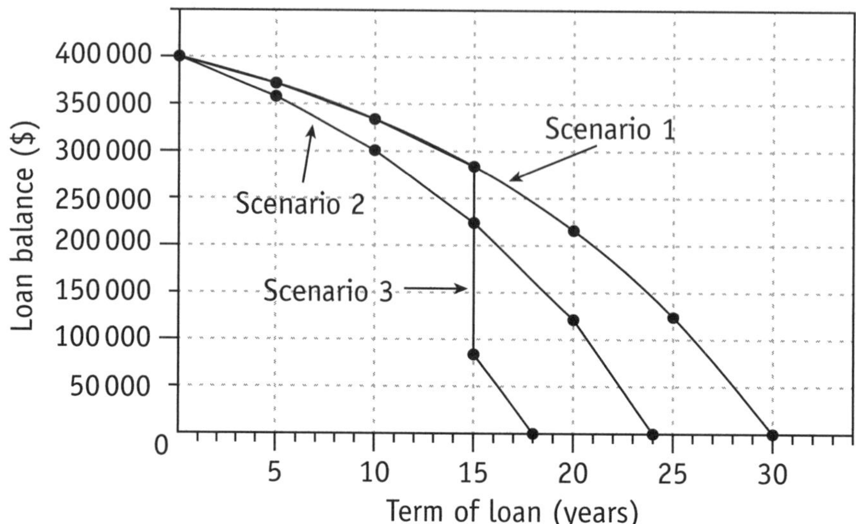

a How much money is still owed on the loan after 15 years? Estimate to the nearest $10 000.

____

b After how many years will half the loan be repaid? ______________

c Calculate the total:

i repayments ______________ **ii** amount of interest ______________

QUESTION 2 Suppose instead, Todd was to make **fortnightly** repayments of $1199. The balance owing throughout this loan is shown on the graph above as **Scenario 2.**

a Approximately how many years earlier will the loan be repaid? Give your answer to the nearest year.

____

____

b Calculate, to the nearest $10 000, the new total:

i repayments **ii** amount of interest

______________ ______________

QUESTION 3 Assume Todd chooses the monthly payment but after 15 years uses an inheritance to pay off some of the balance of his loan as a lump sum. This is shown on the graph as **Scenario 3.**

a Estimate the lump sum amount. ______________

b If he pays the loan out after 18 years:

i find the total amount paid by Todd ______________

ii what is the amount of interest saved by Todd between Scenario 3 and Scenario 1?

____

____

Financial mathematics: Depreciation and loans

Credit cards 1

QUESTION 1 The credit card statement is used to answer the questions below.

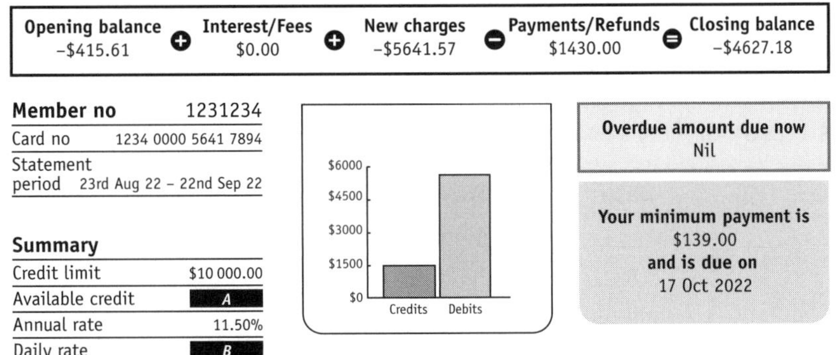

Opening balance		Interest/Fees		New charges		Payments/Refunds		Closing balance
−$415.61	+	$0.00	+	−$5641.57	−	$1430.00	=	−$4627.18

Member no	1231234
Card no	1234 0000 5641 7894
Statement period	23rd Aug 22 – 22nd Sep 22

Summary	
Credit limit	$10 000.00
Available credit	*A*
Annual rate	11.50%
Daily rate	*B*

Overdue amount due now
Nil

Your minimum payment is
$139.00
and is due on
17 Oct 2022

a What is the credit limit on the card? ______

b Find the values of the:

i available credit, *A* ______

ii daily interest rate, *B*. Give your answer correct to five decimal places. ______

c What percentage of the closing balance is the minimum payment, correct to one decimal place?

______

QUESTION 2 Heidi used her credit card to pay her electricity bill valued at $960 on 7 November. She made no other purchases on her credit card account in November. She paid the account in full on 3 December. The credit card has no interest-free period. Compound interest is charged at the rate of 18% p.a., including the date of purchase and the date the account is paid.

a What is the daily interest rate, correct to six decimal places? ______

b For how many days will Heidi be charged interest?

______

c What is the total amount owed?

______

d What is the total amount of interest paid?

______

QUESTION 3 Jack uses a credit card which has no interest-free period and a compound interest rate of 13.2% p.a. including the day of purchase. His credit-card payment is due on the 28th of each month. During May, Jack made the following purchases.

Date	Retailer	Amount
7 May	Supermarket 1	$186.70
15 May	Supermarket 2	$216.80
21 May	Supermarket 3	$187.20
24 May	Supermarket 4	$104.90

a What is the daily interest rate, correct to five decimal places?

______

b How much interest will Jack pay on the transaction from:

i Supermarket 2?

______

______

ii Supermarket 4?

______

______

c What is the total amount due on 28 May? ______

______

______

Credit cards 2

QUESTION 1 A credit card company has the following conditions. There is no charge if the account is paid in full by the due date.

A charge of 0.075 75% compound interest per day accrues on the outstanding balance until it is paid in full.

Below is a copy of Amy's monthly statement.

Date	Transaction	$
	Opening balance	0.00
6.8.22	M Cotton & Co.	42.95
7.8.22	J Electronics	57.60
8.8.22	Recent Fashions	235.65
25.8.22	The Corner	80.20
	Closing balance	

a Calculate the closing balance.

b Calculate the interest charged if Amy repays $200 by the due date and the remainder 18 days later, to the nearest dollar.

QUESTION 2 Steve and Linda each use their credit cards to buy holiday packages to Singapore. The cost of each package is $1900.

a The charge on Steve's credit card is 1.5% interest per month on the unpaid balance. Steve pays $900 after one month and another $900 after the next month. After his second payment, how much does he still owe for his holiday?

b Linda's credit card company charges no interest in the first month and 2% interest per month on the unpaid balance from then on. She pays $900 after one month and another $900 after the next month. How much does she still owe for her holiday after this second payment?

Financial mathematics: Depreciation and loans

TOPIC TEST

SECTION I

Instructions
- This section consists of 5 multiple-choice questions.
- Each question is worth 1 mark.
- Fill in only ONE CIRCLE for each question.

Time allowed: 7 minutes **Total marks: 5**

1 A car is purchased for $26 990. It will depreciate at the rate of 16% p.a. Using the declining-balance method, which of these is closest to the salvage value of the car after 4 years?

Ⓐ $4528
Ⓑ $13 438
Ⓒ $13 552
Ⓓ $21 879

2 A car is purchased but over time its value drops. Which graph best represents the salvage value of the car using the declining-balance method of depreciation over 12 years?

Ⓐ
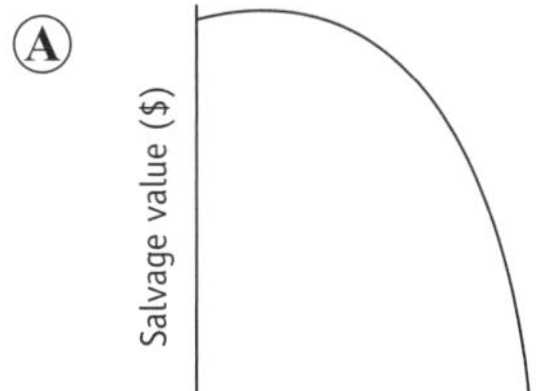

Ⓑ
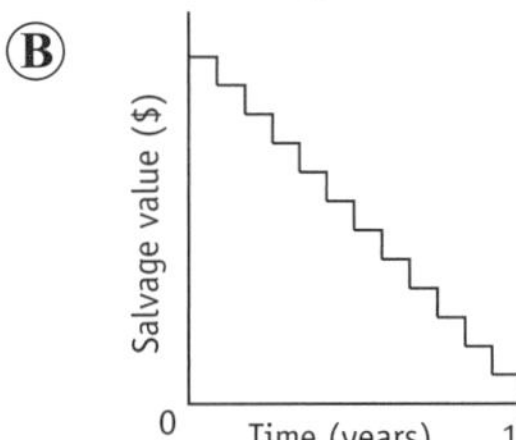

Ⓒ
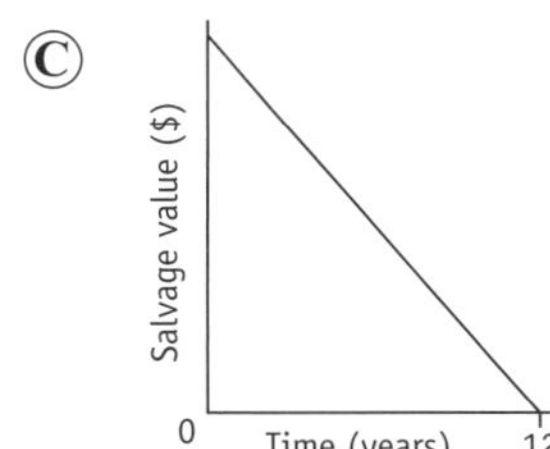

Ⓓ
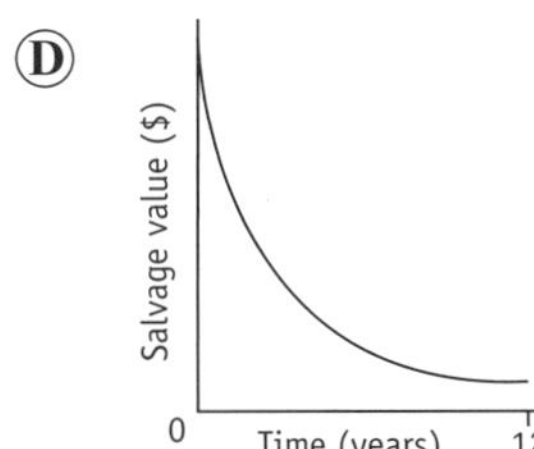

3 Leila used her credit card to purchase a washing machine for $1250 on 18 April. Compound interest is charged at a rate of 16.2% p.a. for purchases on her credit card. There were no other purchases on the credit card account and there was no interest-free period. The period for which interest is charged includes the date of purchase and the date of payment. The account was paid in full on 3 May. Which of these is closest to the amount paid?

Ⓐ $1258.19
Ⓑ $1258.22
Ⓒ $1258.91
Ⓓ $1452.50

4 The price of a new car is $41 250. Rod buys this car on terms of 20% deposit and monthly repayments of $910 over 4 years. How much more does Rod pay by buying the car on terms?

Ⓐ $2430 Ⓑ $8250 Ⓒ $9208 Ⓓ $10 680

5 Ingrid borrows $18 000 to buy a boat and is charged 8% p.a. flat-rate interest over 10 years. Which of these will be her monthly repayment?

Ⓐ $260 Ⓑ $270 Ⓒ $280 Ⓓ $290

TOPIC TEST | SECTION II

Instructions
- This section consists of 4 questions.
- Show all working.

Time allowed: 53 minutes | **Total marks: 35**

6 A spreadsheet is used to present key features of a credit card statement.

Credit card statement							
Credit limit:		$10 000.00		**Date of issue:**	**12/08/22**		
Opening balance:		$1435.00					
Statement duration				Payment summary			
From date:		10/07/22				Due date:	24/08/22
To date:		10/08/22				Balance owing:	$2305.39
Total number of days:		32		Minimum monthly payment:			
Transaction details				Remaining credit available:			$7694.61
Date	Description	Amount	Balance				
29/07/22	Purchase – POS	$125.00	$1560.00	**Interest rates***	**Annual**	**Daily**	
1/08/22	Purchase – POS	$340.00	$1900.00	**Purchases:**	**19.99%**	**0.054 767%**	
4/08/22	Cash advance	$400.00	*A*	**Cash advances:**	**21.49%**	*B*	
	Total interest	$5.39	$2305.39	*** Compound interest rates apply.**			
Total interest charges				**This credit card has no interest-free period.**			
	Purchases:						
	Cash advances:						
	Total:	**$5.39**					

a What is the value of:

i *A*? **1 mark**

____

ii *B*? Give the answer correct to six decimal places. **1 mark**

____

b Calculate the interest charged on the two purchases. **2 marks**

____

____

____

c What is the amount of credit available for the following month? **1 mark**

____

d The minimum monthly payment is calculated as 1% of the balance owing plus the total interest charged. Calculate the minimum monthly payment. **3 marks**

____

____

____

7 Mike takes out a loan of $35 000 and is charged a reducible interest rate of 3.3% p.a. At the end of each month he is charged interest and then makes a repayment of $2140. Mike uses a spreadsheet to calculate the balance owing after each monthly repayment.

Loan table				
	Amount:	**$35 000.00**		
	Annual interest rate:	**3.30%**		
	Monthly repayment:	**$2140.00**		
n	**Principal (*P*)**	**Interest (*I*)**	***P* + *I***	***P* + *I* − *R***
1	$35 000.00	$96.25		***A***
2	***A***	$90.63	$33 046.88	$30 906.88
3	$30 906.88	$84.99	$30 991.87	$28 851.87
4	$28 851.87	$79.34	$28 931.22	$26 791.22
5	$26 791.22	$73.68	$26 864.89	$24 724.89
6	$24 724.89	$67.99	$24 792.89	$22 652.89
7	$22 652.89	$62.30	$22 715.18	$20 575.18
8	$20 575.18	***B***		
9	$18 491.76	$50.85	$18 542.62	$16 402.62
10	$16 402.62	$45.11	$16 447.72	$14 307.72
11	$14 307.72	$39.35	$14 347.07	$12 207.07
12	$12 207.07	$33.57	$12 240.64	$10 100.64
13	$10 100.64	$27.78	$10 128.41	$7988.41
14	$7988.41	$21.97	$8010.38	$5870.38
15	$5870.38	$16.14	$5886.53	$3746.53
16	$3746.53	$10.30	$3756.83	$1616.83
17	$1616.83	$4.45		-$518.72

a Calculate the values of the following:

i *A* **1 mark**

ii *B* **1 mark**

b Calculate the total interest charged by the financial institution in the first 4 months, writing your answer to the nearest dollar. **2 marks**

c What is the meaning of the negative number in the last row of the spreadsheet? **2 marks**

d How much does Mike actually pay as his final repayment? **1 mark**

8 The table below shows the monthly repayments per $1000 on a bank loan for various annual interest rates.

Monthly repayments per $1000								
Term (years)	Interest rate (%)							
	4.5	5	5.5	6	6.5	7	7.5	8
20	$6.326	$6.600	$6.879	$7.164	$7.456	$7.753	$8.056	$8.364
25	$5.558	$5.846	$6.141	$6.443	$6.752	$7.068	$7.390	$7.718
30	$5.067	$5.368	$5.678	$5.996	$6.321	$6.653	$6.992	$7.338

a Calculate the monthly repayment, to the nearest cent, on a loan of $310 000 at 7% p.a. for 25 years. **2 marks**

b What is the interest rate on a loan of $360 000 over 20 years if the monthly repayment is $2476.44? **2 marks**

c Nigel borrows a sum of money and his monthly repayments over 25 years are $2278.78. If he is paying interest of 4.5% p.a., what is the amount Nigel has borrowed? **2 marks**

d Olaf is borrowing $340 000 for 30 years at 6.5% p.a. Find the total amount of monthly repayments made over the life of the loan. **2 marks**

e To the nearest dollar, what is the total amount of interest paid using monthly repayments over the period of the loan of $400 000 for 20 years at 7.5% p.a.? **3 marks**

f Pedro intends to borrow $460 000 over 30 years at a reducible interest rate of 5.5%. How much interest would be saved if his loan had been taken over 20 years at the same interest rate? **3 marks**

g Many years ago Thor borrowed $280 000 for a term of 30 years at an interest rate of 5% p.a. Today, Thor still owes $200 000 and decides to repay the balance with a new loan from a different financial institution which offers Thor 6% p.a. interest for 20 years. What is the difference in the two monthly repayments? **3 marks**

9 For the first 5 years after Cecile's purchase of a new car for $42 500, the annual rate of depreciation is 18%. For the following 5 years the rate is 12%.

Using the declining-balance method of depreciation, what is the value of Cecile's car after:

a 3 years? ____________ **1 mark**

b 10 years? ____________ **2 marks**

CHAPTER 3
Algebra: Simultaneous linear equations

Excel MATHEMATICS STANDARD 1
Ch. 1, pp. 4–5

Graphing linear equations

QUESTION 1 Complete the table of values for each equation.

a $y = 2x$

x	−1	0	1	2
y				

b $y = x + 1$

x	−1	0	1	2
y				

c $y = 2x - 1$

x	−1	0	1	2
y				

d $y = 5 - x$

x	−1	0	1	2
y				

e $y = -2x + 2$

x	−1	0	1	2
y				

f $y = 3x + 3$

x	−2	−1	0	1
y				

QUESTION 2 Sketch the graph of each of the above lines on the number plane below. Clearly label each line.

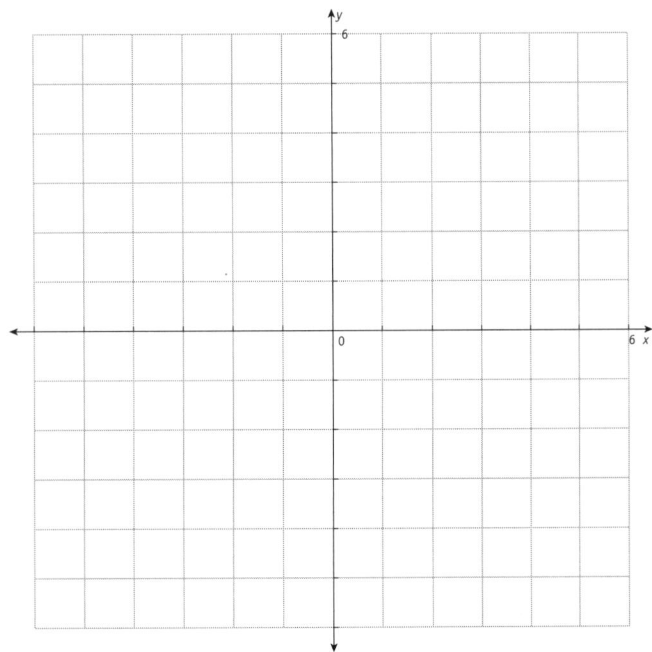

Meaning for gradient and *y*-intercept

Question 1 Liam receives a fixed amount of pocket money each week. In addition, if Liam chooses to help his mother she gives him an extra amount per hour for the time spent. The graph shows the amount of money Liam might receive in pocket money each week.

a What is the intercept on the vertical axis?

b What does the intercept on the vertical axis represent?

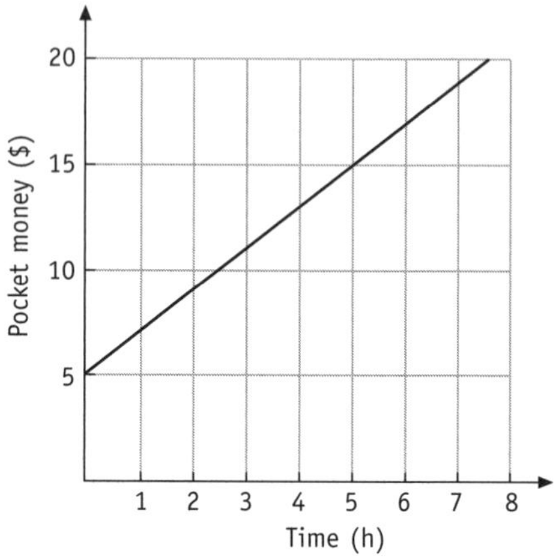

c What is the gradient of this line?

d What does the gradient represent?

Pocket money ($)
20
15
10
5
1 2 3 4 5 6 7 8
Time (h)

Question 2 Dorian intends to ride a bicycle from Aden to Barton to raise money for the local hospital. The graph shows his expected distance from Barton in kilometres over time (in hours).

a What is the intercept on the vertical axis?

b What information does this intercept tell us?

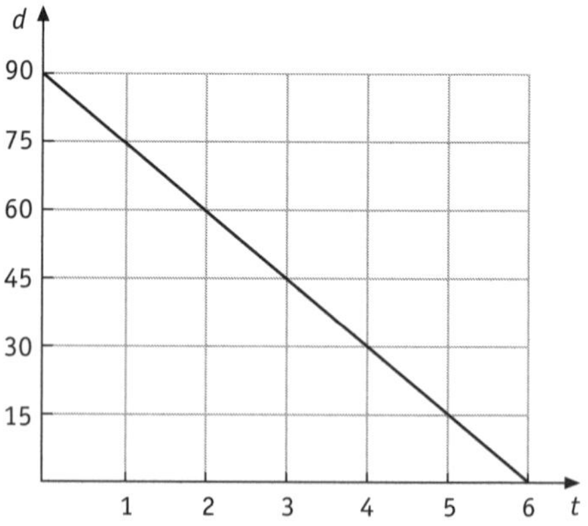

c What is the gradient of the line?

d What information does the gradient tell us?

e What is the equation of the line?

d
90
75
60
45
30
15
1 2 3 4 5 6 t

Solving graphically a pair of simultaneous equations

Question 1

a Complete the tables of values for:

i $y = 20x$

x	0	10	20	30	40
y					

ii $y = 600 - 10x$

x	0	10	20	30	40
y					

b Use the completed tables in part **a** to graph the lines $y = 20x$ and $y = 600 - 10x$ on the number plane.

c What is the point of intersection of the two graphs?

____

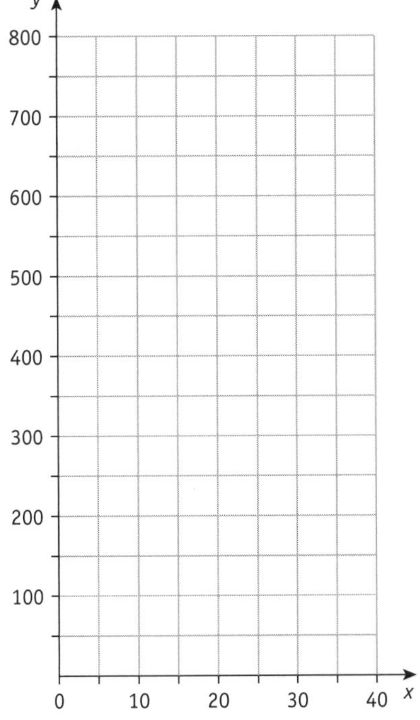

Question 2 Using the values from $x = 0$ to $x = 7$, find the point of intersection of the lines $y = 60 + 20x$ and $y = 300 - 40x$.

$y = 60 + 20x$

x	0	1	2	3	4	5	6	7
y								

$y = 300 - 40x$

x	0	1	2	3	4	5	6	7
y								

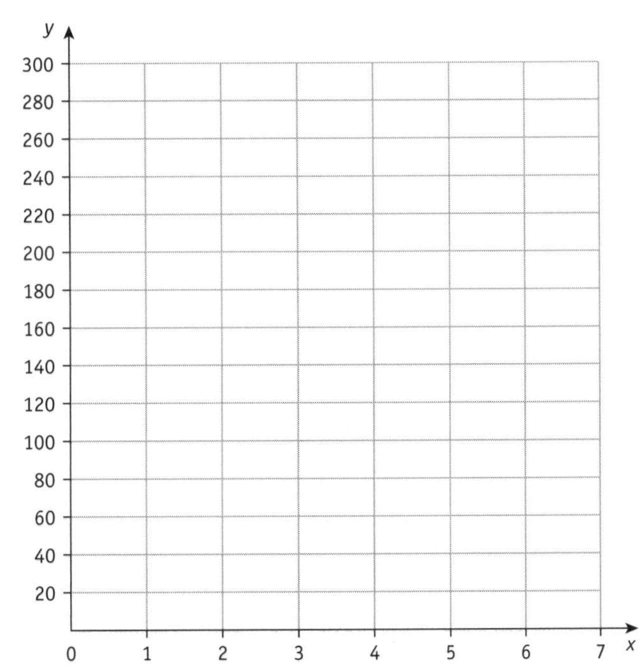

Linear functions and physical phenomena

Question 1 Part of the school oval needs to be returfed. A quote is given from Top Turf stating the cost of the turf as being \$50 for delivery and \$6 per metre.

a Write an equation for the cost of the turf, C.

____

____

b Draw a graph of the cost on the axes provided.

c What is the cost of returfing 100 m of the school oval?

____

____

d Explain why the model cannot be used to calculate the **total** cost of returfing part of the school oval.

____

____

Question 2 A group of Year 12 students are going on an excursion to a chocolate factory. The cost of hiring a bus licensed to seat 54 people involves a \$200 booking fee plus \$5 per person.

a Write an equation to calculate the cost, C, of hiring a bus for n people.

____

____

b Draw a graph for the cost of hiring a bus on the axes provided.

c 45 students and 5 teachers are going on the excursion. What is the cost of hiring the bus?

____

____

d Why is the model not accurate for $n > 54$?

____

____

____

Developing a pair of simultaneous equations

Question 1 Develop an equation to represent the cost (C) of:

a n apples at 60 cents each

b x litres at $7 per litre

Question 2 A taxi driver charges $4 plus $2 per kilometre of each fare.

a Write an equation where C = cost, in dollars, and n = number of kilometres.

b Use the equation to find the cost of a journey of 12 km.

Question 3 The cost of an electrician is a $70 callout fee plus $50 for every half hour or part thereof. If P = price, in dollars, and n = number of half hours, write an equation linking P and n.

Question 4 Over the course of a trip, a train has an average speed of 70 km/h. If d = distance, in kilometres, and t = time in hours:

a write an equation for d in terms of t

b use the equation to find the distance travelled by the train in 6 hours

Question 5 Tom is driving 280 km from his home to his holiday unit. His average speed is 80 km/h.

a Given d = distance (in kilometres) yet to travel and n = number of hours travelled, explain why the formula $d = 280 - 80n$ represents the distance Tom is from his holiday unit.

b After travelling for 2.5 hours, how far does Tom need to travel to reach his destination?

Question 6 It costs Annika $120 plus $8 for every T-shirt she makes. She then sells them for $16 each. Using cost = C, in dollars, sales = S, in dollars, and the number of T-shirts sold = n, write equations for:

a C in terms of n

b S in terms of n

Question 7 Carrie makes small containers of slime for her friends at school, selling them for $2 each. She spent $10 on a box of containers and she estimates that the amount of slime in each container cost her 30 cents to produce. If C = cost in cents, S = sales in cents, P = profit in cents and n = number of containers sold, write equations for:

a C in terms of n

b S in terms of n

c P in terms of n

Question 8 Two women are driving towards each other between Hillcrest and Parklea which are 240 km apart. Keira leaves Hillcrest and drives at an average speed of 60 km/h towards Parklea. Indi leaves Parklea and averages 80 km/h as she drives towards Hillcrest. Find the equation for the distance d, in kilometres:

a Keira is from Hillcrest after time t, in hours

b Indi is from Hillcrest after time t, in hours

Algebra: Simultaneous linear equations

Break-even analysis 1

QUESTION **1** Hayley bakes and decorates cakes and plans to sell them via an online sales page. She works out that her costs (in \$) can be represented by the equation $C = 40x + 250$, where x is the number of cakes produced.

a Graph Hayley's costs on the axes provided.

b What is the gradient of the line?

____

____

c What does it represent?

____

____

d If Hayley wants to break even after 20 cakes, how much does she need to charge per cake?

____

____

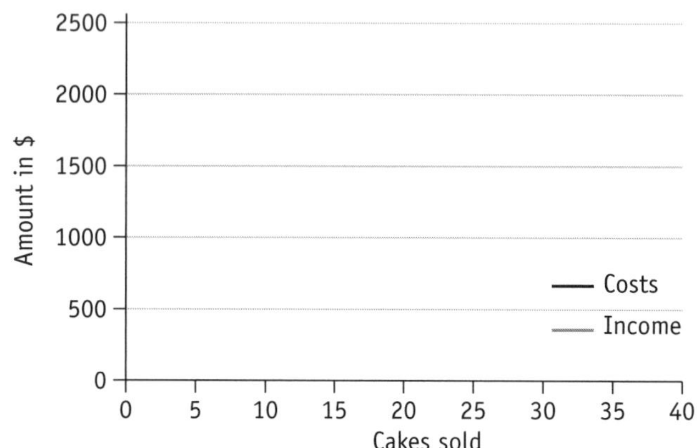

e Graph the line $I = 52.5x$ on the same axes.

f Shade the profit zone indicating where Hayley's income from cake sales exceeds her costs of producing the cakes.

QUESTION **2** Jason plans to sell screen-printed T-shirts at the local markets. He conducts a break-even analysis to evaluate the plan's financial viability, shown to the right.

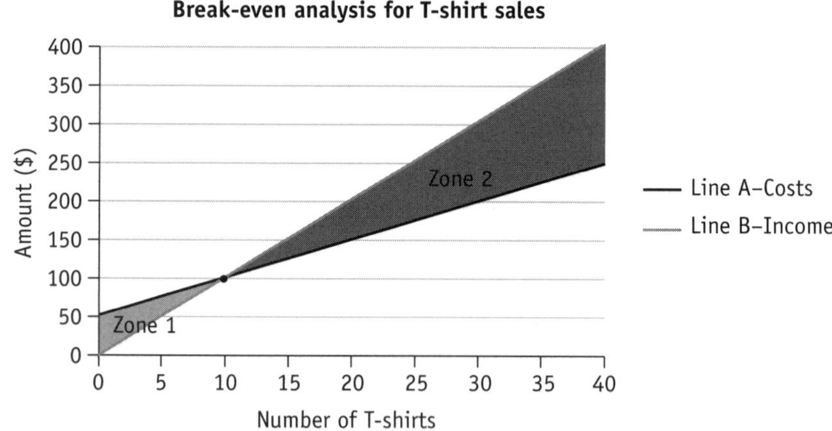

a How many T-shirts sold represent the break-even point for Jason's plan? ______________________________

b Which is the profit zone? ______________________________

c Which is the loss zone? ______________________________

d For Line A, what is represented by: **i** the gradient? **ii** the y-intercept? ______________________________

e For Line B, what is represented by: **i** the gradient? **ii** the y-intercept? ______________________________

f What is the income equation? (Use the form $I = mx$.) ______________________________

g What is the cost equation? (Use the form $C = mx + c$.) ______________________________

Break-even analysis 2

QUESTION 1 Jonny hires a function centre for an MND fundraising dinner, where tickets are sold at \$120 per person. He pays \$2400 to hire the facility and the cost of the meal and drinks will be \$40 per person. The function centre has a capacity of 100 people.

a Given C = total cost of the dinner and n = number of tickets, explain why $C = 2400 + 40n$.

b Given I = income from the dinner and n = number of tickets, explain why $I = 120n$.

c Draw a graph displaying both equations.

d How many tickets will need to be sold for Jonny to break even?

e If 70 tickets were sold, what is the profit made from the dinner?

f Is it possible to make a profit of \$6000?

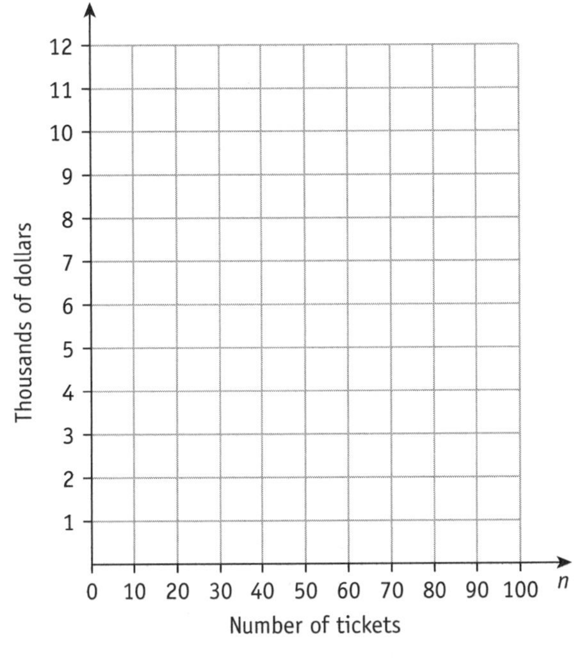

QUESTION 2 The costs of buying and running two types of light globes are recorded in the tables below, where n = the number of hours, H = the total cost of buying and running a halogen globe, and L = the total cost of buying and running an LED globe.

Halogen:

n	0	400	800	1200	1600	2000
H	3	9	15	21	27	33

LED:

n	0	400	800	1200	1600	2000
L	13	15	17	19	21	23

a What is the purchase price of an LED globe?

b What is the hourly cost of using a halogen globe?

c Represent the cost of both globes on the number plane opposite.

d After how many hours is the total cost of the globes identical?

e The life of a halogen globe is 2000 hours while an LED globe lasts 40 000 hours. If the light is turned on for 4000 hours, how much cheaper would it be to use an LED globe compared to a halogen globe (including the cost of the globes)?

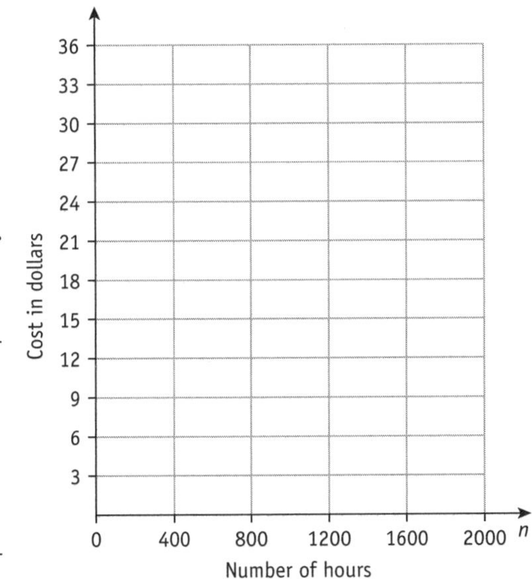

Algebra: Simultaneous linear equations

TOPIC TEST

SECTION I

Instructions
- This section consists of 5 multiple-choice questions.
- Each question is worth 1 mark.
- Fill in only ONE CIRCLE for each question.

Time allowed: 7 minutes **Total marks: 5**

1 A company manufactures widgets. The company's cost equation and income equation are drawn on the same graph. Which region of the graph represents the profit zone?

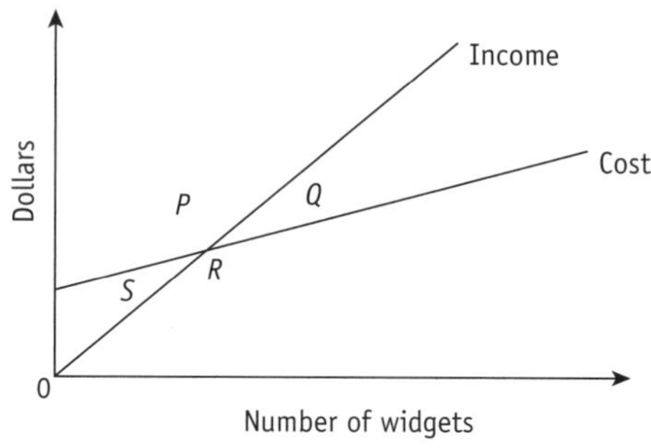

Ⓐ P Ⓑ Q

Ⓒ R Ⓓ S

2 What is the equation of the line k?

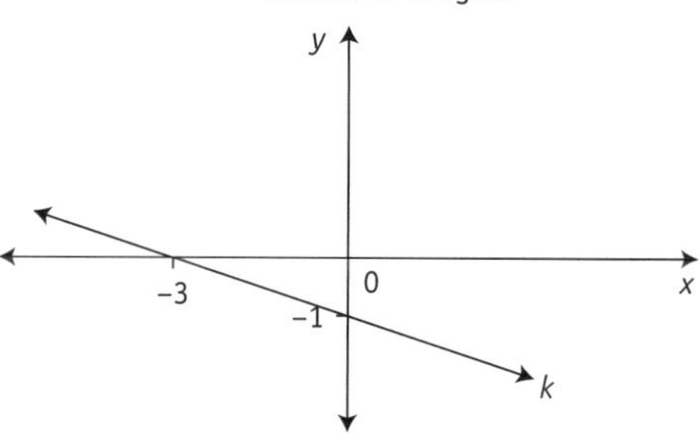

Ⓐ $y = \frac{x}{3} - 1$ Ⓑ $y = 3x - 1$

Ⓒ $y = -\frac{x}{3} - 1$ Ⓓ $y = -3x - 1$

3 Paul and Lauren are working out the cost of hiring tradespeople to build a granny flat on their property. They are told that employing a team of tradespeople will involve a fixed cost of $650 and then an additional $450 per day. What is the correct equation connecting the cost in dollars, C, with the number of days, d, worked by the team of tradespeople?

Ⓐ $C = 650 + 450d$ Ⓑ $C = \frac{450d}{650}$ Ⓒ $C = 650 - 450d$ Ⓓ $C = \frac{650d}{450}$

4 According to the graph, at what value for units produced will production break even?

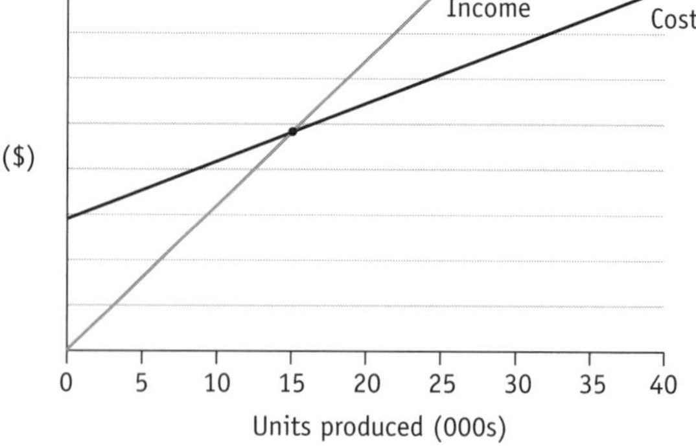

Ⓐ 0 Ⓑ 15

Ⓒ 15 000 Ⓓ none of these

5 Sean drew the graphs of $y = 2x$ and $y = x + 5$ on the same number plane. What are the coordinates of the point of intersection of the two lines?

Ⓐ (2, 4) Ⓑ (3, 8) Ⓒ (4, 9) Ⓓ (5, 10)

TOPIC TEST

SECTION II

Instructions • This section consists of 5 questions.
• Show all working.

Time allowed: 53 minutes **Total marks: 35**

6 **a** Complete the table of values and graph each line on the number plane below.

i $y = 3x - 1$

x	−1	0	1	2
y				

ii $y = x + 5$

x	−1	0	1	2
y				

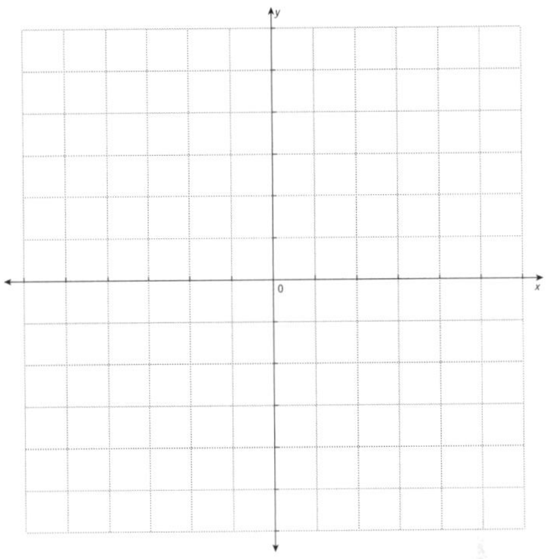

2 marks each

b At what value of x do the two lines intersect? **1 mark**

7 Match each equation with a table of values. **1 mark each**

a

x	−2	−1	0	1
y	−5	−3	−1	1

b

x	−2	−1	0	1
y	−2	1	4	7

c

x	−2	−1	0	1
y	−4	−2	0	2

d

x	−2	−1	0	1
y	9	7	5	3

i $y = 2x$ **ii** $y = 2x - 1$

iii $y = 5 - 2x$ **iv** $y = 3x + 4$

8 Grace makes batches of homemade lemonade which she sells to her friends by the jug. Grace has calculated that the cost of producing the jugs of lemonade is $8 plus $3 for every jug.

a Complete the table of values **2 marks**

Number of jugs	0	4	8	12	16	20
Cost ($)						

b Draw a graph of the cost on the number plane provided. **2 marks**

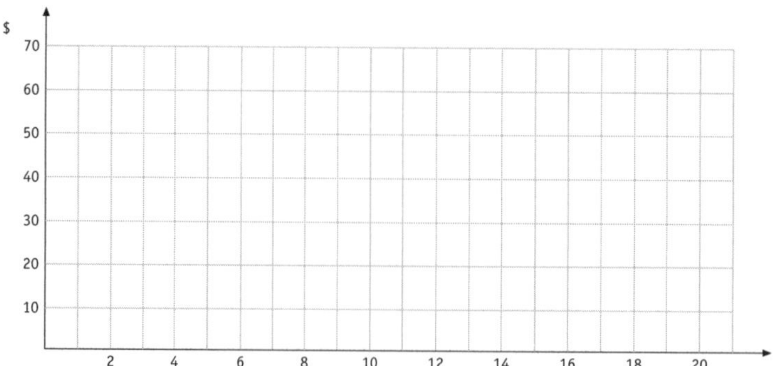

c What is the intercept on the vertical axis? Briefly explain what this represents. **2 marks**

____

____

____

d What is the gradient of the line? Briefly explain what it represents. **2 marks**

____

____

____

e What is the cost of producing 14 jugs of lemonade? **1 mark**

____

f The total cost of a batch Grace made was \$56. How many jugs did this batch contain? **1 mark**

____

g If Grace sells the lemonade for \$4 per jug, draw the graph of her return from sales on the same number plane. **1 mark**

____

h Where do the two lines intersect? Briefly explain what this means. **2 marks**

____

____

____

____

9 Samone runs a small business selling birthday cakes. The graph shows the cost of producing the cakes and the income received from their sale each week.

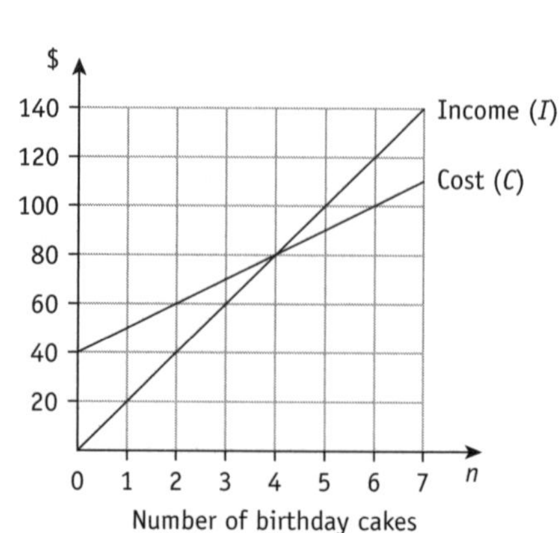

a How many cakes need to be sold to break even? **1 mark**

____

b How much is the profit or loss when Samone sells:

i 2 cakes? **1 mark**

____

ii 7 cakes? **1 mark**

____

c By finding the gradient and vertical intercept, write an equation, in terms of cakes sold (n), for the:

i income (I) **1 mark**

____

ii cost (C) **2 marks**

____

____

10 Harry-Rose makes dresses and sells them at a weekend market. It costs her \$100 plus \$15 for every dress she makes and she sells them for \$40 each.

a Using cost = C, in dollars, sales = S, in dollars, and the number of dresses sold = n, write equations for:

i C in terms of n **1 mark**

____

ii S in terms of n **1 mark**

____

b Use the grid below to draw the graphs of C and S in terms of n. **2 marks**

c How many dresses does Harry-Rose need to sell to break even? **1 mark**

____

____

____

d What is the difference in the profit between Harry-Rose selling 8 dresses and 10 dresses? **2 marks**

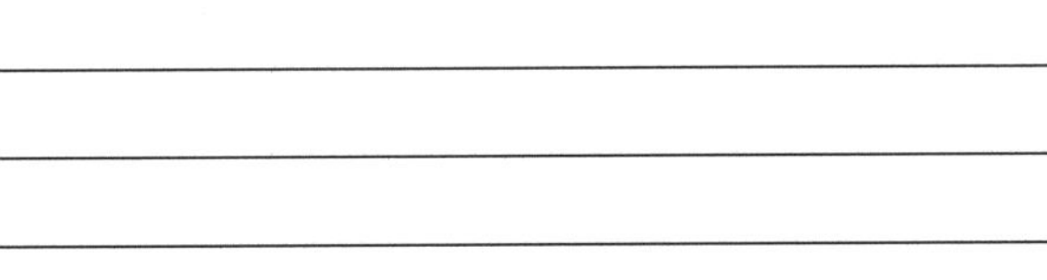

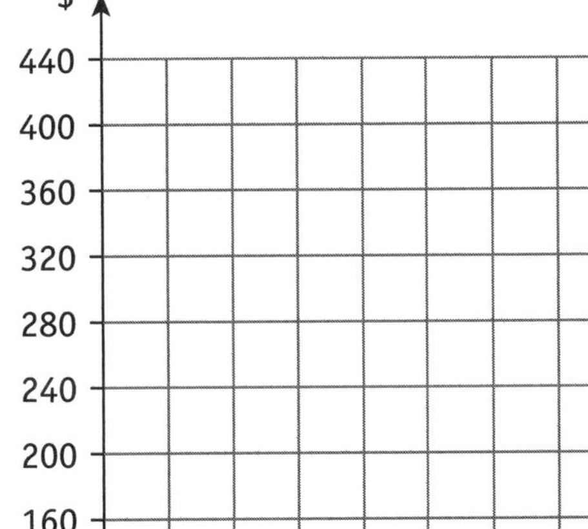

CHAPTER 4
Algebra: Graphs of practical situations

Excel MATHEMATICS STANDARD 1
Ch. 1, p. 11

Sketching a quadratic from a table of values

QUESTION 1 Use the tables of values to sketch the graph of each curve.

a $y = 9 - x^2$

x	0	1	2	3
y	9	8	5	0

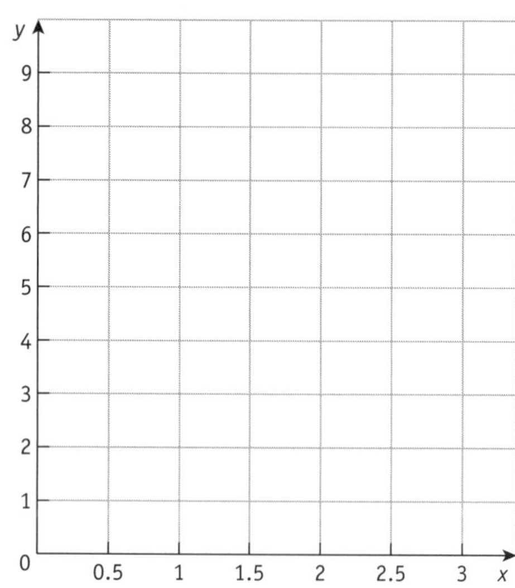

b $h = t^2 - 4t + 4$

t	0	1	2	3	4
h	4	1	0	1	4

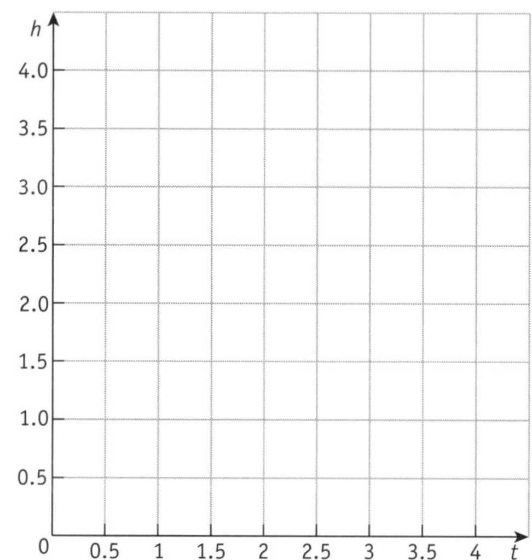

QUESTION 2 Complete the tables and graph each of the curves on the number plane provided.

a $y = 16 - x^2$

x	0	1	2	3	4
y					

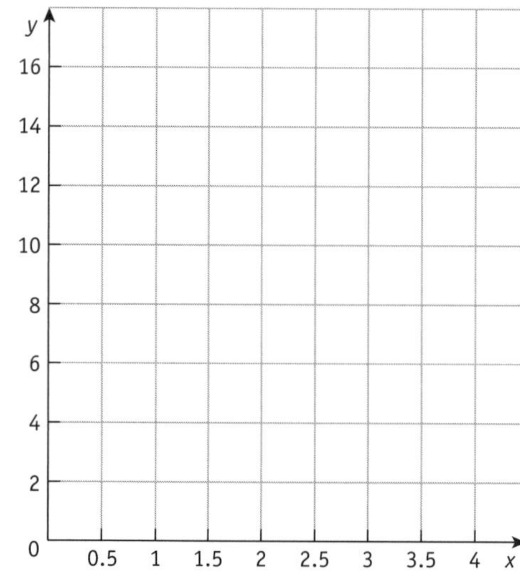

b $h = 8 + 2t - t^2$

t	0	1	2	3	4
h					

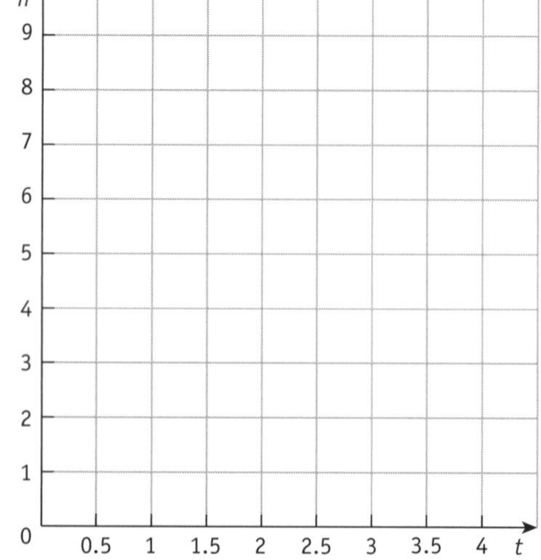

Sketching an exponential from a table of values

QUESTION **1** Use the tables of values to sketch the graph of each curve.

a $y = 3 \times 2^x$

x	0	1	2	3	4
y	3	6	12	24	48

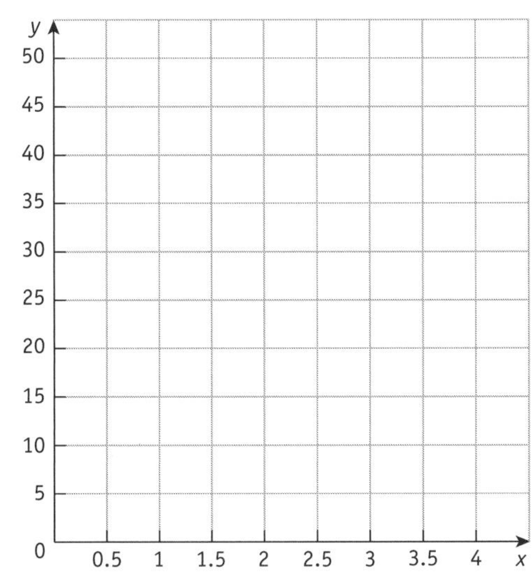

b $P = 10 \times 2.5^t$

t	0	1	2	3
P	10	25	62.5	156.25

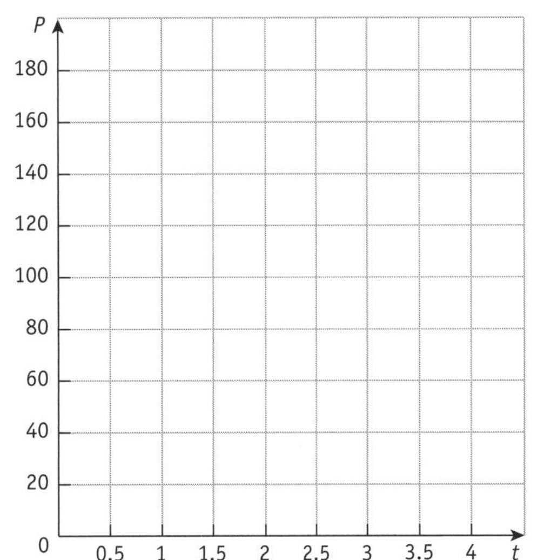

QUESTION **2** Complete the tables of values to sketch the graph of each curve.

a $y = 120 \times 2.5^x$

x	0	1	2	3	4
y					

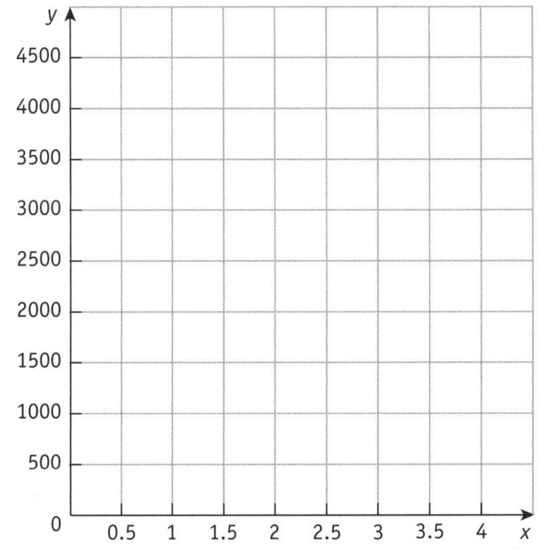

b $P = 20\,000(1.5)^t$

t	0	1	2	3	4
P					

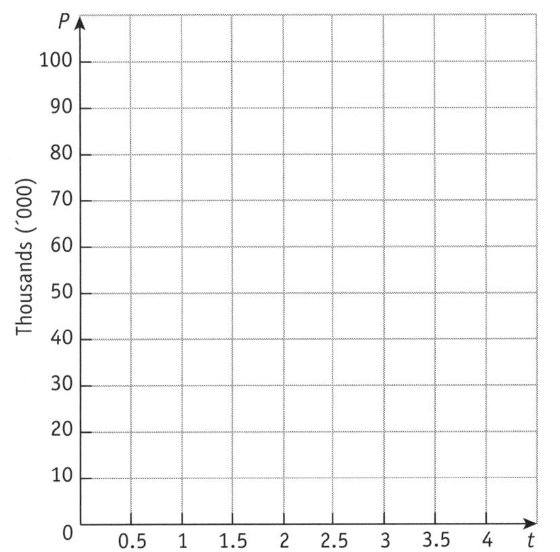

Algebra: Graphs of practical situations

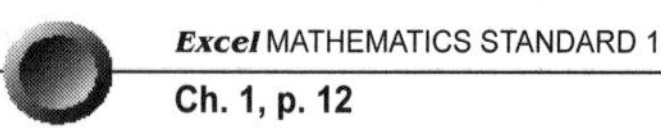

Sketching a graph relating to speed, distance and time

Question 1 A car is travelling at 60 km/h.

a Complete the table showing the distance (d, in km) travelled for different time periods (t, in hours).

t	0	1	2	3	4	5
d						

b Use the table to draw a graph showing the distance travelled.

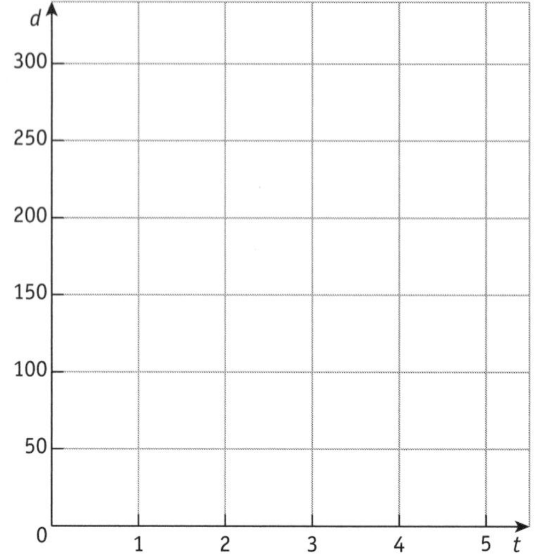

Question 2 A truck is driven for 5 hours.

a Complete the table showing the distance (d, in km) travelled for different speeds (s in km/h).

s	0	50	60	80	100
d		250			

b Use the table to draw a graph showing the distance travelled.

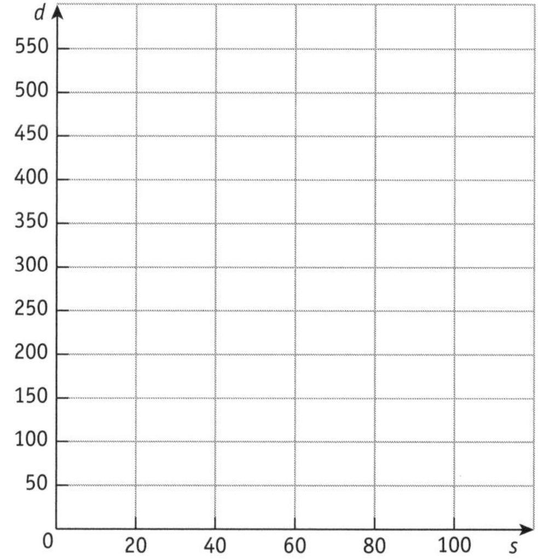

Question 3 A motorcyclist needs to ride a distance of 360 km.

a Complete the table showing the time (t, in hours) taken to ride the distance given the speed of the motorcyclist (s, in km/h).

s	30	40	60	80	90	100
t	12		6			

b Use the table to draw a graph showing the time taken.

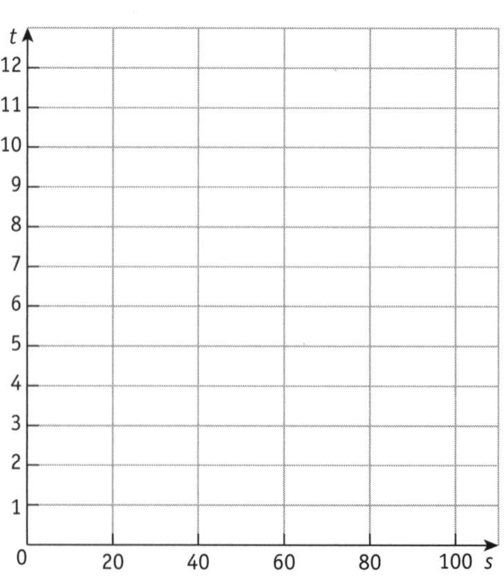

Travel graphs

QUESTION 1 The travel graph displays Joel's car trip along a straight road from home to his parents' farm and back again. The trip has been broken up into five sections: *P*, *Q*, *R*, *S* and *T*.

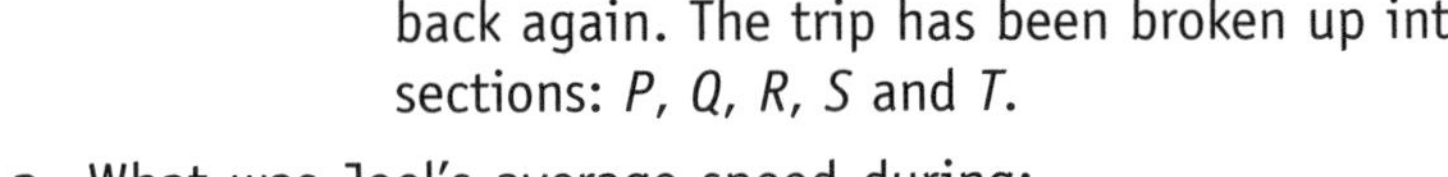

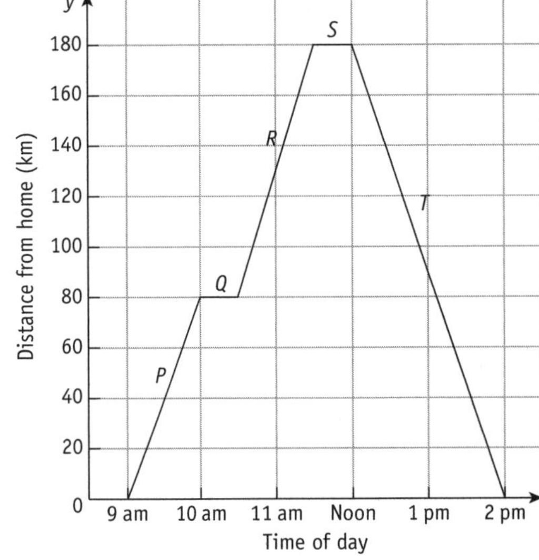

a What was Joel's average speed during:

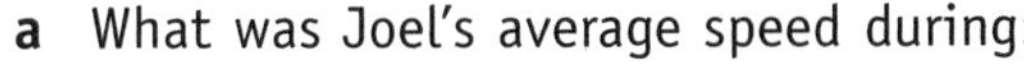

i section *P*? ______________________________

ii section *R*? ______________________________

iii section *T*? ______________________________

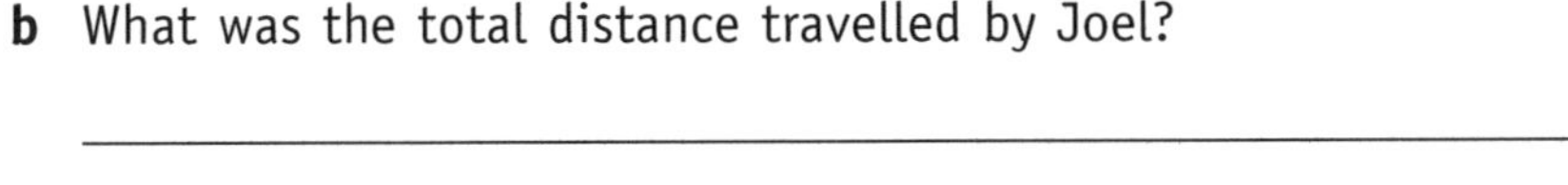

b What was the total distance travelled by Joel?

____

____

c How long was Joel away from home?

____

____

d What was his average speed for the entire day?

____

____

QUESTION 2 A swimming race is held in a harbour. Competitors leave a small beach and swim out to a buoy, before returning to the same beach. Suzy takes 15 minutes to reach the buoy, and maintains the same speed back to the beach. The graph of Suzy's swim is shown.

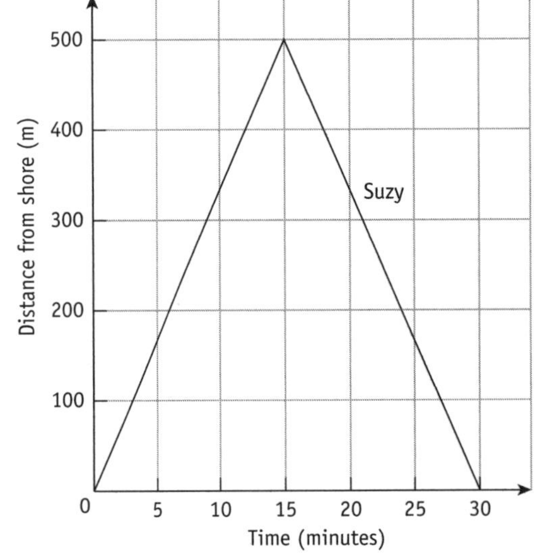

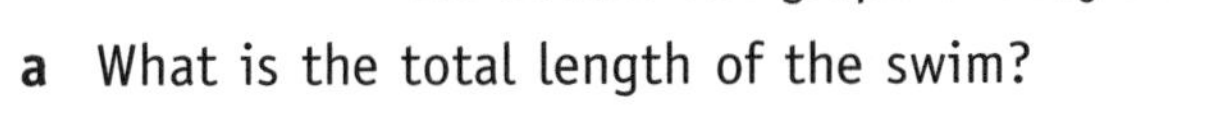

a What is the total length of the swim?

____

____

b What was the average speed of Suzy's swim?

____

____

Jodie also competed in the same race. She completed the race in 20 minutes, maintaining the same speed through the race.

c Draw the graph of Jodie's swim on the same diagram.

d Suzy was still swimming when Jodie finished the race. Estimate the distance that Suzy still had to swim.

____

____

____

Containers and graphs

QUESTION 1 Water is poured into each of these containers at a steady rate. Draw a graph showing the depth of the water in each container over time.

a

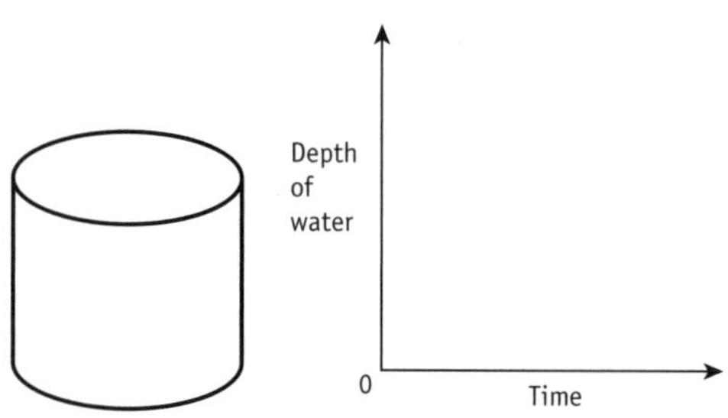

b

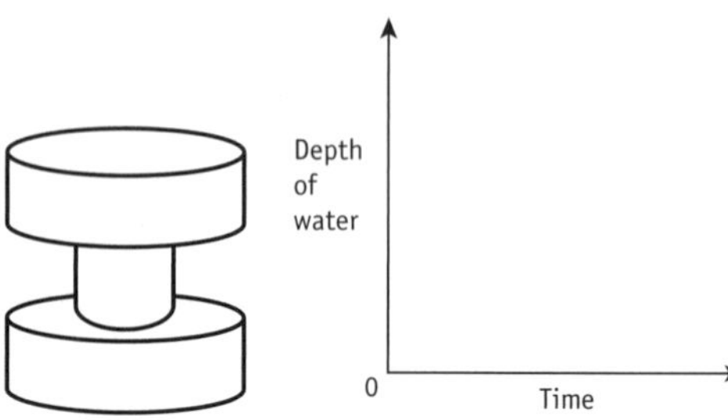

c

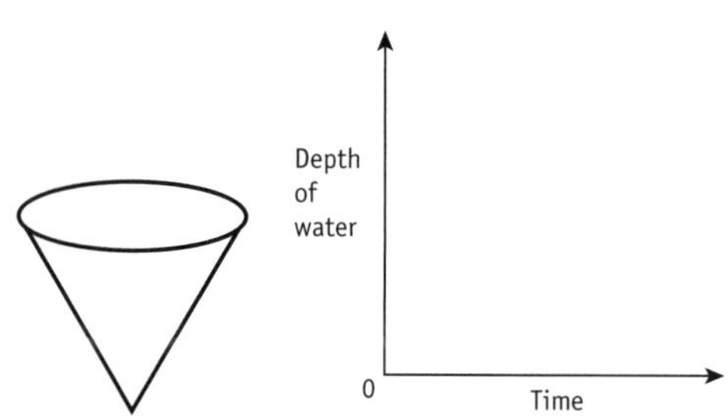

d

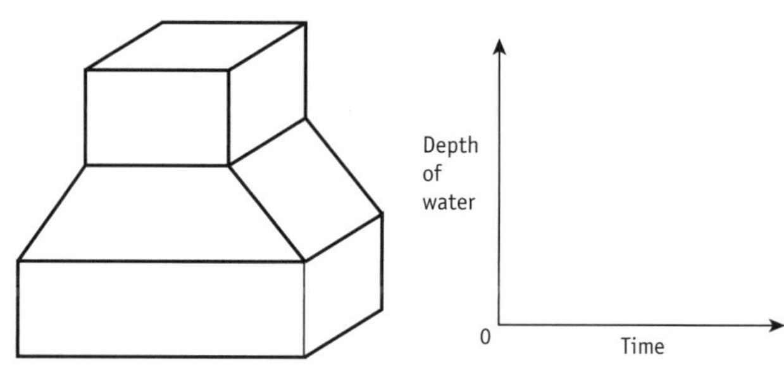

QUESTION 2 Water is poured into four containers at a constant rate and the following graphs have been drawn to show the increase in the depth of water. Draw a sketch of the container that is related to each graph.

a

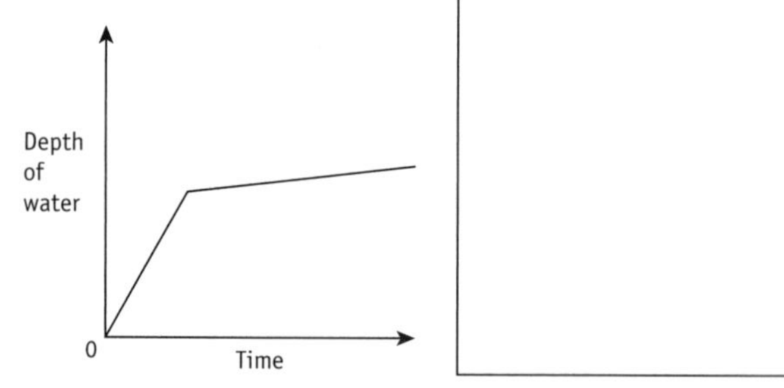

b

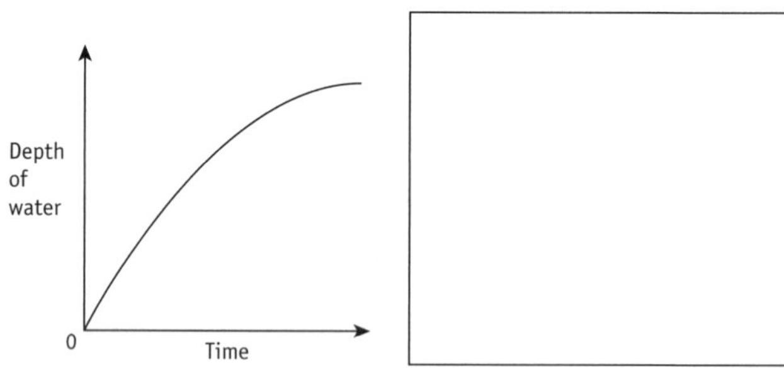

c

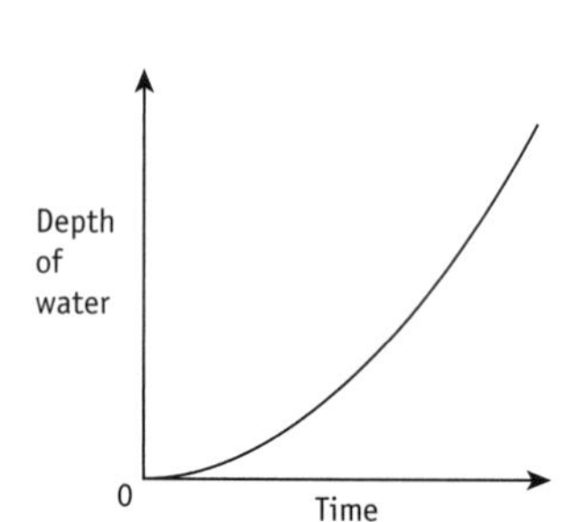

d

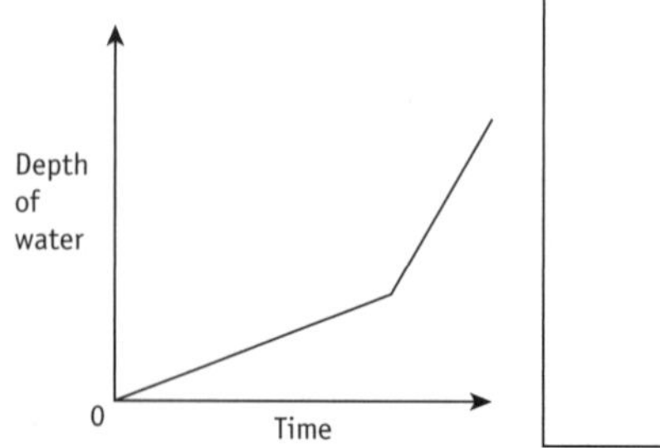

Algebra: Graphs of practical situations

TOPIC TEST

SECTION I

Instructions
- This section consists of 5 multiple-choice questions.
- Each question is worth 1 mark.
- Fill in only ONE CIRCLE for each question.

Time allowed: 7 minutes **Total marks: 5**

1 Water was poured into a container at a constant rate. The graph shows the depth of the water as time passed. Which of these is a sketch of the container?

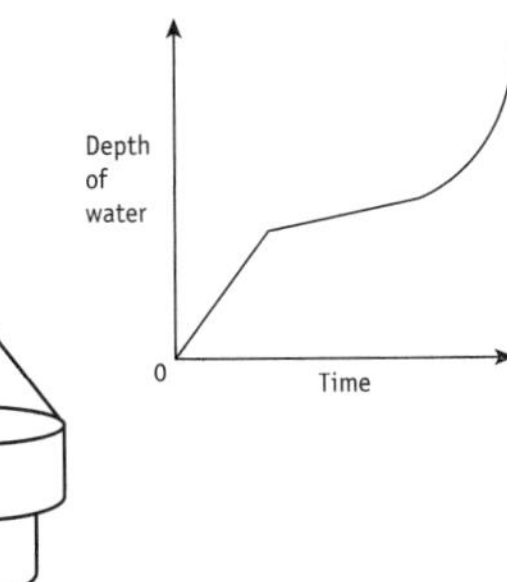

Ⓐ 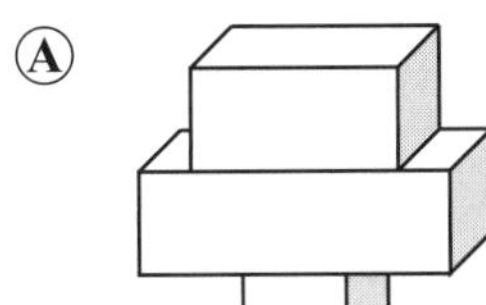

Ⓑ 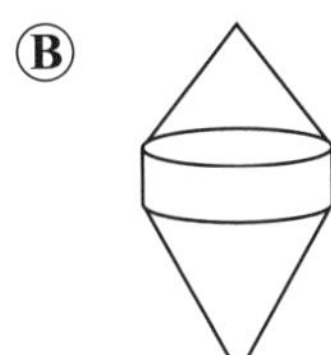

Ⓒ 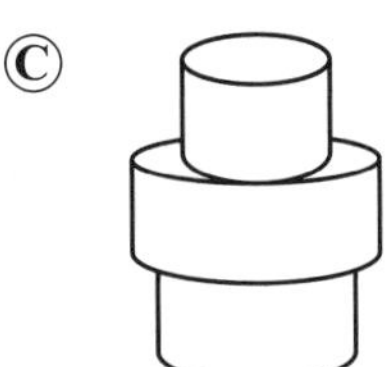

Ⓓ 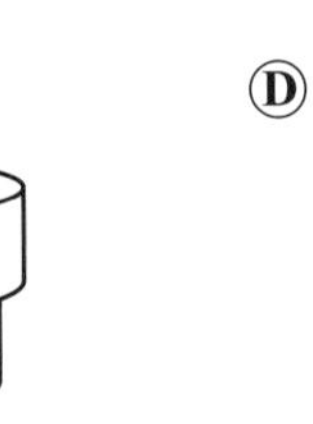

2 The equation $P = 200(1.06)^t$ can be used to determine the population of a penguin colony after t months. A table is shown and is used to graph the curve representing the equation.

t	0	1	2	3	4
P	200	a	225	238	252

Which of these is the value of a?

Ⓐ 206 Ⓑ 212 Ⓒ 213 Ⓓ 216

3 Water is poured into a container at a constant rate. The container is shown. Which of these is a graph of the height of the water in the container?

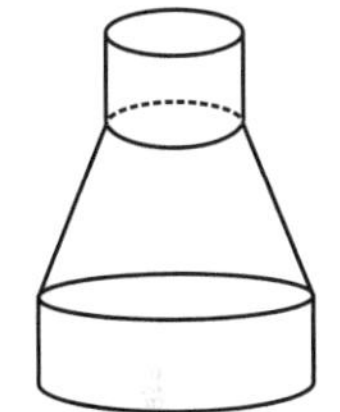

Ⓐ 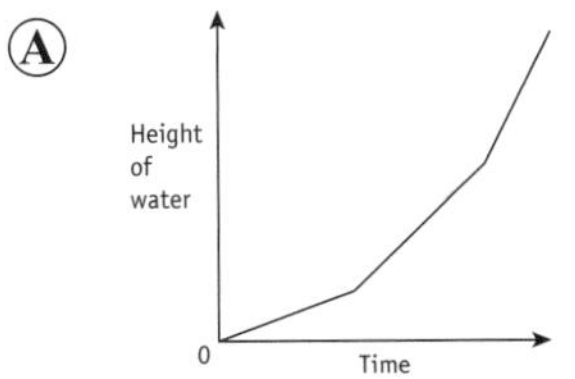

Ⓑ 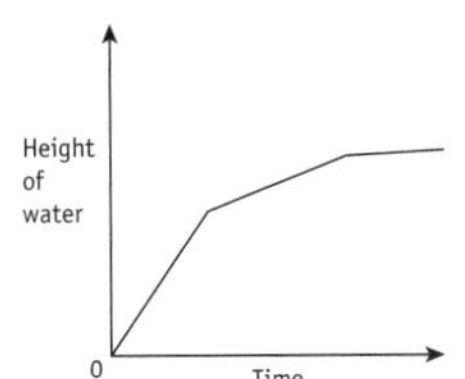

Ⓒ 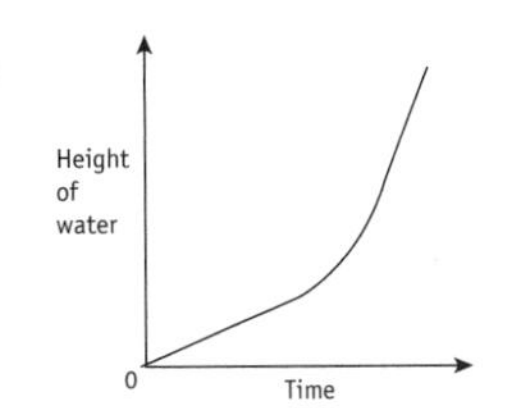

Ⓓ 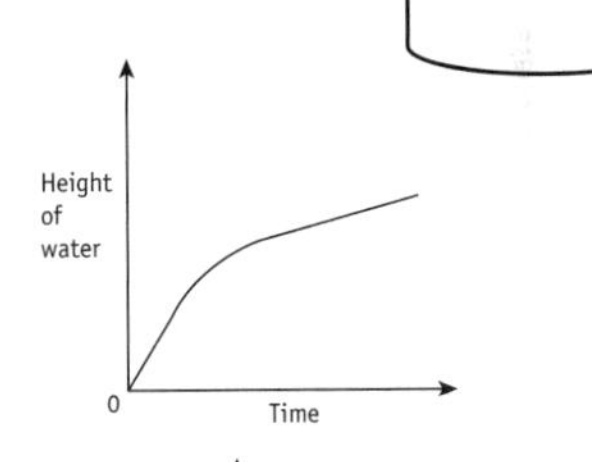

4 Which type of relationship is represented by the graph?

Ⓐ linear Ⓑ hyperbolic

Ⓒ inverse Ⓓ exponential

y, 2, 0, x

5 The graph shows the distance travelled by a car over a period of 4 hours. Which of these is the average speed of the car?

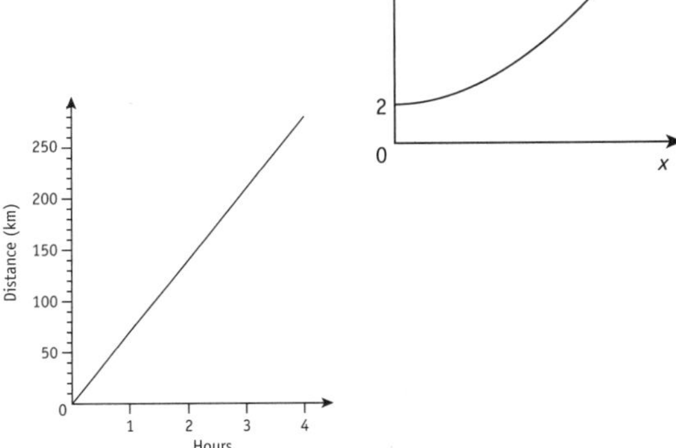

Ⓐ 56 km/h Ⓑ 70 km/h

Ⓒ 92 km/h Ⓓ 100 km/h

TOPIC TEST

SECTION II

Instructions
- This section consists of 6 questions.
- Show all working.

Time allowed: 53 minutes **Total marks: 35**

6 The travel graph displays Chloe's car trip along a straight road from home to her cousin's apartment and back again.

The trip has been broken into seven separate sections: *A*, *B*, *C*, *D*, *E*, *F* and *G*.

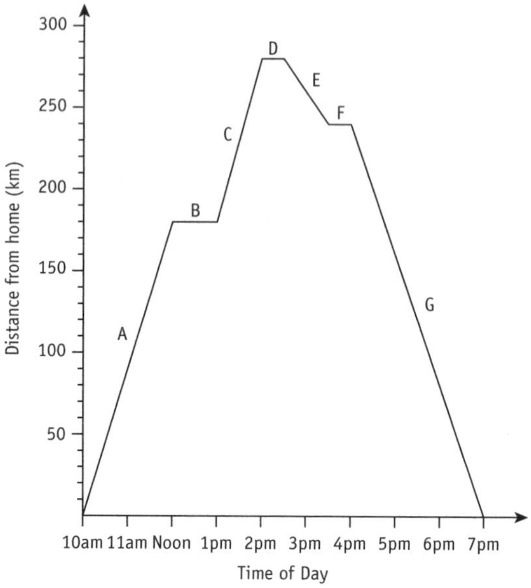

a How far did Chloe travel in total? **1 mark**

____

b What was the total amount of time Chloe's car was stationary? **1 mark**

____

c Find the total time Chloe was driving her car. **1 mark**

____

____

d Calculate Chloe's average speed represented by:

i section *C* **1 mark**

____

____

ii section *E* **1 mark**

____

____

e In which section was Chloe's average speed the highest? **1 mark**

____

____

7 The equation $y = 25 - x^2$ is to be graphed.

a Complete the table for $y = 25 - x^2$. **2 marks**

x	0	1	2	3	4	5
y						

b Use the table to graph $y = 25 - x^2$ on the number plane. **2 marks**

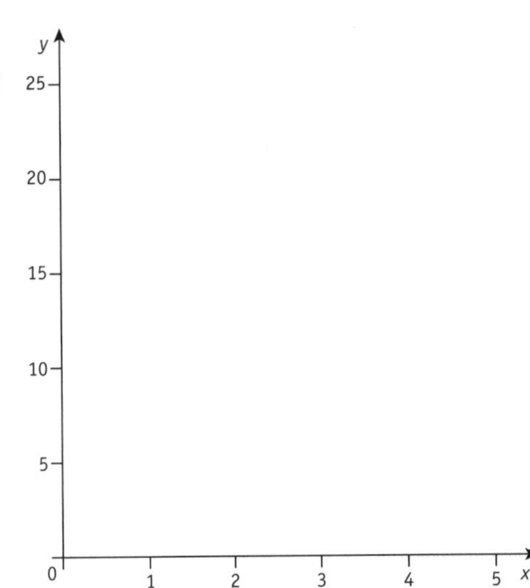

8 The graph shows the height above the ground of a ball t seconds after it was thrown into the air from the top of a building.

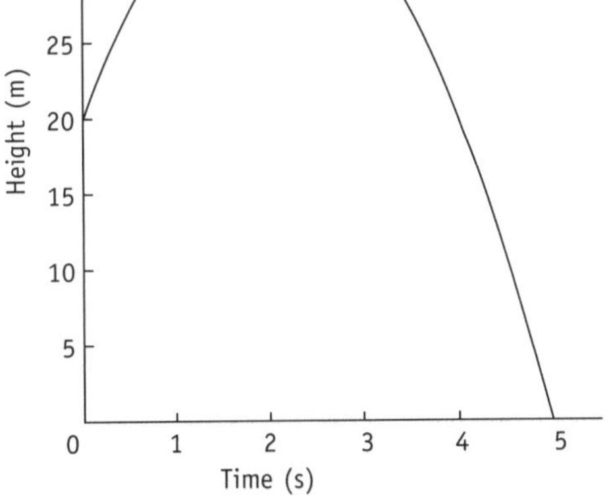

a What was the maximum height reached by the ball above the ground? **1 mark**

____

b How high was the building? **1 mark**

____

c How long was it before the ball hit the ground? **1 mark**

____

9 On a property the population of mice doubles every month. Initially there were 10 mice transported to the farm on a truck.

a Complete the table where t is the number of months and m is the number of mice. **2 marks**

t	0	1	2	3	4	5	6
m	10					320	

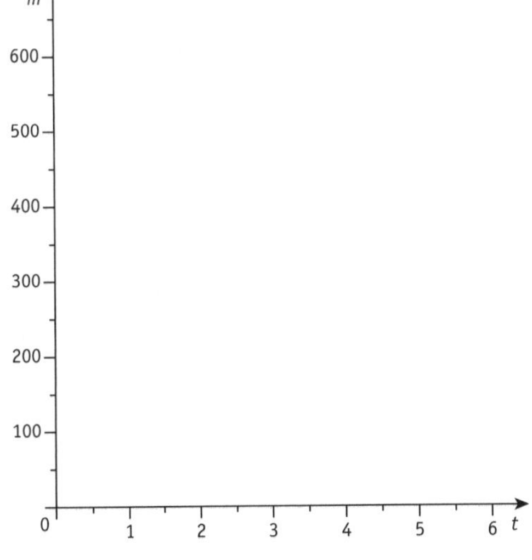

b Use the number plane to draw a graph representing the population of mice over the six-month period. **2 marks**

c The formula $m = 10 \times 2^t$ can be used to estimate the population of mice.

What will the population of mice be after 1 year? **1 mark**

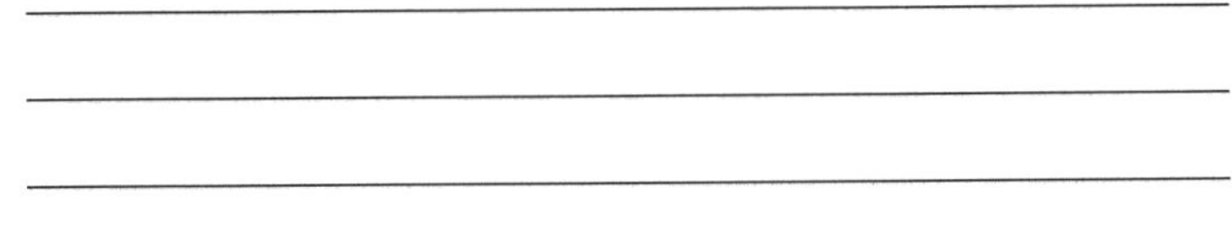

10 Four containers are to be filled with water at a constant rate.

Four graphs are also shown representing the height of the water level in each of the containers.

Draw lines to match the containers with the correct graph. **4 marks**

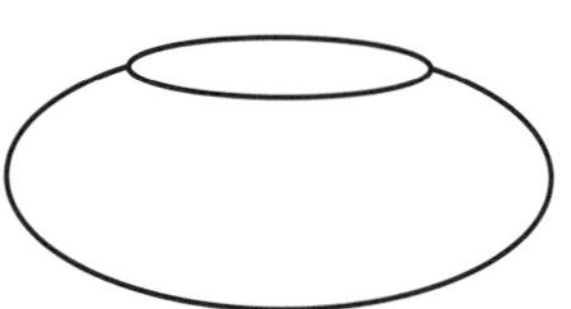

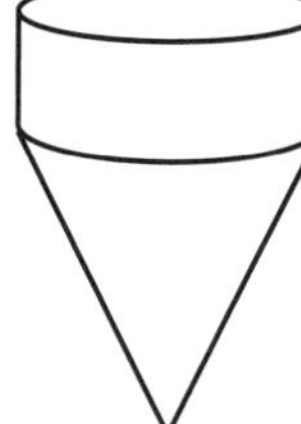

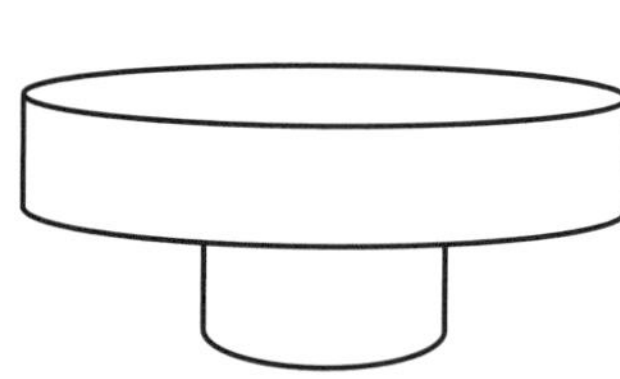

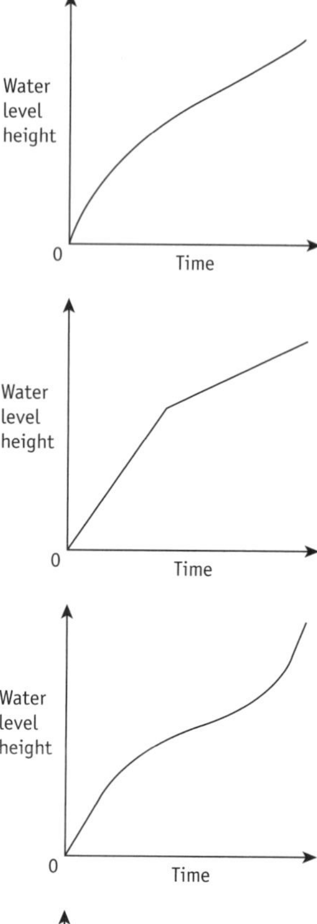

11 Kara threw a stone from an open window into the air.

The stone landed on the ground.

The movement of the stone forms part of a quadratic function.

Graphing software has been used to represent the height h, in metres, of the stone over time, in seconds.

a The path of the stone is represented by a part of the graph drawn. With the use of a highlighter, or pen, draw on the graph the section that represents the path of the stone. **1 mark**

b How high was the window from where Kara threw the stone? **1 mark**

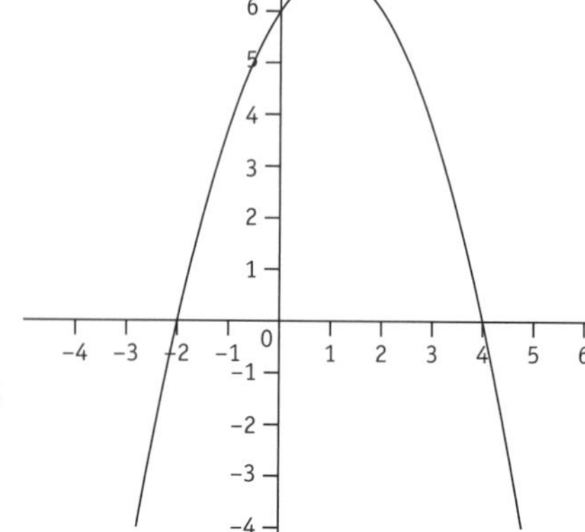

c How long was the stone in the air? **1 mark**

---

d What was the maximum height achieved by the stone? **1 mark**

---

12 The population of a city is modelled using the equation $P = 24\,000 \times 1.08^t$ where P is the population and t is the time in years.

a Complete the table of values for $P = 24\ 000 \times 1.08^t$. **2 marks**

t	0	1	2	3	4	5
P	24 000		27 994		32 652	35 264

b Use the table above to graph the equation on the number plane. **2 marks**

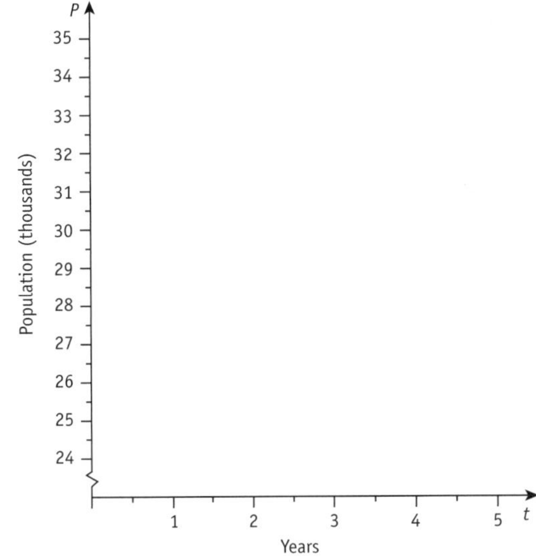

13 George leaves his home at 7:00 am to drive to Bathurst.

At the beginning, he drives 160 km which takes him 2 hours.

He then stops and visits his mother for one hour.

George then continues his journey towards Bathurst.

He drives for an hour at an average speed of 90 km/h and arrives in Bathurst.

George has lunch with some friends and has a dentist appointment.

At 2 pm he leaves Bathurst and then takes 3 hours to drive back on the same road to his home.

a Represent George's trip to and from Bathurst by drawing a graph. **2 marks**

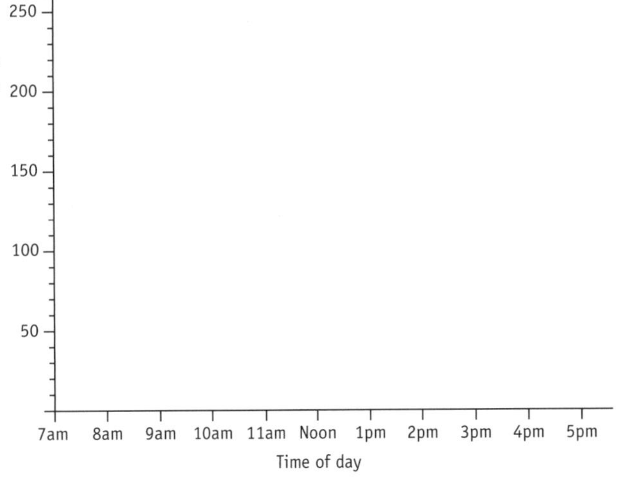

b What was the total distance travelled by George? **1 mark**

---

c What was the average speed travelled on the journey from Bathurst to home? **2 marks**

CHAPTER 5
Measurement: Right-angled triangles

Excel MATHEMATICS STANDARD 1
Ch. 2, pp. 38–39

Using Pythagoras' theorem

QUESTION 1 Find the value of the pronumeral, correct to two decimal places:

a

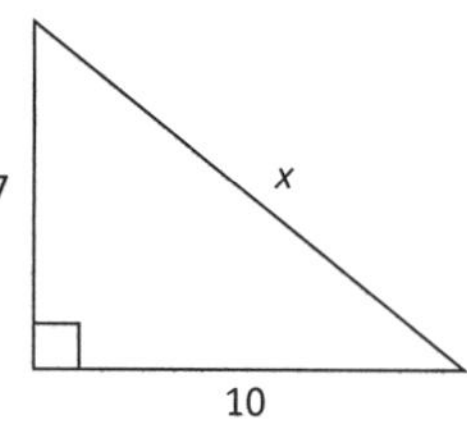

b

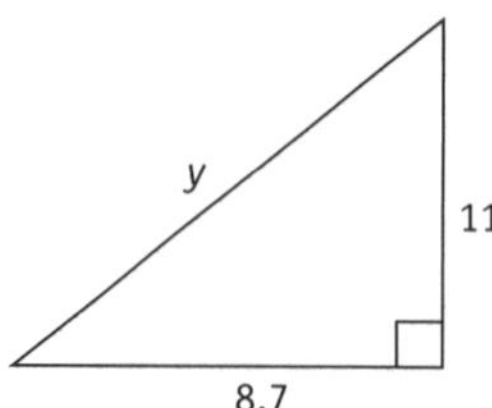

c

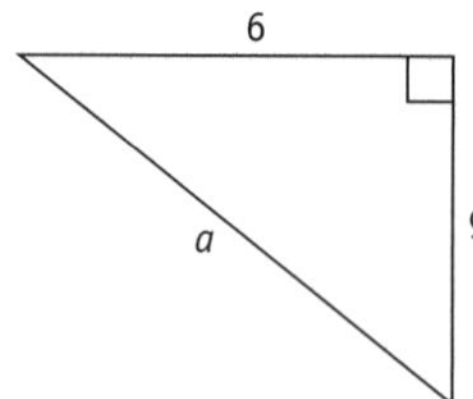

QUESTION 2 Find the value of the pronumeral, correct to two decimal places:

a 2

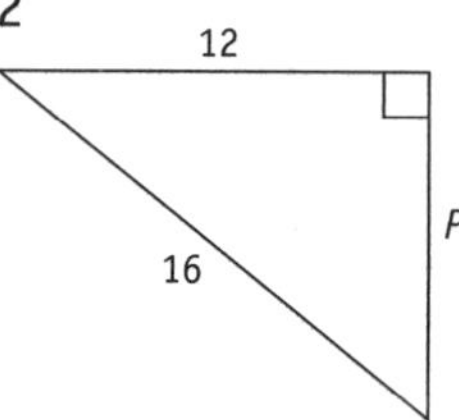

b

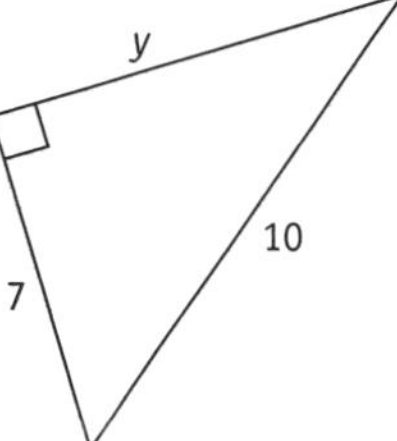

c

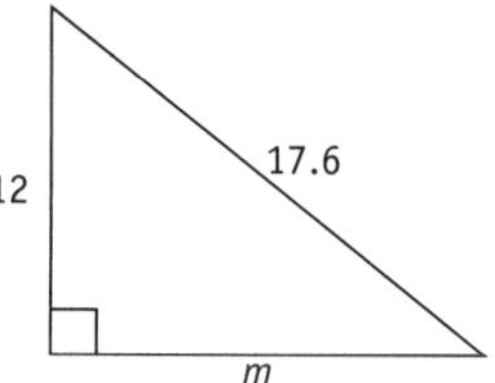

QUESTION 3 Find the width of the rectangle with a length of 40 cm and a diagonal of 56 cm. Leave your answer correct to one decimal place.

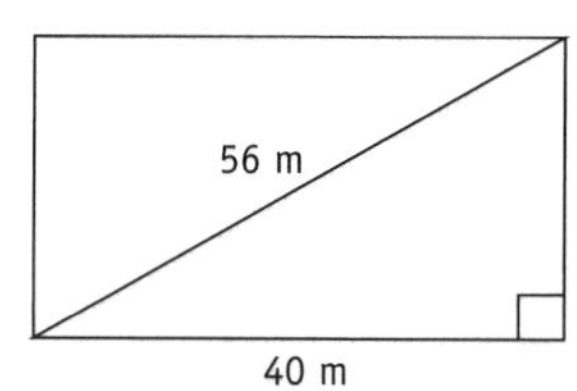

QUESTION 4 Three towers *A*, *B* and *C*, are positioned where *A* is 20 km north of *B* and *C* is 31 km east of *B*. Find the distance from *A* to *C*, to the nearest kilometre.

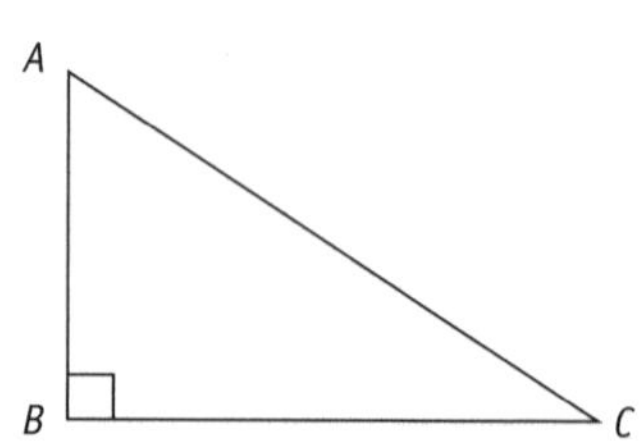

QUESTION 5 Find the area of the triangle *PQR*.

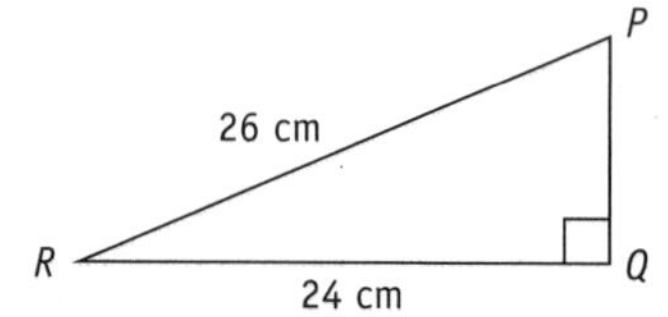

Measurement: Right-angled triangles

The trigonometric ratios

Question 1 For each of the triangles, complete the trig ratios, expressing as a fraction:

a

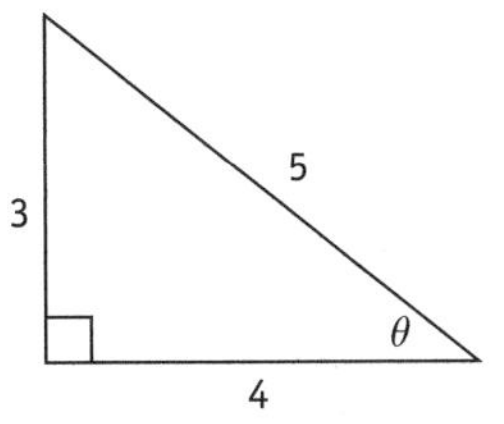

$\sin\theta$ = ______________

$\cos\theta$ = ______________

$\tan\theta$ = ______________

b

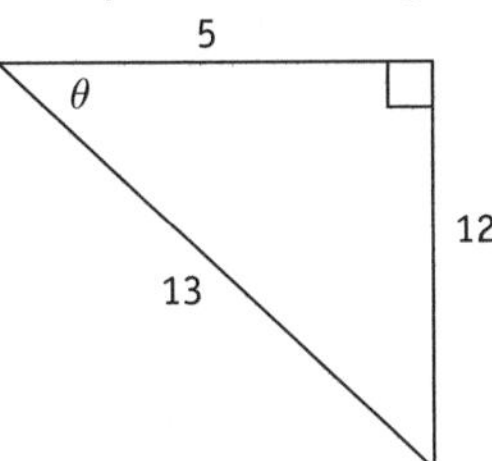

$\sin\theta$ = ______________

$\cos\theta$ = ______________

$\tan\theta$ = ______________

c

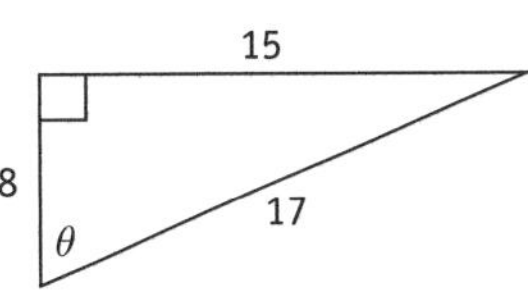

$\sin\theta$ = ______________

$\cos\theta$ = ______________

$\tan\theta$ = ______________

Question 2 Calculate, correct to four decimal places:

a $\cos 45° =$ ______________ **b** $\tan 19° =$ ______________ **c** $\sin 68° =$ ______________

Question 3 Calculate, correct to four decimal places:

a $\tan 36°10' =$ ______________ **b** $\sin 68°56' =$ ______________ **c** $\cos 25°8' =$ ______________

Question 4 Calculate, correct to two decimal places:

a $12 \times \sin 58° =$ ______________ **b** $38 \times \tan 27° =$ ______________

c $53 \times \cos 71° =$ ______________ **d** $17 \times \cos 71°24' =$ ______________

e $59 \times \tan 39°48' =$ ______________ **f** $35.6 \times \sin 83°46' =$ ______________

Question 5 Calculate, correct to two decimal places:

a $\frac{15}{\tan 62°} =$ ______________ **b** $\frac{38}{\sin 73°} =$ ______________

c $\frac{31.7}{\cos 17°} =$ ______________ **d** $\frac{11}{\sin 39°15'} =$ ______________

e $\frac{18}{\cos 11°47'} =$ ______________ **f** $\frac{29.8}{\tan 37°8'} =$ ______________

Question 6 Find the value of acute angle θ, to the nearest degree:

a $\sin\theta = 0.4562$ **b** $\tan\theta = 1.3589$ **c** $\cos\theta = 0.8589$

______________ ______________ ______________

Question 7 Find the value of acute angle θ, to the nearest minute:

a $\cos\theta = 0.6708$ **b** $\sin\theta = 0.3461$ **c** $\tan\theta = 2.0358$

______________ ______________ ______________

Using trigonometry to find a side

QUESTION **1** Find the length of the unknown side, correct to two decimal places.

a

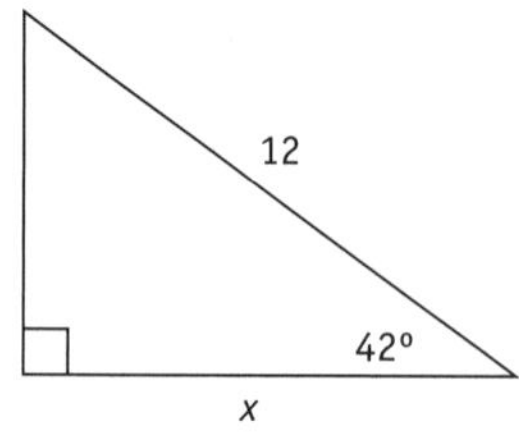

b

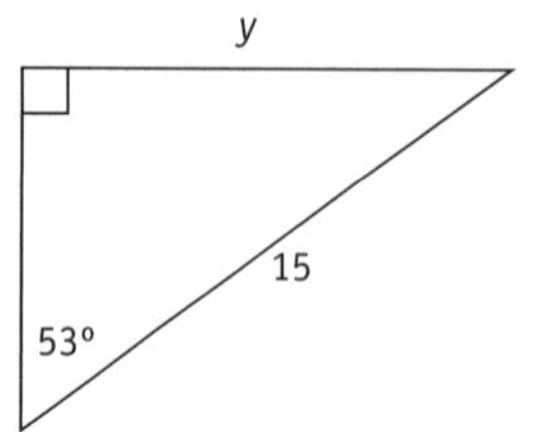

c

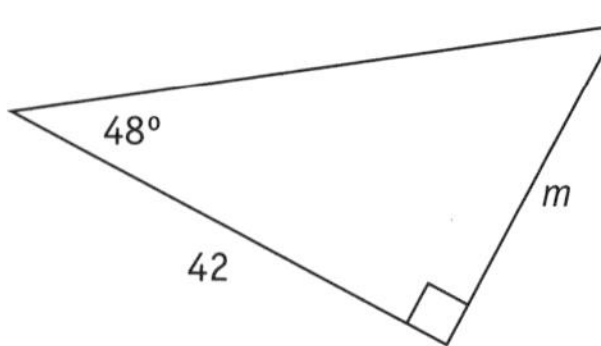

d

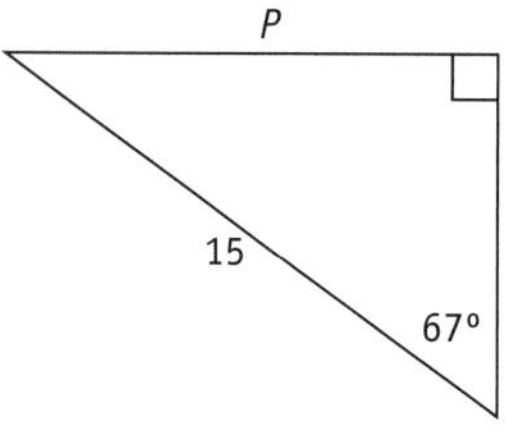

e

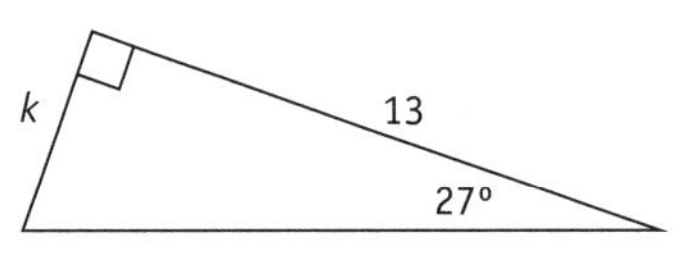

f

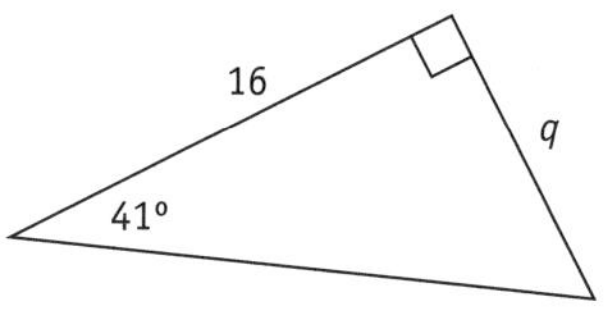

g

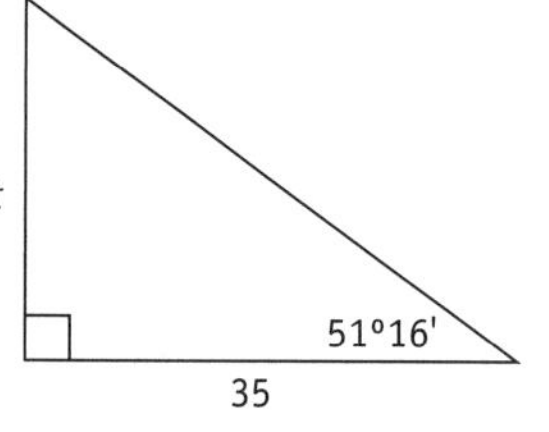

h

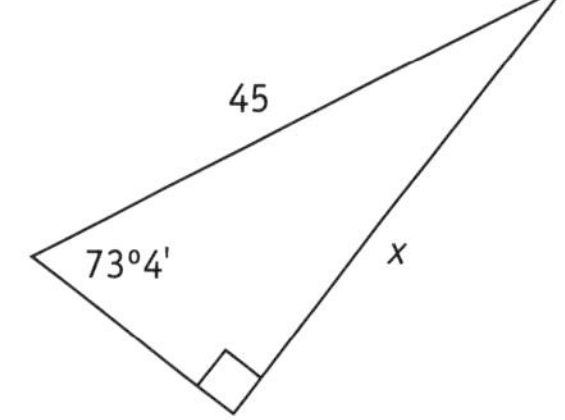

i

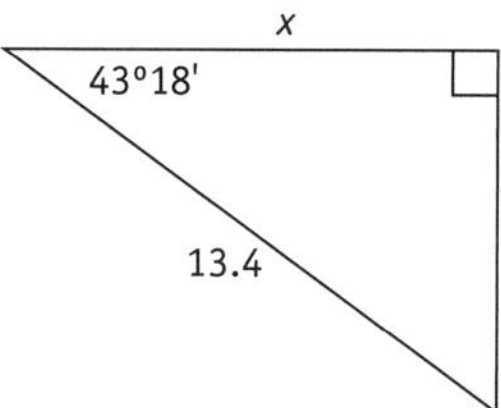

QUESTION **2** Find the length of the unknown side, correct to two decimal places.

a

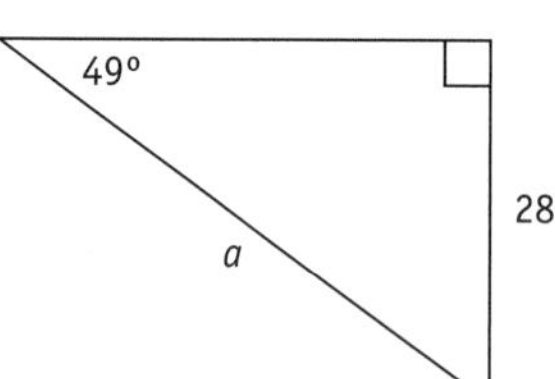

b

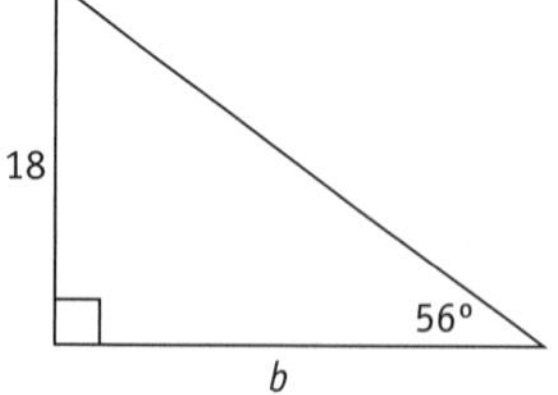

c

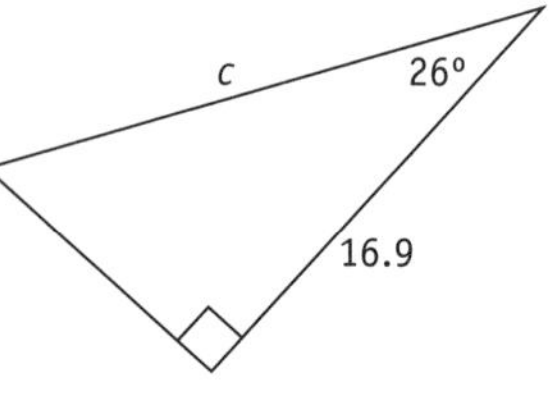

Using trigonometry to find an angle

Question 1 Find the size of the angle θ, to the nearest degree.

a

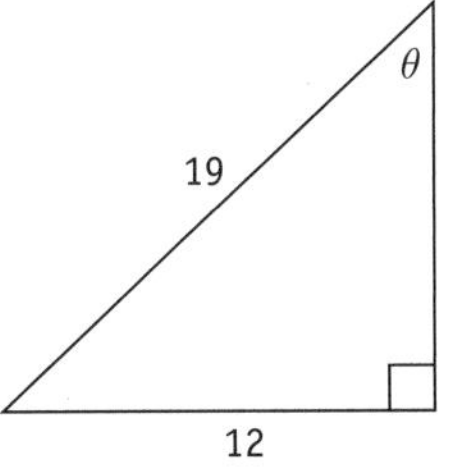

b

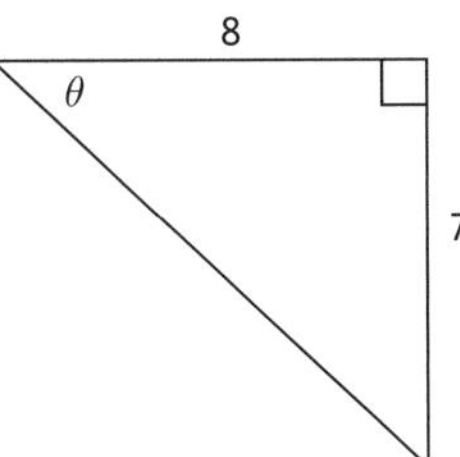

c

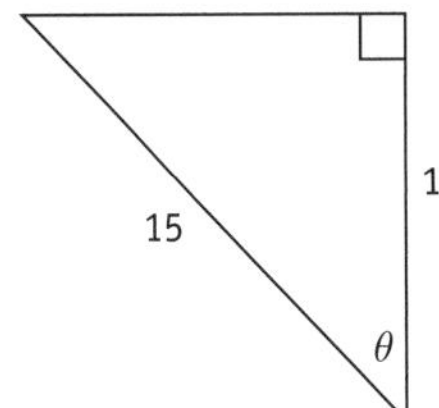

d

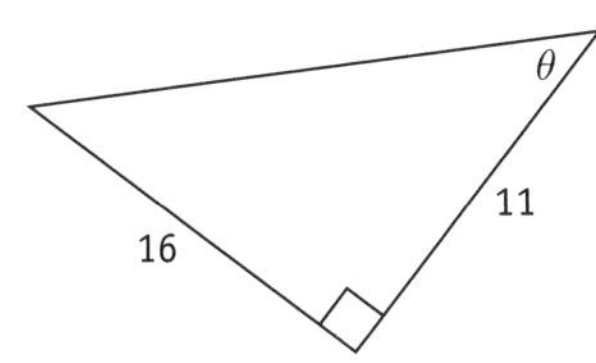

e

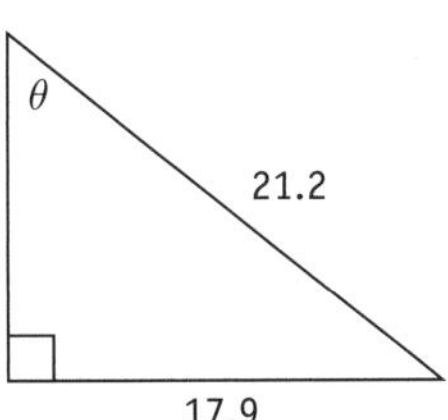

f

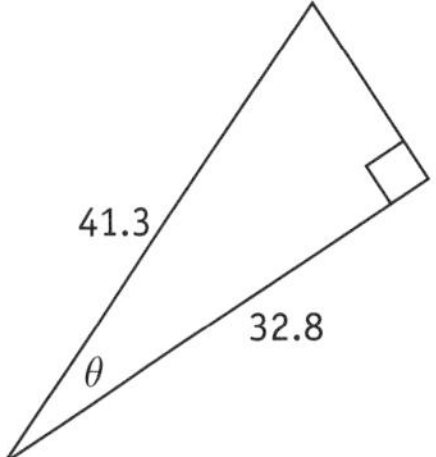

Question 2 Find the size of the angle θ, to the nearest minute.

a

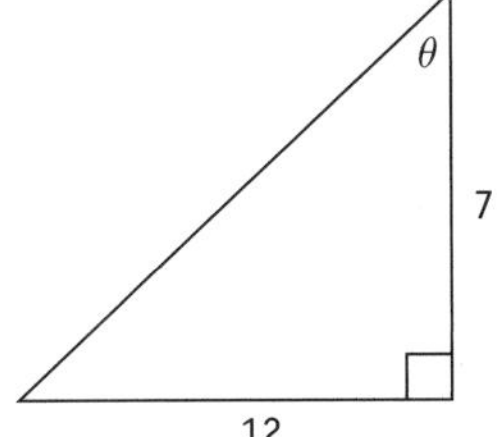

b

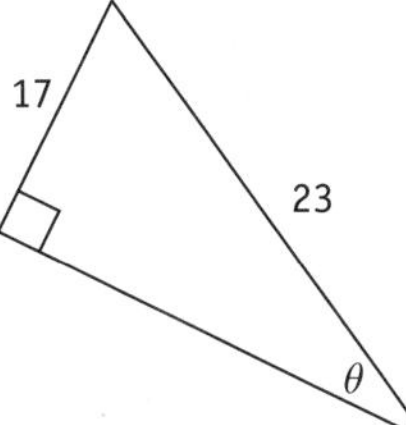

c

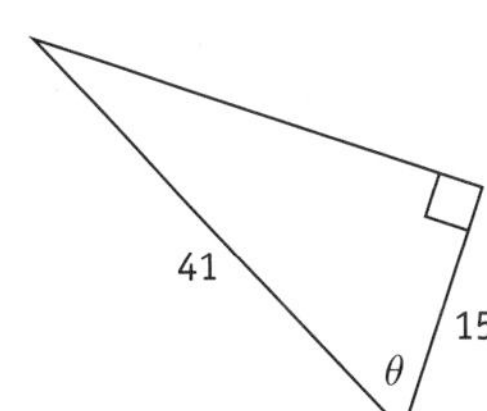

Measurement: Right-angled triangles

Angles of elevation and depression

QUESTION 1 Write the given angles on the two diagrams below.

a The angle of elevation of the top of a tree from P is 48°.

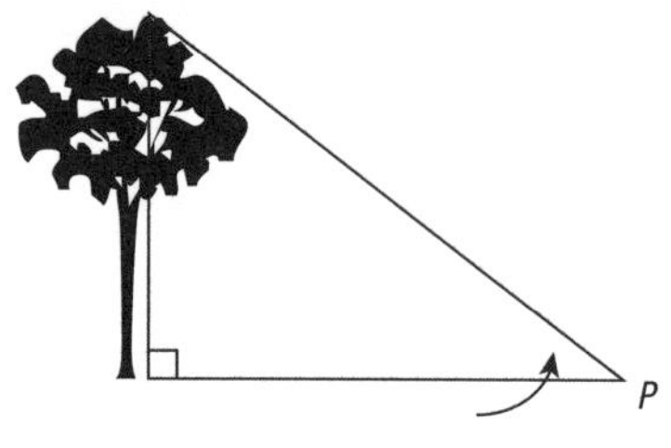

b The angle of depression from the top of a wall to P is 35°.

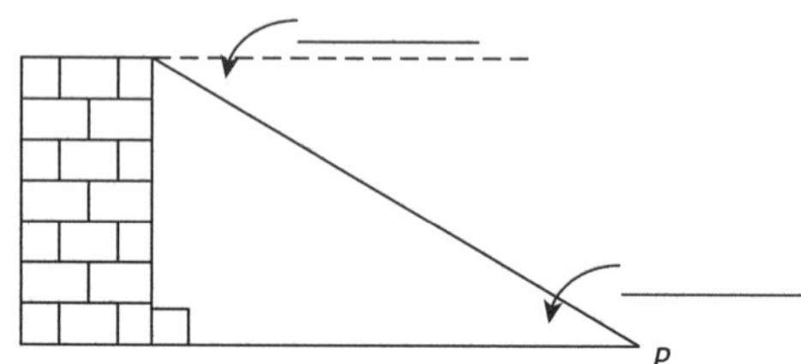

QUESTION 2

a The angle of elevation from A to the top of a tower is 24°. If A is 40 m from the base of the tower, find the height of the tower, to the nearest metre.

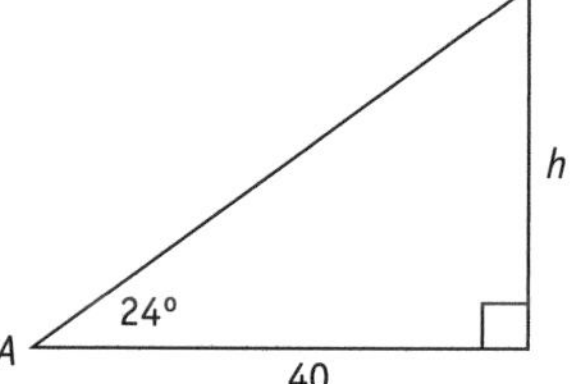

____

____

____

b A tree is 8 m high. The angle of elevation of the top of the tree from B is 54°. How far is B from the tree, correct to two decimal places?

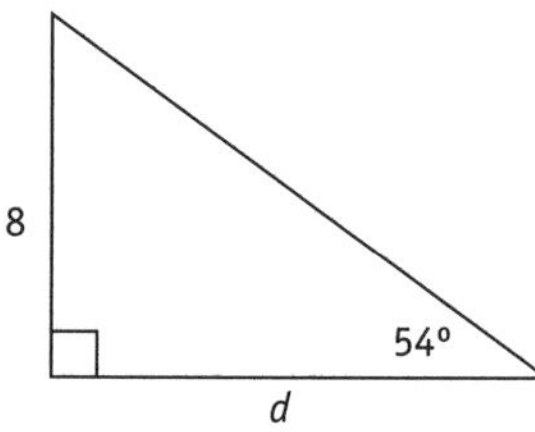

____

____

____

QUESTION 3

a Chris is standing on the top of a 20-m cliff and looks down to a swimmer. The angle of depression from Chris to the swimmer is 54°. How far is the swimmer from the cliff, correct to two decimal places?

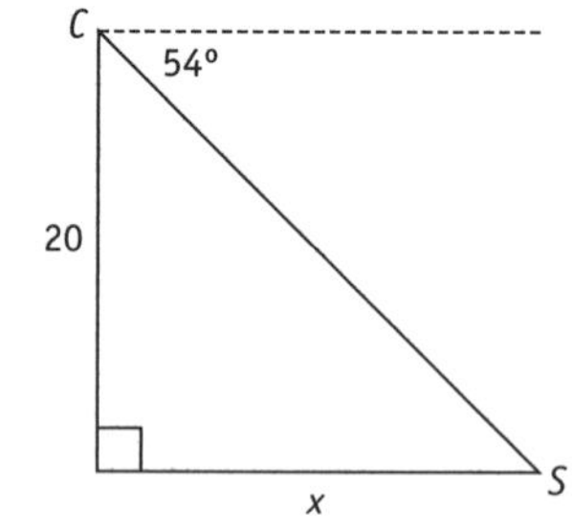

____

____

____

b Dave is cutting a branch from a tall tree and sees a child walking nearby. The angle of depression from Dave to the child is 12°. If Dave is working at a height of 4 m, how far is the child from the tree, to the nearest metre?

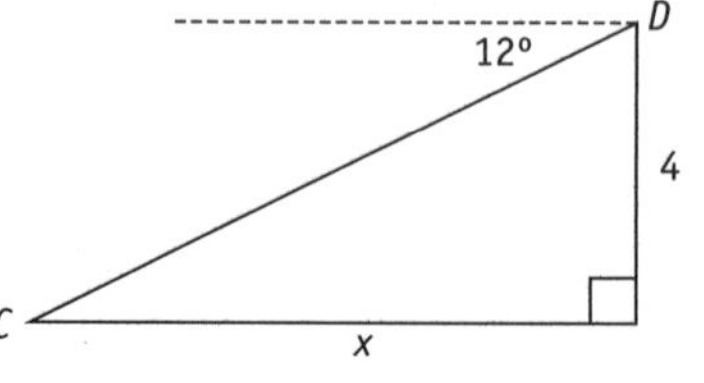

____

____

Measurement: Right-angled triangles

Solving problems with right-angled triangles 1

QUESTION 1

a Find the length of DC.

b Find the length of CB.

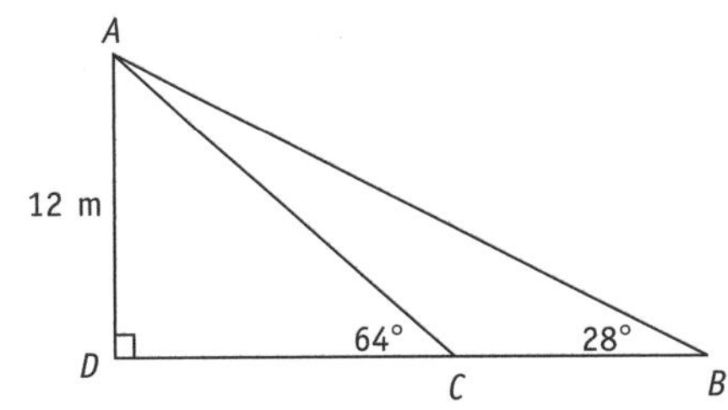

QUESTION 2

a The angle of elevation of the top of a tower AB is 65° from a point C on the ground at a distance of 30 m from the base of the tower. Calculate the height of the tower to the nearest metre.

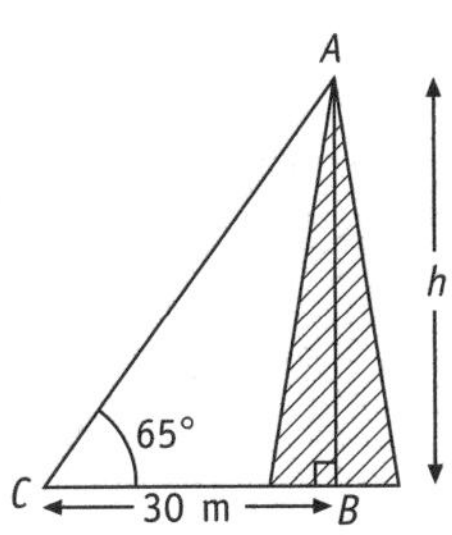

b A man 1.65 m tall is 28 m away from a tower 38 m high. What is the angle of elevation of the top of the tower from his eyes, to the nearest degree?

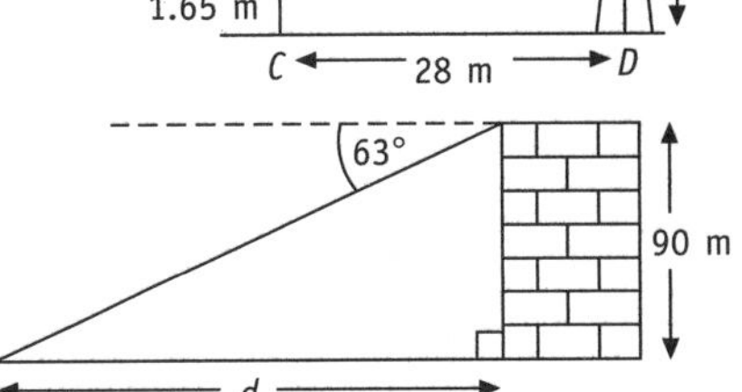

c From the top of a building 90 m high, the angle of depression of a car parked on the ground is 63°. Find the distance of the car from the base of the building. (Answer correct to two decimal places.)

QUESTION 3 Give each answer to the nearest centimetre.

a

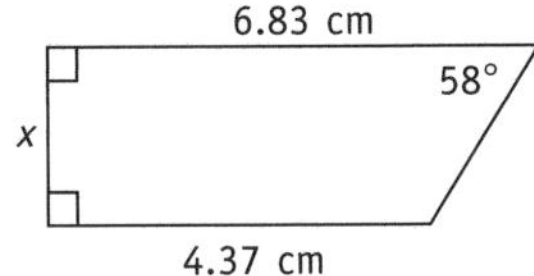

Find x.

b

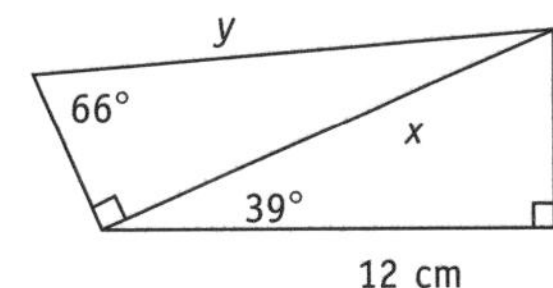

i Find x.

ii Find y.

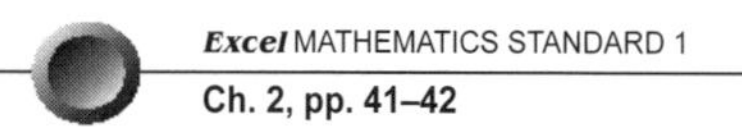

Solving problems with right-angled triangles 2

Question 1 An 18-m ladder standing on level ground reaches 14 m up a vertical wall. The angle that the ladder makes with the ground is θ.

a Draw a diagram to represent this information.

b Find θ to the nearest degree.

Question 2 *ABCD* is a rectangle with sides *AD* = 5 cm and *AB* = 12 cm.

a Draw a diagram to represent this information.

b Find the length of *AC*, the diagonal, using Pythagoras' theorem.

c Find angle *ACD* correct to the nearest minute.

Question 3 A pole 15 m tall casts a shadow 20 m long. The angle of elevation of the top of the pole from the tip of the shadow is θ.

a Draw a diagram to represent this information.

b Find θ, the angle of elevation (to the nearest degree).

c Find the distance from the tip of the shadow to the top of the pole.

Bearings 1

Question 1 For each diagram, write down the **true bearing** of *Q* from *P*.

a

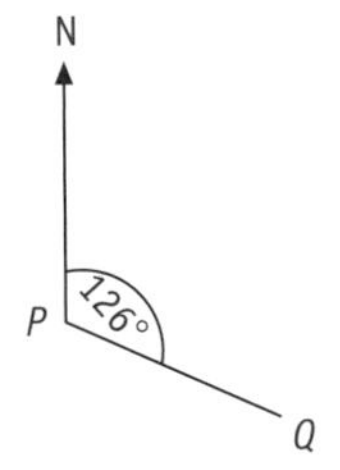

b

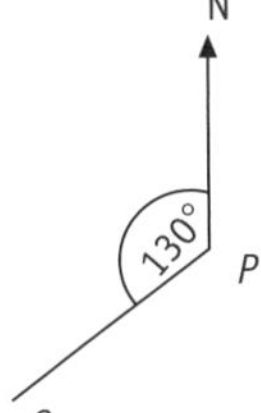

c

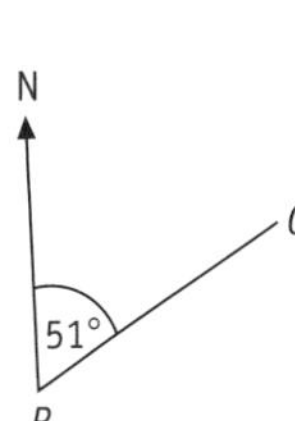

d

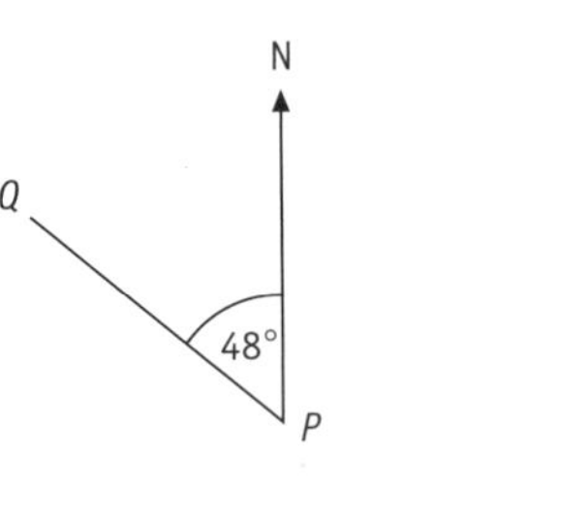

e

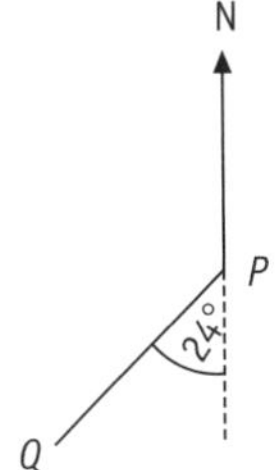

f

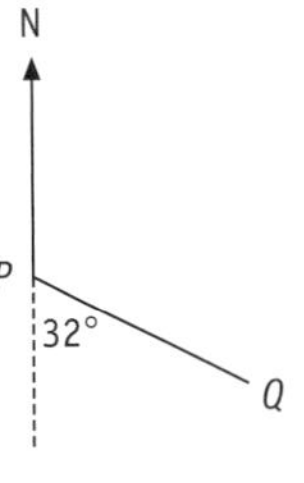

Question 2 For each diagram, write down the **compass bearing** of *B* from *A*.

a

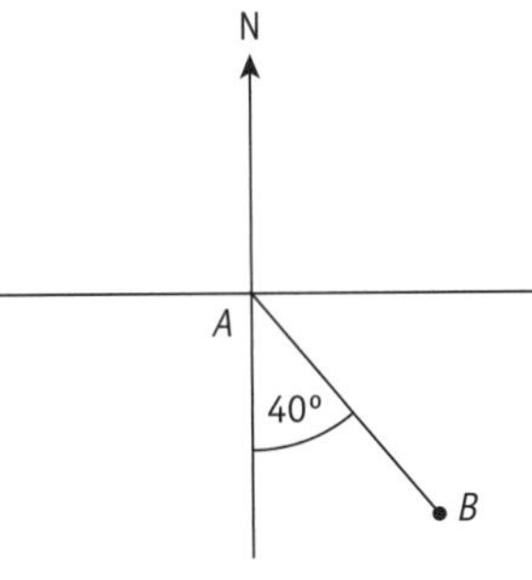

b

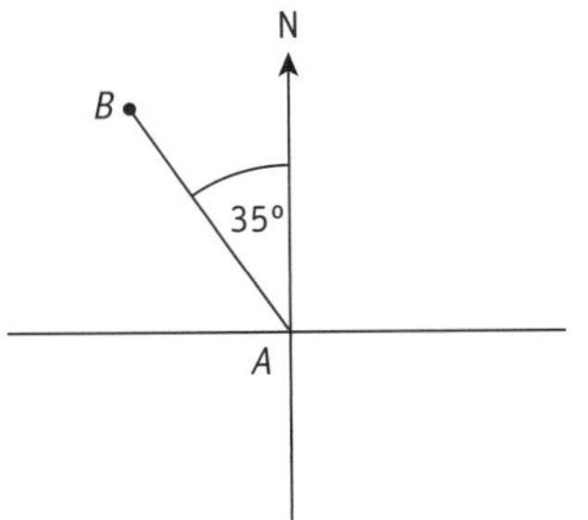

c

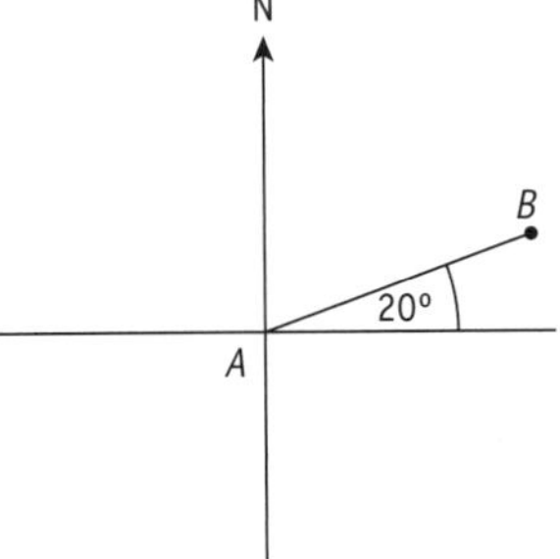

d

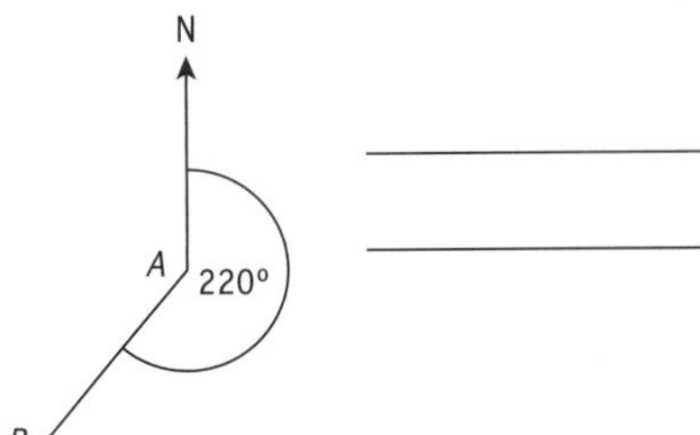

e

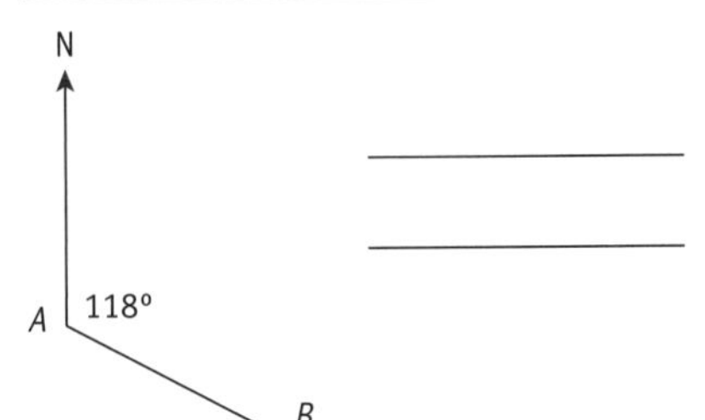

f

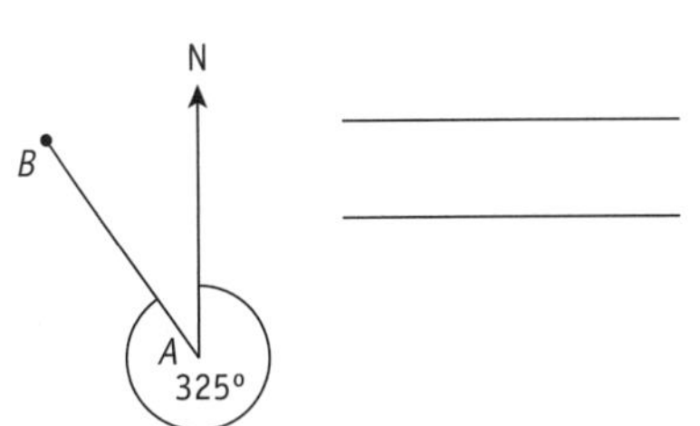

Question 3 Express each of these compass bearings as true bearings:

a S 80° E

b N 60° W

c S 75° W

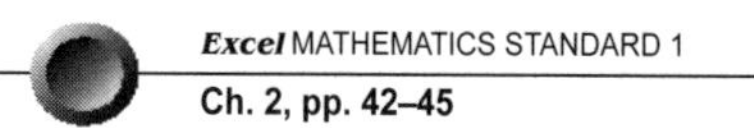

Bearings 2

QUESTION 1 A ship sets out from a point A and sails due north to a point B, a distance of 150 km. It then sails due east to a point C. If the bearing of C from A is 048°, find:

a the distance BC

b the distance AC

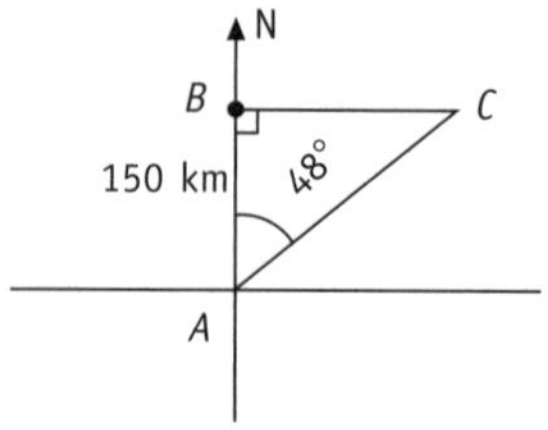

QUESTION 2 A ship leaves port for a destination 80 km east and 70 km north. On what compass bearing should it sail, to the nearest degree?

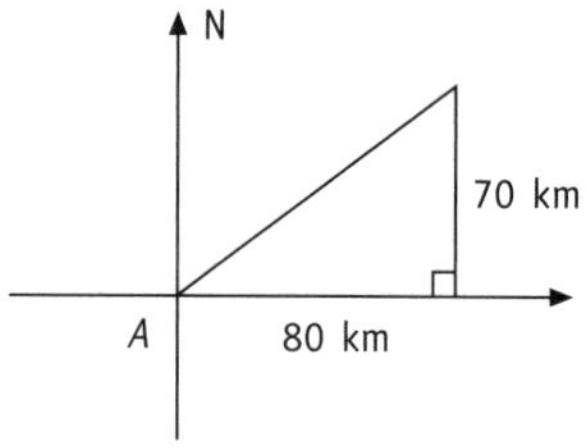

QUESTION 3 A ship starts from a port P, sails 226°T for a distance of 120 km. Find:

a how far south of P it is

b how far west of P it is

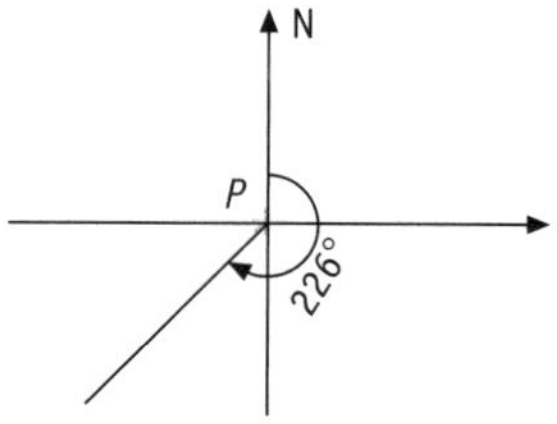

QUESTION 4 A ship sailed 12 km north and then 20 km west. Find its bearing (to the nearest degree) from the starting point.

QUESTION 5 A man walked due south and then turned and walked due east. He was then 3 km and S 50° E from his starting point. How far (to the nearest metre) was he south of his initial position?

QUESTION 6 A lighthouse is 10 km north-east of a ship. How far is the ship west of the lighthouse (correct to two decimal places)?

Measurement: Right-angled triangles

TOPIC TEST

SECTION I

Instructions
- This section consists of 5 multiple-choice questions.
- Each question is worth 1 mark.
- Fill in only ONE CIRCLE for each question.

Time allowed: 7 minutes **Total marks: 5**

1 In which triangle is $\sin\theta = \frac{5}{8}$?

Ⓐ
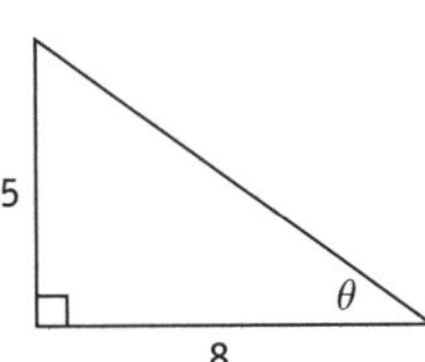

Ⓑ
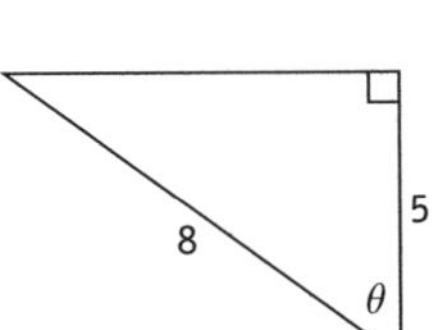

Ⓒ
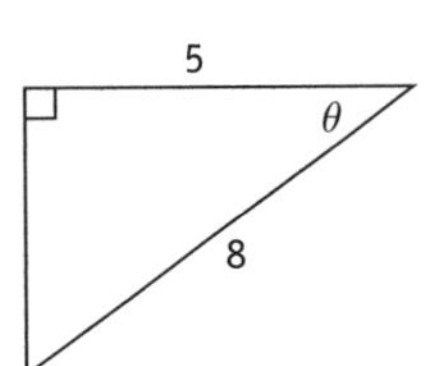

Ⓓ
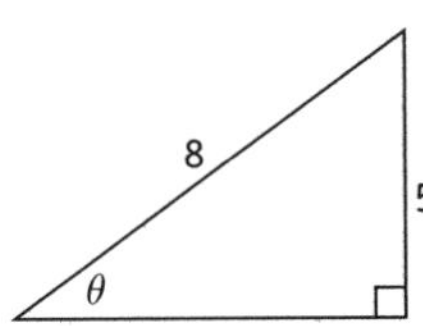

2 Which of the following expresses S 30° E as a true bearing?

Ⓐ 120° Ⓑ 150° Ⓒ 210° Ⓓ 330°

3 Sam leaves *A* and walks on a bearing of 070° to *B*.
He then turns and walks on a bearing of 150° to *C*.
What is the size of the obtuse angle *ABC*?

Ⓐ 95° Ⓑ 100°

Ⓒ 120° Ⓓ 130°

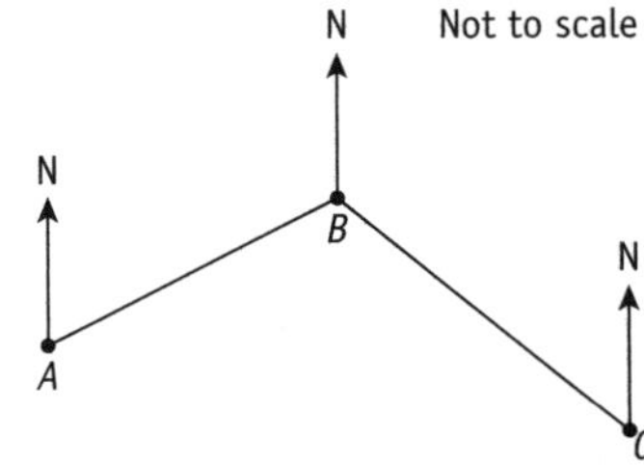

4 What is the area of triangle *PQR*?

Ⓐ 200 cm^2 Ⓑ 300 cm^2

Ⓒ 400 cm^2 Ⓓ 800 cm^2

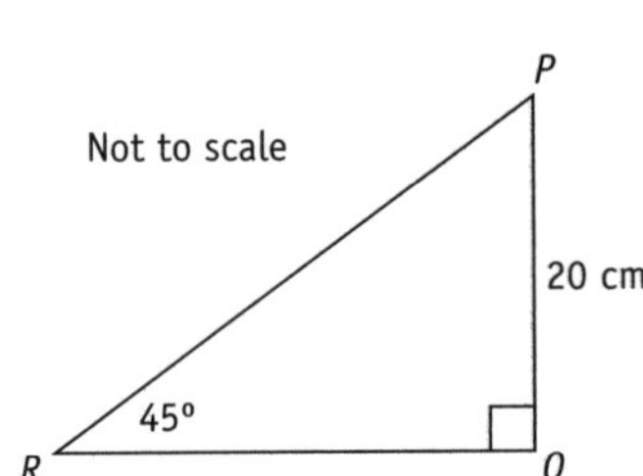

5 The bearing of *Y* from *X* is 130°.
Which of these is the bearing of *X* from *Y*?

Ⓐ 330° Ⓑ 300°

Ⓒ 310° Ⓓ 340°

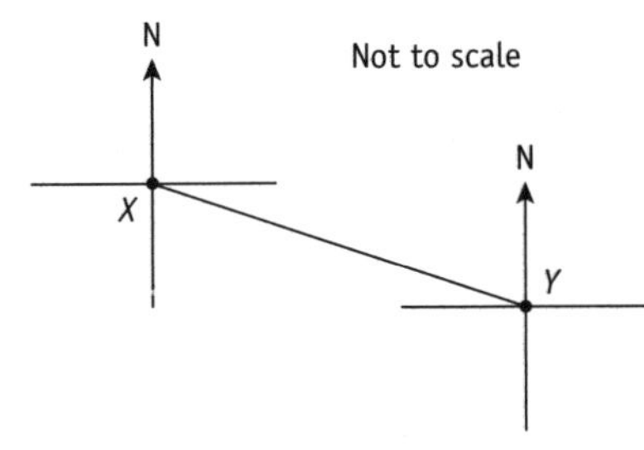

TOPIC TEST

SECTION II

Instructions
- This section consists of 11 questions.
- Show all working.

Time allowed: 53 minutes **Total marks: 35**

6 Calculate the value of x, correct to two decimal places. **2 marks**

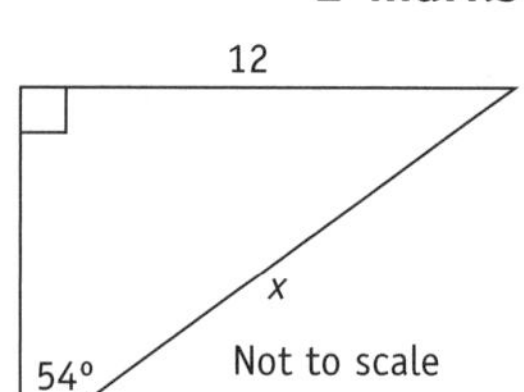

7 Each school day Mae walks 1.4 km on a bearing of 235° to Ella's home.
Both girls then walk in an easterly direction to school.
How far does Mae walk each morning from her home to school?

Give your answer to the nearest metre. **3 marks**

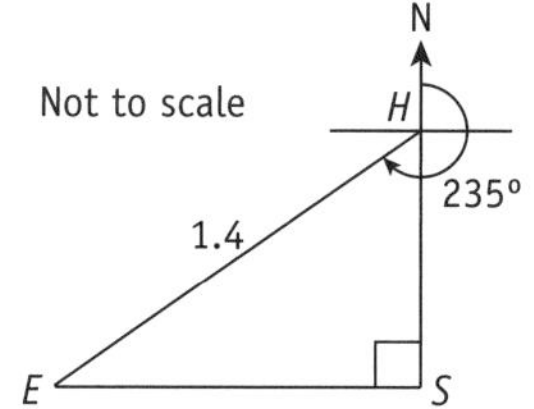

8 **a** Find the perimeter of the triangle ABC, correct to two decimal places. **3 marks**

b Find the size of $\angle ACD$, to the nearest minute. **2 marks**

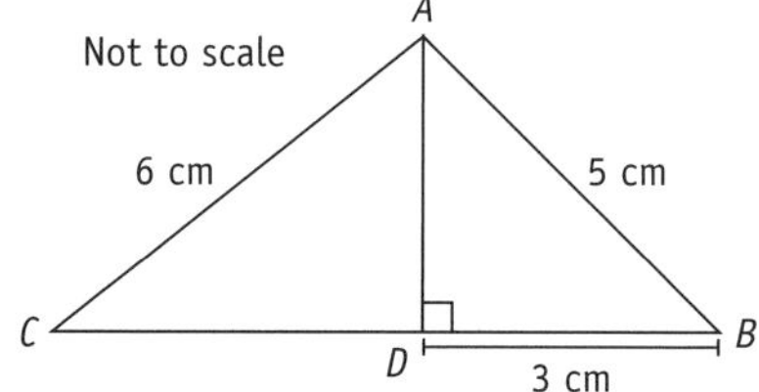

9 From the top of a cliff 36 m above sea level, the angle of depression of a small boat is 48°. How far is the boat from the base of the cliff, to the nearest metre? **2 marks**

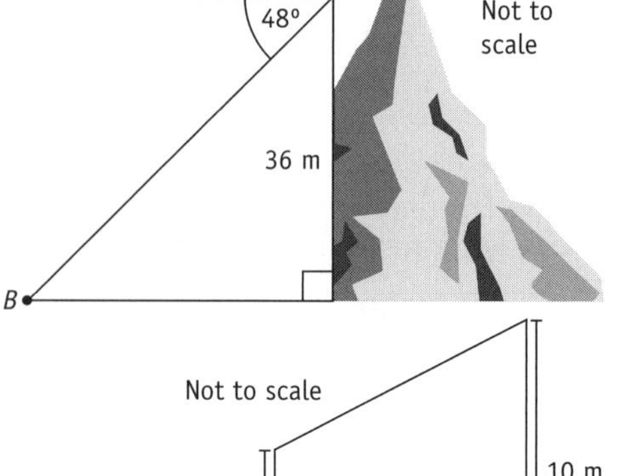

10 Two poles on level ground, 8 m apart, are joined by a cable.
The height of the two poles are 6 m and 10 m.

Not to scale

6 m

10 m

8 m

What is the length of the cable between the two poles, correct to two decimal places? **3 marks**

---

11 The angle of elevation from A to the top of a building is 42°. The distance from A to the base of the building is 16 m.

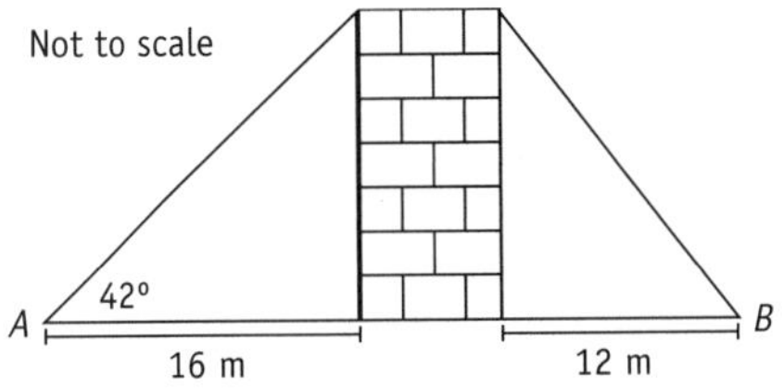

Not to scale

A 42° 16 m 12 m B

a Show that the height of the building is 14.41 m (correct to two decimal places). **2 marks**

---

b If B is 12 m from the base of the building, find the angle of elevation from B to the top of the building. Give your answer to the nearest degree. **2 marks**

---

12 Farzan is in the local swimming pool.

From where Farzan is swimming the angle of elevation to the top of a diving tower is 38°.

If the height of the tower is 10 m, how far is Farzan from the base of the tower, correct to two decimal places? **2 marks**

T Not to scale 10 38° F x

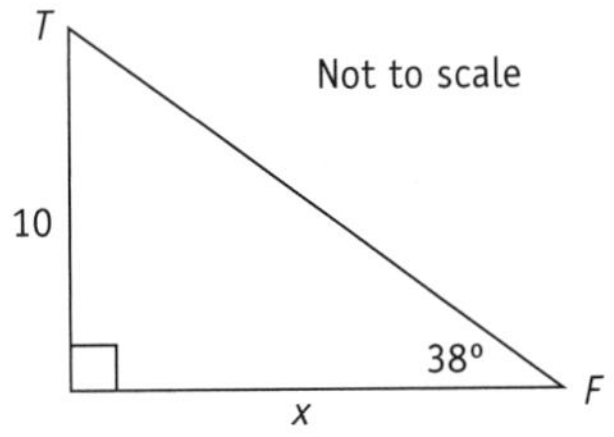

13 What is the size of the smallest angle in the triangle?

Leave your answer to the nearest degree. **2 marks**

Not to scale 10 cm 24 cm 26 cm

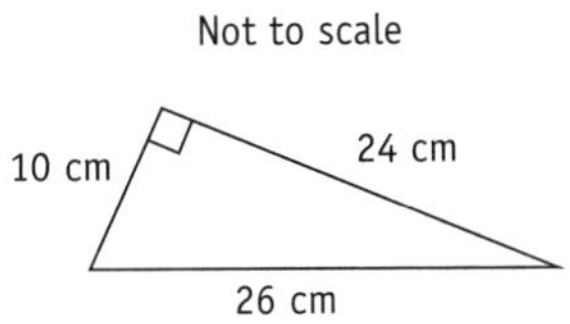

14 Symon leaves P and flies 32 km south to Q before turning and flying east a distance of 28 km to R.

What is the bearing of R from P? **2 marks**

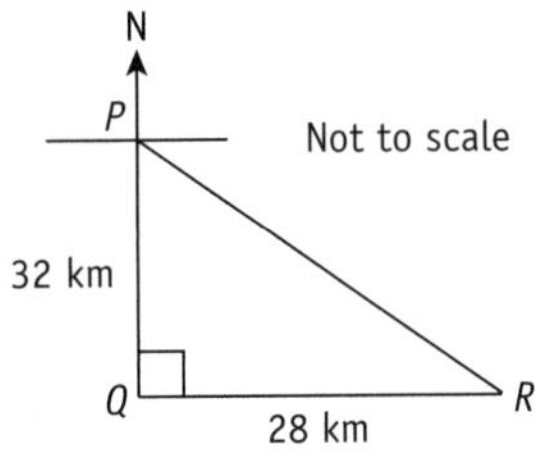

N P Not to scale 32 km Q R 28 km

---

15 The diagram shows the three towns P, Q and R.

Town R is due west of Town Q.

The bearing of Town Q from Town P is S 35° E and the bearing of Town R from Town P is S 55° W.

The distance between Town R and Town Q is 400 km.

A plane flies between the three towns.

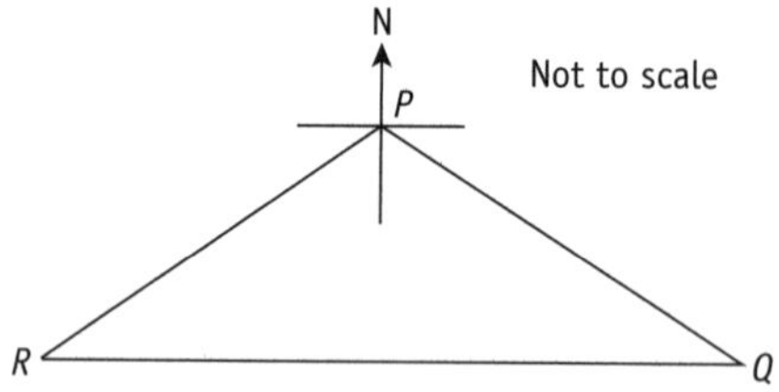

N

P

Not to scale

R

Q

a Mark the given information on the diagram and explain why $\angle RPQ$ is 90°. **2 marks**

---

b Find the distance between Town P and Town Q to the nearest kilometre. **2 marks**

---

c If the speed of the plane is 480 km/h, how long does the flight take from Town P to Town Q, in minutes? **2 marks**

---

16 Ainsley is in an office building.

The distance from Ainsley to the ground is 12 m.

She looks out of the window at her two friends Belinda and Chloe who are standing 4 m apart.

The angle of depression from Ainsley to Belinda is 24°.

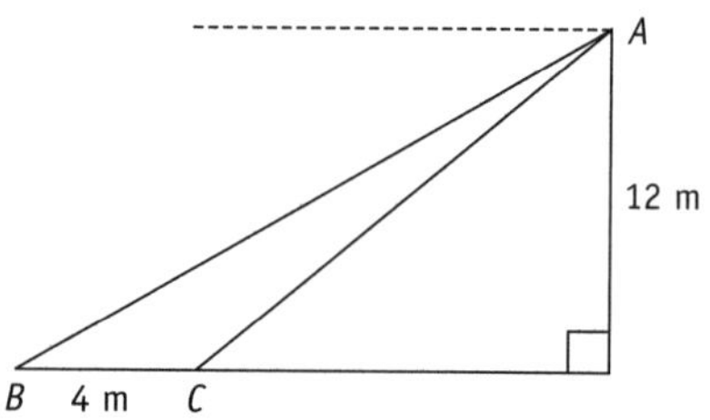

A

12 m

B 4 m C

a Show that Belinda is 26.95 m from the base of the building. **2 marks**

---

b What is the angle of depression from Ainsley to Chloe, to the nearest degree? **2 marks**

CHAPTER 6
Measurement: Rates

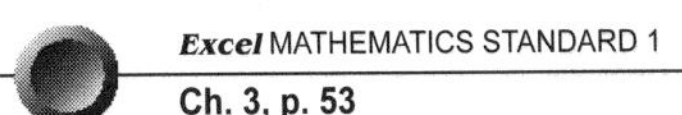
Excel MATHEMATICS STANDARD 1
Ch. 3, p. 53

Converting between units of rates

QUESTION 1 Complete the following speed conversions, correct to two decimal places if necessary:

a 72 km/h = ________ m/h = ________ m/minute = ________ m/s

b 60 km/h = ________ m/h = ________ m/minute = ________ m/s

c 110 km/h = ________ m/h = ________ m/minute = ________ m/s

QUESTION 2 Complete the following speed conversions, correct to two decimal places if necessary:

a 10 m/s = ________ m/min = ________ m/h = ________ km/h

b 8 m/s = ________ m/min = ________ m/h = ________ km/h

c 25 m/s = ________ m/min = ________ m/h = ________ km/h

QUESTION 3

a Ben walks at a speed of 6 km/h. Express this speed in metres per second. ________

b A greyhound has a top speed of 68 km/h. What is this speed in metres per second?

____

c A cyclist in the Tour de France averaged 46 km/h for one stage. Find this speed in metres per second.

____

d A honey-bee flies at a speed of 6 m/s. What is this speed in kilometres per hour?

____

e A snail averages 0.6 cm/second. Find this speed in metres per hour. ________

f Zoe walks at a speed of 100 m/minute. Express this speed in kilometres per hour.

____

QUESTION 4 Complete the following conversions, correct to two decimal places if necessary:

a 6 mL/minute = ________ mL/h = ________ L/h

b 15 mL/minute = ________ mL/h = ________ L/h

c 600 mL/minute = ________ mL/h = ________ L/h

QUESTION 5 Complete the following conversions, correct to two decimal places if necessary:

a 0.45 L/h = ________ mL/h = ________ mL/min

b 0.8 L/h = ________ mL/h = ________ mL/min

c 23 L/h = ________ mL/h = ________ mL/min

QUESTION 6 A tap drips 0.25 mL every 5 seconds. How much water is wasted:

a every hour? ________

b every day? ________

c every year? ________

QUESTION 7 Complete the following conversions, correct to two decimal places if necessary:

a 12 L/minute = ________ L/h = ________ kL/h

b 26 L/minute = ________ L/h = ________ kL/h

c 165 L/minute = ________ L/h = ________ kL/h

QUESTION 8 Complete the following conversions, correct to two decimal places if necessary:

a 1.8 kL/h = ________ L/h = ________ L/min

b 3.2 kL/h = ________ L/h = ________ L/min

c 0.08 kL/h = ________ L/h = ________ L/min

Using unit pricing

QUESTION 1 Find the cost per 100 g for the following products, correct to four decimal places if necessary:

a 440-g packet of spaghetti at $2.35

b 175-g packet of chocolate biscuits at $3.45

c 640-g box of cereal at $5.40

d 420-g can of crushed tomatoes at $1.60

QUESTION 2 Find the cost per 100 mL for the following products, correct to four decimal places:

a 600-mL container of thickened cream at $3.40

b 510-mL bottle of barbecue sauce at $2.80

c 390-mL bottle of shampoo at $6.80

d 1.25 L of soft drink at $2.18

QUESTION 3 Macca surveyed the price of vegemite available in different sized jars in different stores and recorded the results in the table below. Complete the table by calculating each price/100 g, correct to two decimal places.

Store	Size	Price	Price/100 g
A	560 g	$7.90	
B	453 g	$6.45	
C	380 g	$5.80	
D	280 g	$4.90	
E	220 g	$3.75	

QUESTION 4 Arrange these packets of toilet rolls from the most to least expensive: 6 rolls for $4.05, 12 rolls for $8.40, 8 rolls for $5.80 and 24 rolls for $14.60.

QUESTION 5 Arrange these nuts from the least to most expensive: brazil nuts at $5.49/100 g, cashews at $7.49/250 g, pecans at $2.35/50 g, hazelnuts at $9.29/300 g.

QUESTION 6 Kiwifruit is sold in different ways: 4 for $2.85, 6 for $4.40, 12 for $8.65 and 3 for $2.15. Which product gives the cheapest way to buy one dozen kiwifruit?

Hourly charges and call-out fees

QUESTION 1 Steve's Appliances charge a call-out fee of $154 which includes the first 30 minutes of labour. If the repair is extended, the company charges an additional $25 per 15 minutes or part thereof. What is the cost of a repair that takes:

a 20 minutes?

b 1 hour?

c 1 hour 10 minutes?

QUESTION 2 Kel's Computers offers an in-home repair service. The company charges a call-out fee of $65 plus $12 per 10-minute block or part thereof.
A repairer arrived at Josh's home at 10:20 am and charged $125 for their services.
What is the latest the repairer had completed their work?

QUESTION 3 A taxi company has the following charges: hire charge $3.60, distance rate $2.19/km, waiting time $0.944/minute.
What is the cost of the following journeys?

a a distance of 12.6 km, with no waiting time

b a distance of 18.8 km, with 5 minutes waiting time

QUESTION 4 An emergency plumber charges a call-out fee of $80 plus an hourly rate. For a job on Tuesday the plumber charged $365 for three hours work.

a What is the hourly rate?

b How much will be charged if a job takes five hours?

c How long does the plumber work if the charge is $792.50?

QUESTION 5 A ride-share company has a fee that includes a fixed cost and a cost that depends on both the time spent travelling, in minutes, and the distance travelled, in kilometres.
The fixed cost is $2.75, the distance cost is $1.20/km and the time cost is $0.35/minute.
Find the cost of travelling:

a 8 km in 12 minutes

b 30 km at an average speed of 50 km/h

QUESTION 6 Many people are eligible for free ambulance transport in emergencies. For those not eligible the cost in a certain state is $392 call-out plus $3.54/km. If the total cost of one journey was $512.36, what was the distance travelled?

Measurement: Rates

Speed

QUESTION 1 Find the average speed of a:

a truck that travels 230 km in 2 hours 30 minutes

b car that travels 360 km in 3 hours 20 minutes

c plane that travels 1330 km in 1 hour 45 minutes

QUESTION 2 What is the distance travelled by a:

a cyclist riding at 38 km/h for 30 minutes?

b bus travelling at 87 km/h for 1 hour 20 minutes?

c motorbike after two and a half hours averaging 85 km/h?

QUESTION 3 How long will it take a:

a motorist to drive 250 km if they average 75 km/h?

b jogger to run 3 km if they are running at 6 km/h?

c plane travelling at 680 km/h to fly 7310 km?

QUESTION 4 A certain tortoise can walk at 60 m/h and a certain hare can run at 36 km/h.

a How far, in metres, can both animals travel in a minute?

b The hare and the tortoise plan to have a race over a distance of 100 m. The hare is so confident he gives the tortoise a 99.5-m head start. Which animal is likely to win the race? Justify your answer with calculations.

QUESTION 5 Daniel left Albay at 9:20 am and drove to Bichero averaging 90 km/h. He arrived at his destination at 11:10 am.

a What is the distance from Albay to Bichero?

b Daniel drove the same route home. If he left Bichero at 4:25 pm and averaged 75 km/h, at what time would he arrive in Albay?

QUESTION 6 The Mecano highway has a 200-km straight section of road. The midpoint of the straight section is the town of Cronkers Creek. At 2:30 pm, Bill leaves Cronkers Creek and drives east at an average speed of 100 km/h. Jill leaves from the same spot and at the same time as Bill and drives west at an average speed of 90 km/h.

a How far are they apart at 3 pm?

b After what time will the distance between them exceed 114 km?

QUESTION 7 When Juliet is driving her car, her reaction time before she brakes is 0.95 seconds. Find her reaction-time distance, to the nearest metre, if she is travelling at:

a 40 km/h **b** 90 km/h **c** 110 km/h

Distance-time graphs

QUESTION **1** The distance-time graph for Logan's trip is shown.

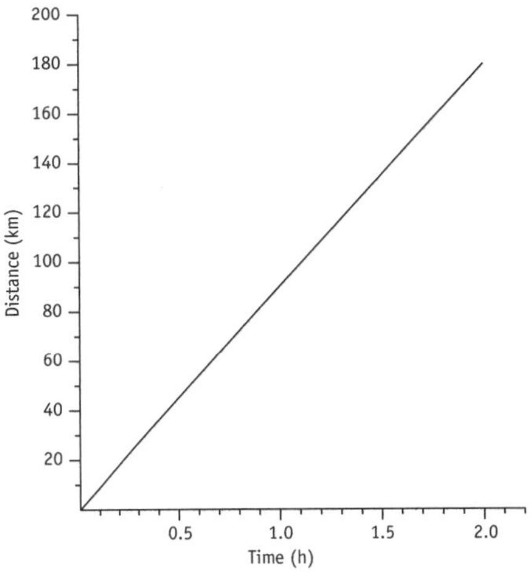

a Find his average speed.

b If he maintained this speed, how far would Logan travel in 3 hours 30 minutes?

QUESTION **2** The distance-time graph represents two stages of a truck's journey.

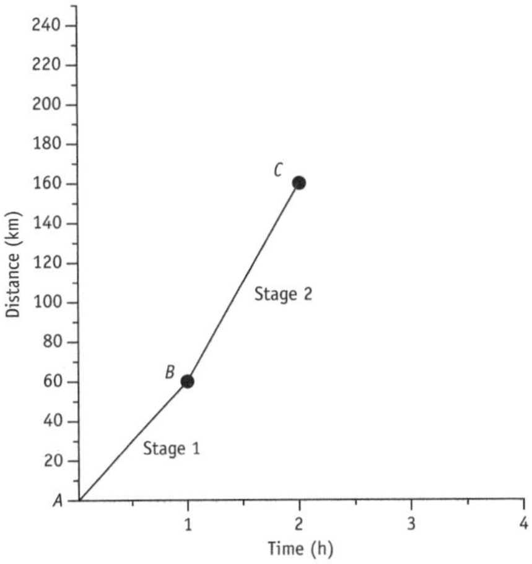

a What is the average speed of the truck in:

i stage 1 (from *A* to *B*)?

ii stage 2 (from *B* to *C*)?

b In stage 3 of the journey, from *C* to *D*, the truck had an average speed of 40 km/h for one-and-a-half hours.

i Represent stage 3 of the truck's journey by drawing a line segment on the graph.

ii What is the total distance travelled by the truck?

QUESTION **3** The distance-time graph for two moving objects is shown.

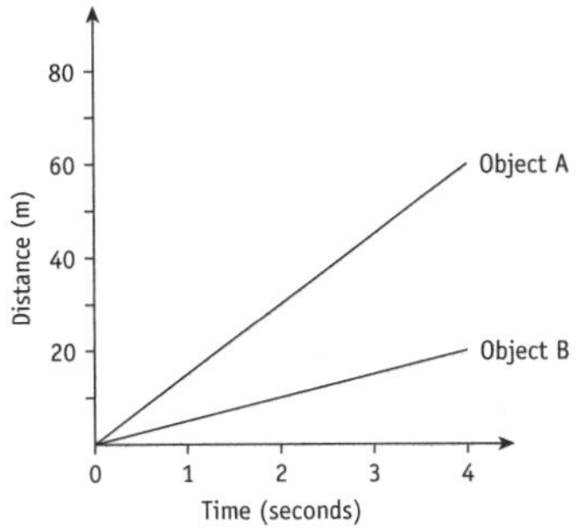

a What is the speed of object *A*, in kilometres/hour?

b How much further has object *A* travelled than object *B* after 2 hours?

QUESTION **4** Laura cycles for 20 minutes at 7.5 m/s and then for a further 10 minutes at 6 m/s.

a Find the total distance Laura has cycled, in kilometres.

b Use the diagram to draw a distance-time graph for the 30 minutes of cycling.

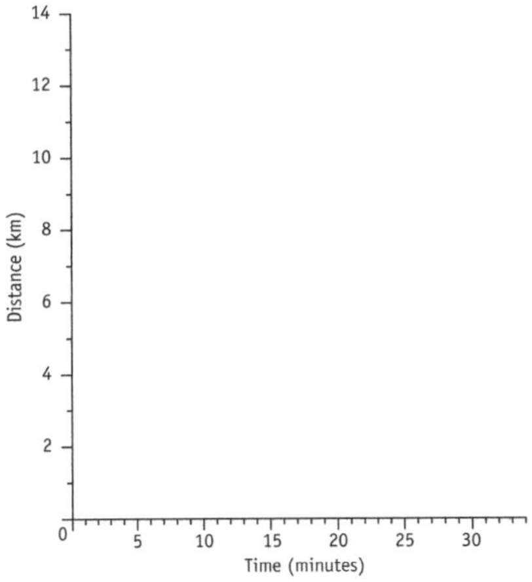

Fuel consumption

Question 1 Find the fuel consumption (in L/100 km) for a car, correct to one decimal place, that:

a uses 45 L of fuel on a journey of 650 km

b travels 586 km and uses 38 L of fuel

Question 2 On a trip Andrew's car has a fuel consumption rate of 7.6 L/100 km. How much fuel has been used on a trip of:

a 380 km?

b 515 km?

Question 3 If a car uses fuel at the rate of 8.3 L/100 km, how far will it travel, to the nearest kilometre, using:

a 45 litres of fuel?

b $60 of fuel if fuel costs 172.9 c/L?

Question 4 Luke buys fuel at 166.9 cents/L. If his car has a fuel consumption rate of 8.1 L/100 km, what is the cost of fuel, to the nearest cent, for a journey of:

a 180 km?

b 435 km?

Question 5 Michael and Jenny are both driving their own vehicles from Gloucester to Broadbeach, a distance of 605 km. Michael's car has a fuel consumption rate of 7.3 L/100 km while Jenny's car averages 5.9 L/100 km. If both drivers buy fuel at a price of 179.9 c/L, what is the difference in the amount of money they pay for fuel, correct to the nearest cent?

Question 6 The table shows the price per litre of three types of petrol.

Cost of petrol	
Fuel type	**Cost/L**
91-octane	143.9
95-octane	154.9
98-octane	161.9

a Connor spent $45 on 95-octane petrol. If his car averages 6.8 L/100 km, how far will he travel on his purchase, to the nearest kilometre?

b Eliana's car has an average fuel consumption rate of 7.9 L/100 km. What is the cost of a trip of 290 km, to the nearest cent, if she uses:

i 91-octane petrol?

ii 98-octane petrol?

c Ryan drove 480 km and his car averaged 6.9 L/100 km. How much would he have saved on petrol if he had used 91-octane petrol rather than 95-octane petrol, to the nearest cent?

d Lincoln and Sarah both paid $60 for their petrol. Lincoln used 91-octane petrol and his car averaged 8.3 L/100 km while Sarah used 98-octane and her car averaged 7.7 L/100 km. Which person drove the greater distance on the petrol purchase? Justify your answer with calculations.

Measurement: Rates

Heart rates

QUESTION **1** Cedric measures his heart rate by counting his pulse for 30 seconds. What is his heart rate in beats per minute (bpm) if he counts:

a 32 beats? **b** 38 beats?

QUESTION **2** The table shows the resting heart rate (RHR) for women of different age ranges across four fitness levels.

Resting heart rate (RHR) for women (bpm)					
	Age range				
Fitness level	**18 to 25**	**26 to 35**	**36 to 45**	**46 to 55**	**56 to 65**
Athlete	54–60	54–59	54–59	54–59	54–59
Good	66–69	65–68	65–69	66–69	65–68
Average	74–78	73–76	74–78	74–77	73–76
Poor	85+	83+	85+	84+	84+

a Jessica is 42 years old and her resting heart rate is 76 bpm. What is her likely fitness level?

b Melissa is 27 years old and represents NSW in netball. What is a likely range of Melissa's resting heart rate?

c The resting heart rate for a man is about 3 beats per minute less than for a woman of a similar age and fitness level. Estimate the range of RHR for 59-year-old Bob with a 'good' fitness level.

QUESTION **3** The suggested maximum heart rate is 220 bpm minus your age. What is the maximum heart rate for a person aged:

a 25? **b** 45? **c** 75?

QUESTION **4** An alternative method to determine a person's maximum heart rate (MHR), in beats per minute (bpm), is to use the formula MHR = 208 − 0.7a, where a is the person's age in years. Use the formula to estimate the maximum heart rate, to the nearest bpm, of a person who is:

a 30 years old **b** 60 years old **c** 86 years old

QUESTION **5**
Generally, the target heart rate during moderate intensity activity is 50–70% of the maximum heart rate, while during vigorous physical activity the rate is 70–85% of the maximum heart rate. Use this information to complete the table.

Heart rates after activity based on age			
	Target heart rate zone (bpm)		
	Moderate activity	**Intense activity**	**Maximum heart rate**
Age	**50–70% max.**	**70–85% max.**	**(bpm)**
20	100–140	140–170	200
30	95–133	133–162	190
40	90–126		180
50		119–145	170
60	80–112	112–136	160
70	75–105	105–127.5	150

Blood pressure

QUESTION 1 Olivia's blood pressure was recorded as 115/76. This meant that her systolic pressure was 115 mm/Hg and her diastolic pressure was 76 mm/Hg.
If Henry's blood pressure was reported as 119/73, what was his:

a systolic pressure? ______________________ **b** diastolic pressure? ______________________

QUESTION 2 The table below shows a number of blood pressure categories.

Blood pressure stages			
Blood pressure category	**Systolic (mm/Hg)**		**Diastolic (mm/Hg)**
Optimal	less than 120	and	less than 80
Normal	120–129	and/or	80–84
High to normal	130–139	and/or	85–89
Mild hypertension	140–159	and/or	90–99
Moderate hypertension	160–179	and/or	100–109
Severe hypertension	more than or equal to 180	and/or	more than or equal to 110

Determine the blood pressure categories for the following people:

a Isla with a blood pressure of 126/82

____

b Hannah with a blood pressure of 147/96

____

c Alexander with a blood pressure of 131/88

____

QUESTION 3 Ethan is recovering in hospital from surgery and his blood pressure is recorded every 3 hours in the table and on the graph below. Some data is missing.

Time of day	Systolic (mm/Hg)	Diastolic (mm/Hg)
03:00		
06:00	130	100
09:00	120	95
12:00	125	90
15:00		
18:00	120	85
21:00	115	90
00:00	110	95

Complete the table **and** the graph with the missing data.

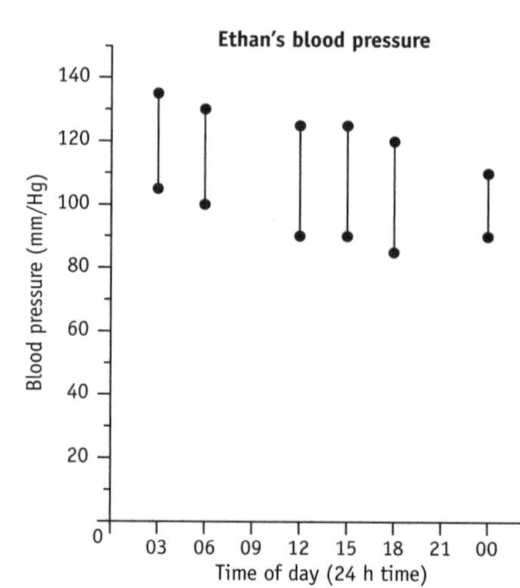

Measurement: Rates

TOPIC TEST

SECTION I

Instructions
- This section consists of 5 multiple-choice questions.
- Each question is worth 1 mark.
- Fill in only ONE CIRCLE for each question.

Time allowed: 7 minutes **Total marks: 5**

1 Vegemite is available in different-sized jars. Which of the following is the best buy?

Ⓐ 150 g vegemite jar for $3.50 Ⓑ 220 g vegemite jar for $5.30

Ⓒ 380 g vegemite jar for $7.80 Ⓓ 560 g vegemite jar for $11.30

2 A car travels 540 km on 40 L of petrol. What is its fuel consumption?

Ⓐ 5 L/100 km Ⓑ 7.4 L/100 km

Ⓒ 8.6 L/100 km Ⓓ 13.5 L/100 km

3 A car is travelling at 94 km/h. How far will it travel in 1 hour and 45 minutes?

Ⓐ 136.3 km Ⓑ 154.3 km

Ⓒ 164.5 km Ⓓ 186.2 km

4 Herman's systolic blood pressure decreased from 128 to 116.
Which of these is closest to the percentage decrease?

Ⓐ 9% Ⓑ 10%

Ⓒ 12% Ⓓ 14%

5 The diagram shows a distance-time graph for a moving object.
What is the speed of the object in **kilometres per hour**?

Ⓐ 10 km/h

Ⓑ 36 km/h

Ⓒ 40 km/h

Ⓓ 144 km/h

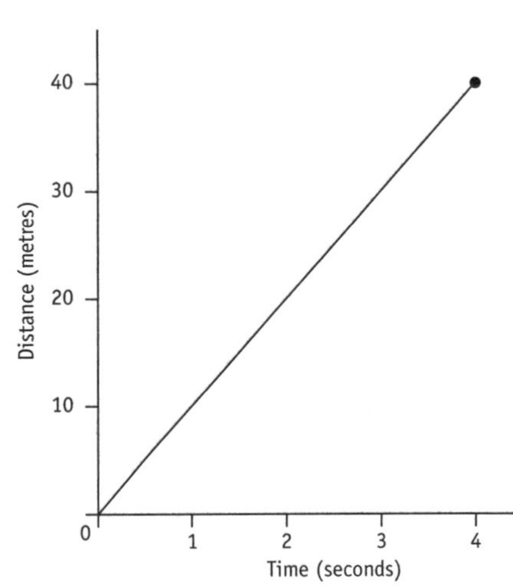

TOPIC TEST — SECTION II

Instructions
- This section consists of 12 questions.
- Show all working.

Time allowed: 53 minutes **Total marks: 35**

6 Ethan's car uses fuel at the rate of 6.8 L per 100 km for long-distance driving and 8.4 L per 100 km for short-distance driving.
He used the car to make a journey of 480 km, which included 110 km of short-distance driving.
Approximately how much fuel did Ethan's car use on the journey?
Give your answer to the nearest litre. **2 marks**

7 A train departs from Carguille at 9:30 am to travel to Peakfield.
Its average speed for the journey is 80 km/h and it arrives at 12:30 pm.
A second train departs from Carguille at 1:00 pm and arrives at Peakfield at 5:00 pm.

a What is the distance from Carguille to Peakfield? **1 mark**

b What is the average speed of the second train? **1 mark**

8 Adam travels to and from work 5 days each week.
He can ride his scooter or catch a bus.
His scooter uses 1 L of fuel for every 20 km travelled.
The cost of fuel is $1.60/L and the distance from his home to his place of work is 30 km.
If he catches the bus the cost of each trip is $3.20.
Which mode of travel is cheaper per week and by how much?
Support your answer with calculations. **3 marks**

9 The costs for membership at a fitness centre are shown below.

Hunter Gym and Swim Centre			
Membership costs			
	Gym only ($)	**Swim only ($)**	**Gym + swim ($)**
12 months (monthly direct debit)	26	15	30
1 month (direct debit)	30	18	36
Casual (per visit)	6	4	8
Family discount: One full-paying, second family member 50% discount			

a Thomas buys a gym + swim membership for 3 months.
How much will the membership cost? **1 mark**

b Alex buys a 12-month gym-only membership for both himself and his wife.
What will be the total cost? **2 marks**

c Shelby plans to use the fitness centre's gym for 10 months.
She compares buying a 12-month membership and a monthly membership for 10 months.
Which provides the cheaper option and by how much?
Support your answer with calculations. **2 marks**

10 A car driven 540 km uses a total of 46.44 L of petrol.
The cost of petrol is 159.9 cents/litre.

a Explain why the petrol consumption is 8.6 L/100 km. **1 mark**

b Find the cost of petrol used on the 540-km trip. **1 mark**

11 Oliver earns $33.80 per hour as a train driver.
He works 36 hours a week at normal time and 8 hours at double time.
What is his total pay for the work? **2 marks**

12 Olivia leaves her grandpa's home at 10:45 am.
She drives 260 km to her home, arriving at 2:15 pm the same day.
What was Olivia's average speed?
Leave your answer correct to one decimal place. **2 marks**

13 The diagram shows the distance-time graphs for two motorists *A* and *B* who are travelling the same route.

a What was the difference in the average speeds of the two motorists? **3 marks**

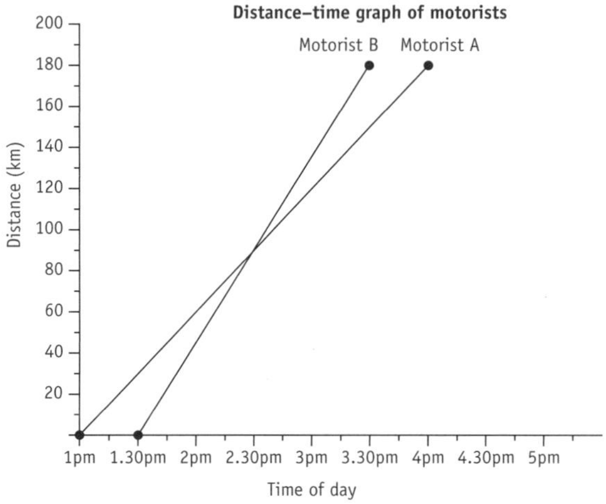

b How far was motorist *A* from the destination when passed by motorist *B*? **1 mark**

c Motorist *C* completes the same route, leaving at 2:30 pm and averaging 72 km/h.

Draw a line on the graph to represent the journey of motorist *C*. **2 marks**

14 Amelia drives to and from work every day from Monday to Friday each week.
The distance from home to work is 28 km.
Her car uses petrol at the rate of 9.6 L/100 km and petrol costs 172.9 cents per litre.
What is the total cost of her driving to and from work each fortnight? **3 marks**

15 Twelve people measured their heart rates. The results in beats per minute (bpm) are recorded below.

65 75 68 74 67 68 70 71 74 70 74 71

Use a five-number summary to draw a box plot on the scale below and find the median bpm. **3 marks**

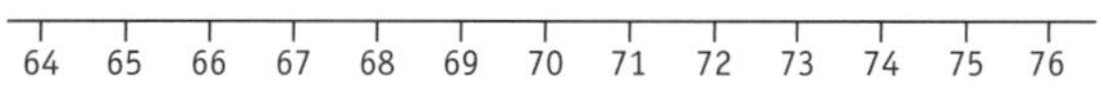

16 The distance-time graph shows Chloe's journey from her apartment to her friend's farm.

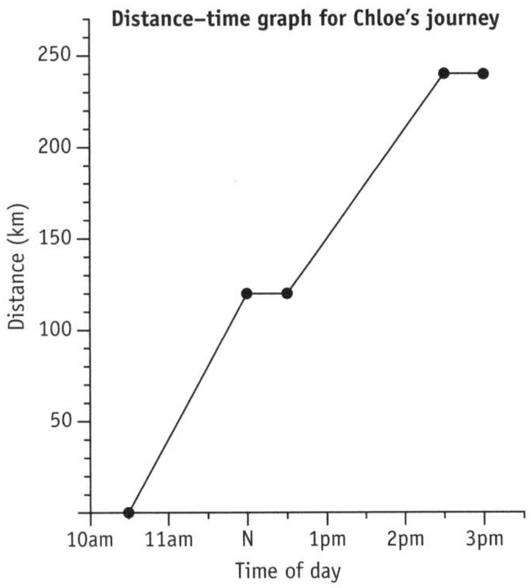

a What was Chloe's average speed between 10:30 am and noon? **1 mark**

____

____

b At 3 pm she left the farm to drive back to her apartment.

If she made no stops and averaged 100 km/h, at what time did Chloe arrive home? **2 marks**

____

____

____

17 A ride-sharing company has the following charges: hire charge $3.80, distance rate $2.14/km, waiting time $0.85/minute.

What is the cost of a journey of 24.3 km with a 3-minute waiting time? **2 marks**

____

____

____

Chapter 7
Measurement: Scale drawings

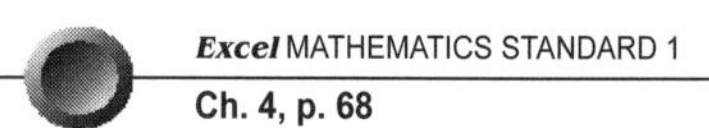

Excel MATHEMATICS STANDARD 1
Ch. 4, p. 68

Simplifying ratios

Question 1 Simplify the ratio of:

a 12:15

b 20:60

c 12:48

d 64:16

e 5:15:25

f 12:120:240

Question 2 Simplify:

a \$4:\$28

b 50c:\$2

c 3 km:10 km

d 600 mL:1 L

e 100 g:10 kg

f 5 cm:25 mm

Question 3 Of the 60 passengers on a bus, 24 are women, 16 are men and the remainder are children. Write the simplified ratio of:

a men to women

b women to children

c men to children to women

Question 4 A painting is purchased for \$6500 and 4 years later is sold for a profit of \$3500. Write the simplified ratio of:

a purchase price to profit

b profit to sales price

c cost price to sales price

Question 5 A bag contains 40 balls. There are 12 blue balls, 8 green balls and 10 red balls. The remaining balls are purple. Write the simplified ratio of:

a blue balls to red balls

b red balls to purple balls

Question 6 Each week Mia is paid a net wage of \$450. She saves \$120 and spends the remainder. Express the following in simplest form:

a the ratio of money saved to money spent

b the ratio of money spent to money earned

Question 7 An apartment was purchased in 2017 for \$1.25 million. Last year it was sold at a loss of 5%. Write the simplified ratio of:

a purchase price to loss

b loss to sales price

Problems involving ratio

QUESTION **1** A rectangle has its sides in the ratio of 2:3. Calculate the length of the:

a shorter side if the longer side is 18 cm

b perimeter if the shorter side is 24 cm

QUESTION **2** Eucalypts and acacias are planted along a roadside in the ratio of 3:7. If there are 210 acacias, how many eucalypts are there?

QUESTION **3** James and Merrill are employed at a restaurant to provide background music. James plays the piano and Merrill sings. They decide to split any tips they receive in their tip-jar in the ratio of 3:4.

a If Merrill's share of the tips is $60, how much will James receive?

b If James's share of the tips is $36, what was the total amount collected?

QUESTION **4** In a horse race the first three placegetters receive prize money in the ratio 8:5:3. The prize for second place is $45 000.

a What is the total prize money for the three placegetters?

b How much more does the first placegetter receive than the third placegetter?

QUESTION **5** A concrete mix uses gravel, cement and sand in the ratio of 3:1:2. If 7.2 kg of sand is used in the mix, what will be the mass of gravel?

QUESTION **6** At a party the ratio of girls to boys is 3:4. An hour later the number of girls increased from 12 to 18, while the number of boys remained the same. What is the new ratio of girls to boys at the party?

QUESTION **7**

a Steve mixes petrol and oil in the ratio of 25:1 to make fuel for his brush cutter. How much oil does Steve add to a container of 4.75 L of petrol?

b How much fuel does the container now hold? ____________

c If Steve wants the ratio to change to 60:1 how much petrol would he now need to add to the container?

Measurement: Scale drawings

Dividing a quantity in a given ratio

Question 1 Divide the following amounts in the ratio given:

a \$20 in the ratio of 3 : 5

b 1.2 kg in the ratio of 2 : 3 : 1

c 8 hours in the ratio of 3 : 5 : 2

Question 2 A 2-L bottle of soft drink is to be shared between three boys in the ratio of 1 : 4 : 5. How many millilitres will each boy drink?

Question 3 Grandma gives \$20 000 to her three granddaughters in the ratio of their ages. If Olivia is 16, Ava is 14 and Sophia is 10 years old how much money will each girl receive?

Question 4 Concrete is mixed using cement, sand and gravel in the ratio of 1 : 3 : 6. If Davo wants to make 120 kg of concrete, how much of each product will he need?

Question 5 Barb and Ernie together bought a \$10 lottery ticket. Barb contributed \$6 and Ernie the remainder. If they share in a win of \$120 000 in the ratio of their contributions, how much will Ernie receive?

Question 6 The side lengths of a triangle are in the ratio of 3 : 5 : 7. What is the length of the longest side if the perimeter is 75 cm?

Question 7 The ratio of the populations of town A and town B is 1 : 2, and the populations of town B to town C is 4 : 5.

a What is the ratio of the population of town A to town B to town C? ____________

b If the total population of the three towns is 11 000, what is the population of town C?

Question 8 A rectangular paddock had a perimeter of 480 m. If the ratio of its length to width is 7 : 5, find the area of the paddock, in hectares.

Measurement: Scale drawings

Converting map scales

Question 1 Write each of the following scales in simplest ratio form:

a 1 mm to 1 m

b 1 cm to 1 m

c 1 cm to 100 m

d 10 cm to 1 km

e 4 mm to 1 m

f 5 cm to 1 m

g 1 mm to 20 m

h 20 cm to 1 m

i 1 mm to 6 m

Question 2 Using a scale of 1:100, what length, in metres, is represented by:

a 1 cm?

b 3 cm?

c 5 cm?

d 8 mm?

e 6 mm?

f 12 m?

Question 3 Using a scale of 1:1000, what is the real length represented by each of the following?

a 8 mm

b 5 cm

c 6 m

d 9.5 cm

e 8.3 m

f 63.25 m

Question 4 The distance between two points in real life is given. What is this distance on a scale drawing with scale 1 cm to 100 m?

a 500 m

b 400 m

c 1260 m

d 80 m

e 3000 m

f 2835 m

Question 5 A map has a scale of 1:100 000.

a A distance of 1 cm on the map will represent a real distance of how many kilometres?

b The distance between two towns on the map is 8.4 cm. How far apart are the two towns?

c The real straight-line distance between *A* and *B* is 26 km. How long will this distance be on the map?

Similar figures

Question 1 In each pair of shapes, find the scale factor and the value of the pronumeral:

a

Not to scale

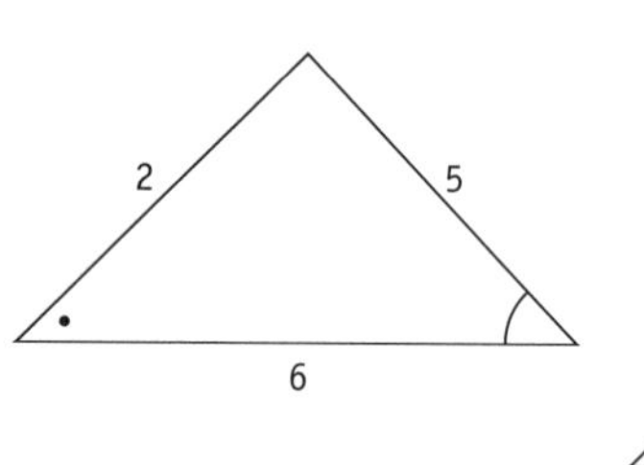

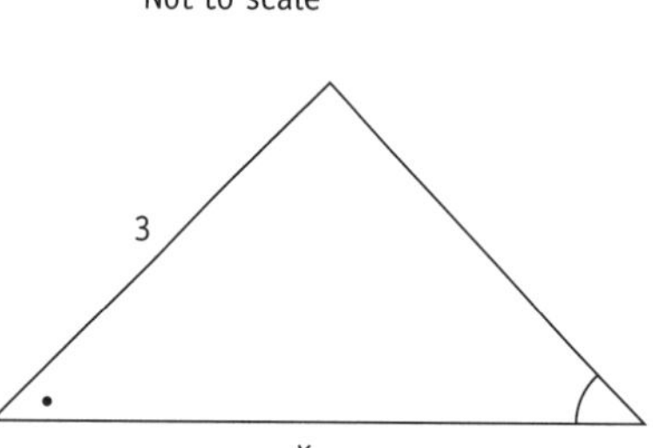

b

Not to scale

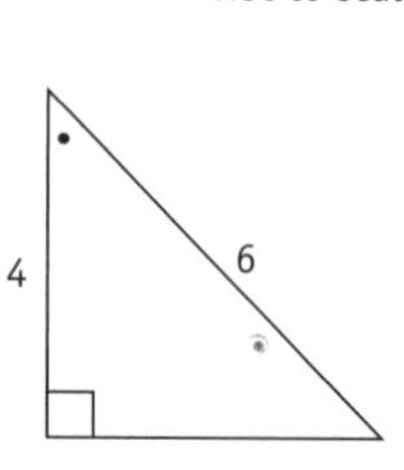

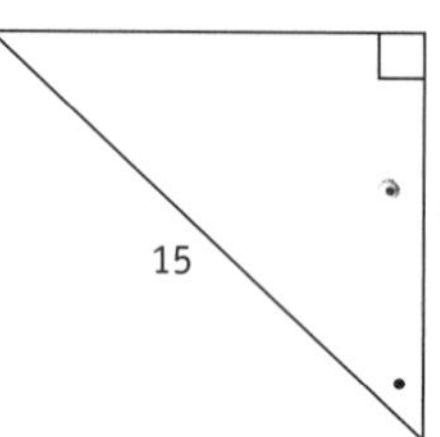

Question 2 The diagram shows similar triangles ABC and PBQ. Also, $AC = 12$ cm, $PQ = 8$ cm and $QB = 6$ cm. Find the:

Not to scale

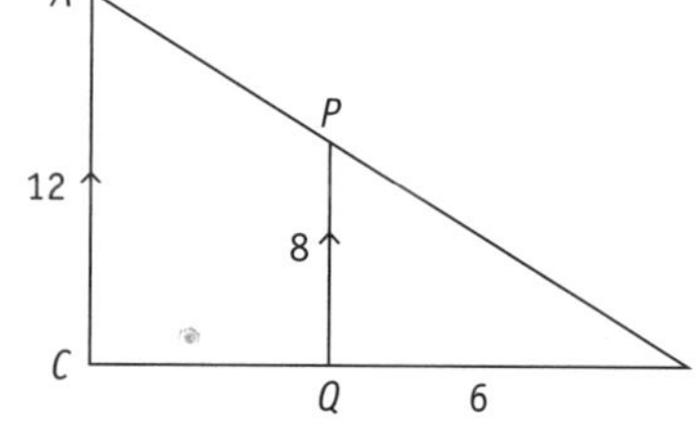

a scale factor for triangle PBQ to triangle ABC

b length of BC

c length of CQ

Question 3 Find the value of the pronumeral:

a

Not to scale

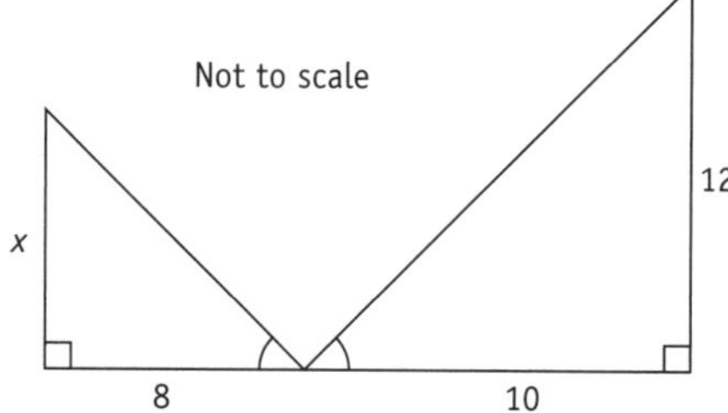

b

Not to scale

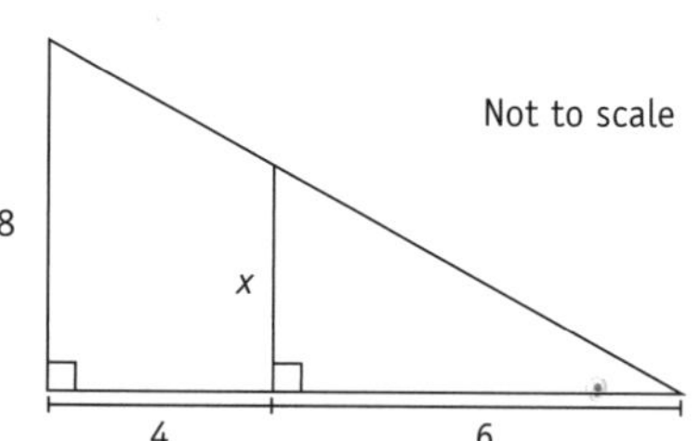

Question 4 The sun makes a shadow of 9 m from a building which is 6 m tall. At the same moment a pole of height 2.8 m makes a shadow of length d m. What is the value of d?

Not to scale

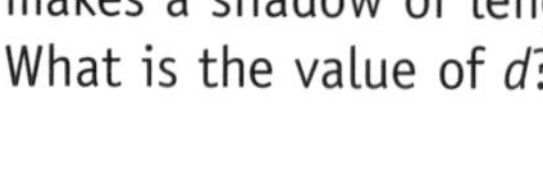

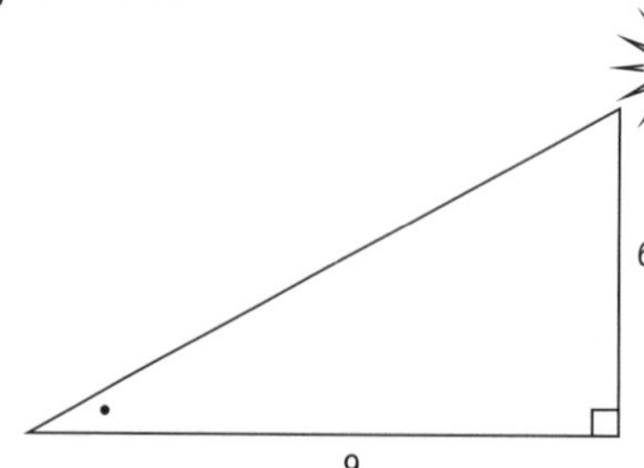

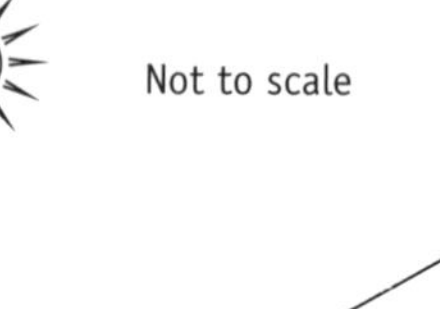

Measurement: Scale drawings

Scale drawings 1

QUESTION 1 A block of land, along with the proposed building, has been drawn using a scale of 1:500. By measurement and calculation find:

a the width of the block

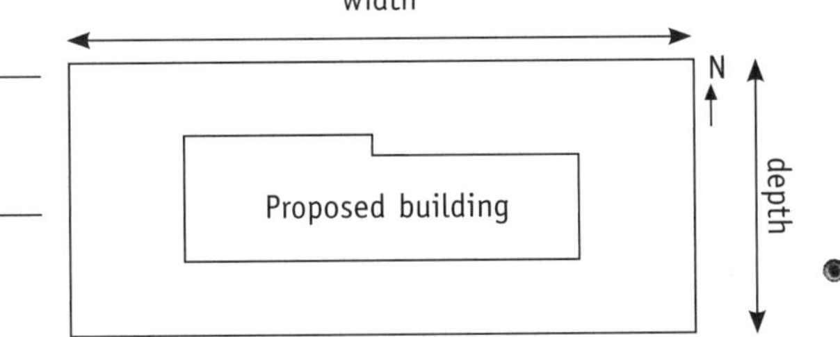

b the depth of the block

c the area of the block

d how far the proposed building is from the southern boundary

e the area of the proposed building

QUESTION 2 The diagram shows a scale drawing of a cross-section of a pipe. The outer diameter of the pipe is 1.44 m.

a By measurement and calculation find the scale used for the drawing.

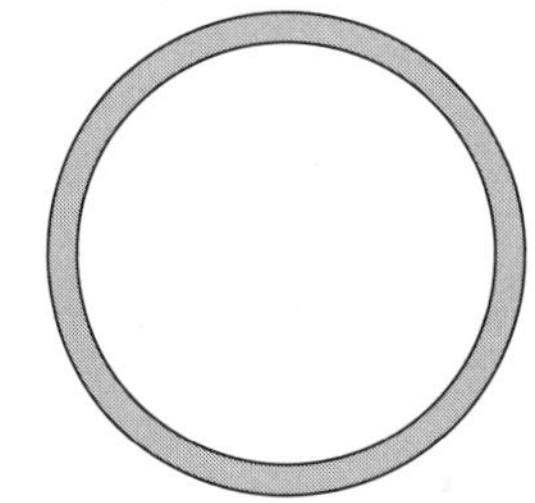

b What is the inside diameter of the pipe?

QUESTION 3 Toby has made a rough sketch of a block of land he is considering buying.

a Make a scale drawing of the block (below) using a scale of 1:400.

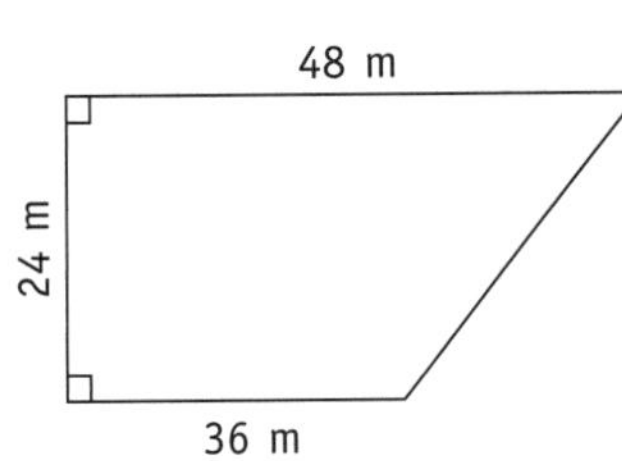

b What is the perimeter of the block to the nearest metre? ______

Scale drawings 2

Question 1 Use the scale to determine the length and width of this swimming pool.

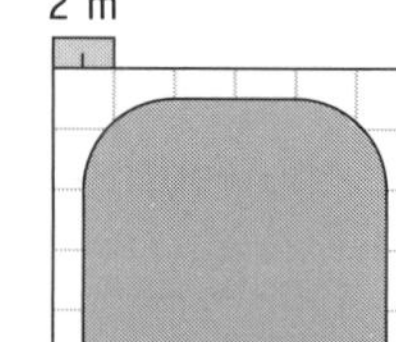

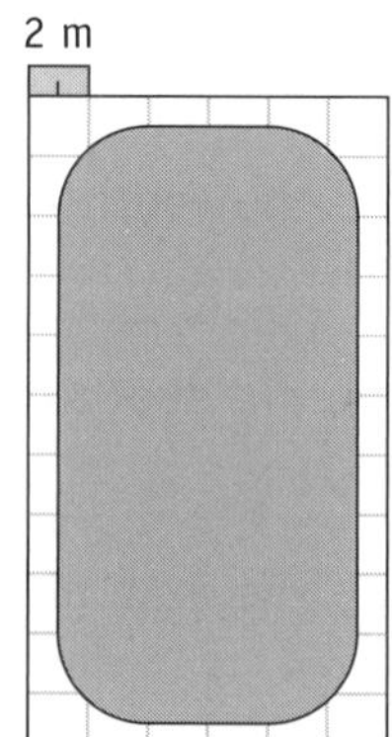

Question 2 *Edmontosaurus* was a large, duck-billed dinosaur of the late Cretaceous Period.

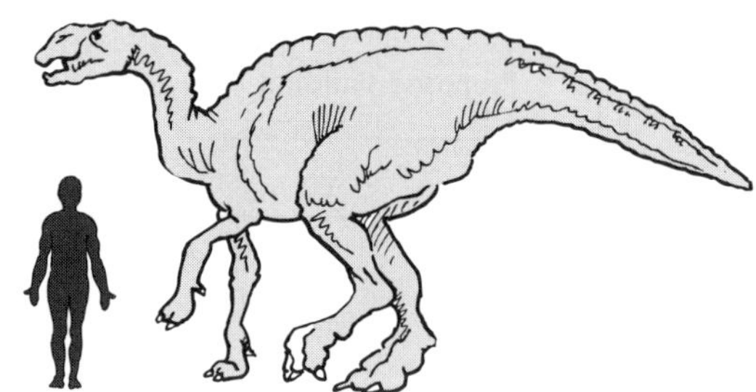

Assuming the man stands 1.8 m tall, how long is the dinosaur?

Question 3 Calculate the perimeter of this land.

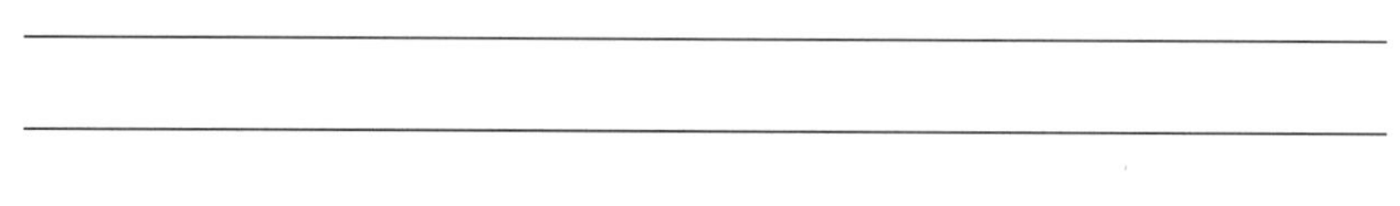

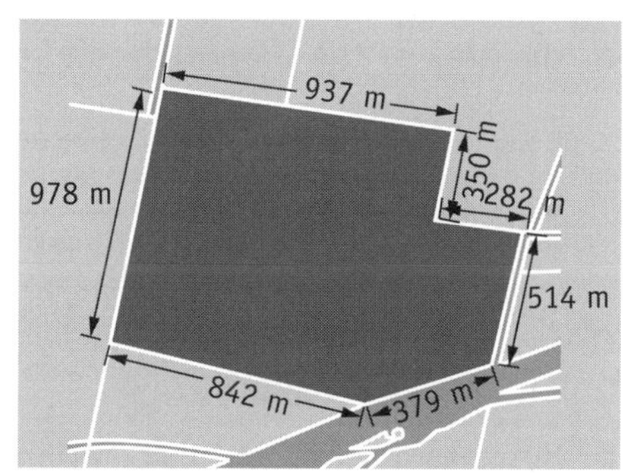

Question 4 The diagram shows some land area with a lake inside it. A scale is provided.

a Use the scale to estimate the perimeter of the land.

b Estimate the perimeter of the lake.

c What is the area of each of the small squares on the grid?

d By counting the number of these small squares, estimate the area of land (including lake). Write the answer both in square metres and hectares (1 ha = 10 000 m^2).

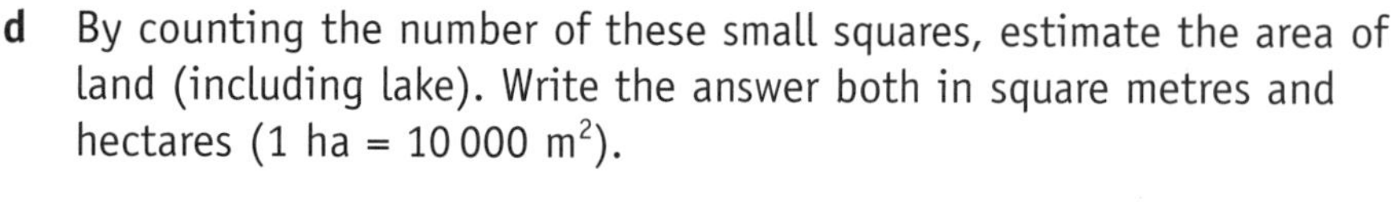

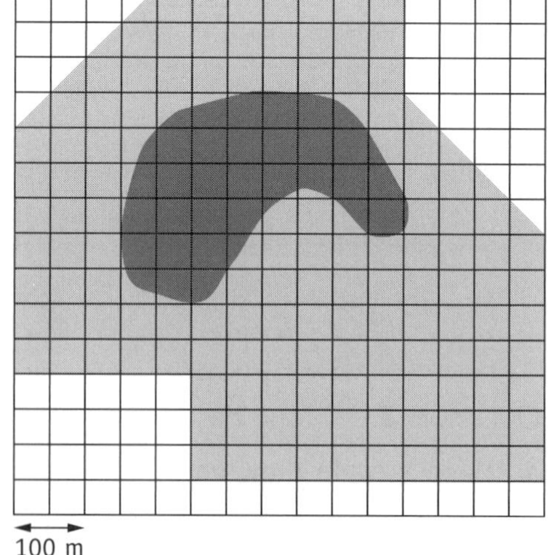

e Estimate the area of the lake.

f What percentage of the land area is taken up by the lake?

Interpreting floor plans

QUESTION 1 The diagram shows a floor plan of a house.

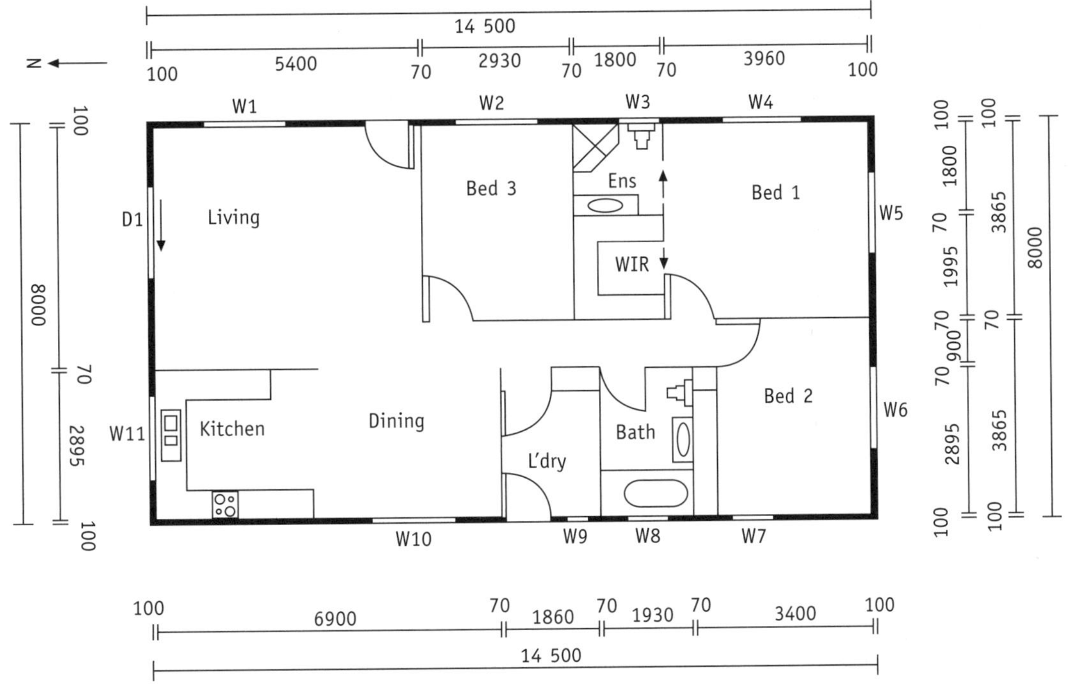

a What is the width of the house? ____________________

b What is the feature marked D1? ____________________

c What is the feature labelled WIR? ____________________

d What are the dimensions of bedroom 3? ____________________

e What is the width of each internal wall? ____________________

f What is the width of the external walls? ____________________

g In which elevation is there not a door? ____________________

h One of the measurements for the living room is missing. What should it be?

____

i If building costs are $1525 per square metre, how much will it cost to build this house?

____

____

____

____

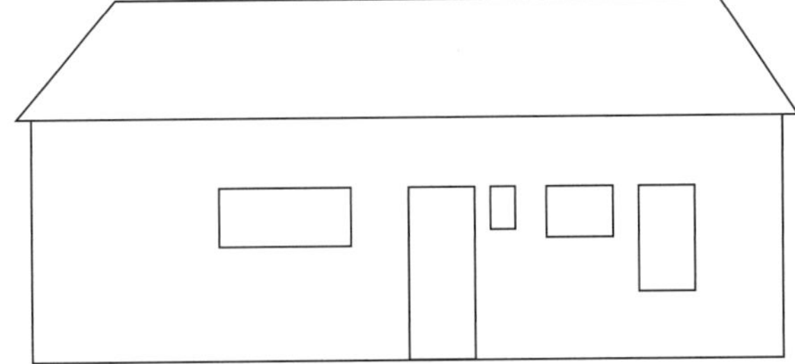

j Lisa has drawn a sketch, not to scale, of one side of the house. Which elevation did she sketch?

____

Measurement: Scale drawings

Excel MATHEMATICS STANDARD 1
Ch. 4, pp. 75–81

Land areas and lengths

QUESTION **1** The island of Hispaniola is shared by Haiti and the Dominican Republic.

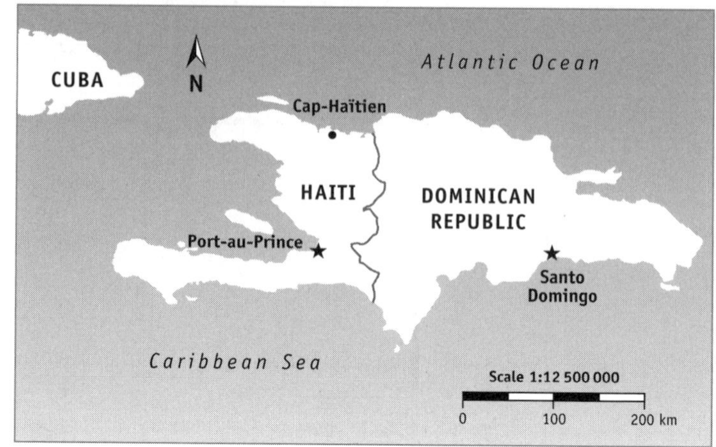

a Using North–South distances, how wide is the widest part of the island?

____

b How far, in a straight line, is Port-au-Prince from Santo Domingo?

____

c Calculate the closest distance from Haiti to Cuba.

____

____

d Given that the area of Haiti is 27 750 km^2, which of the following is closest to the area of the Dominican Republic?

(A) 30 000 km^2 (B) 40 000 km^2 (C) 50 000 km^2 (D) 60 000 km^2

QUESTION **2** A land area is being subdivided to build houses. The diagram is drawn to scale.

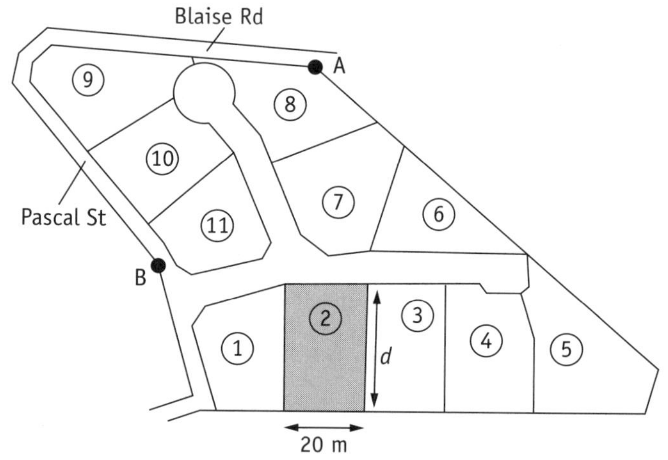

a Block number 2 has a 20-m frontage. Calculate the length of the dimension marked.

____

b Calculate the area for block number 2.

____

c What is the combined length of Blaise Road and Pascal Street (measured between points A and B) shown in the diagram?

____

____

QUESTION **3** The diagram shows the water catchment area for a river.

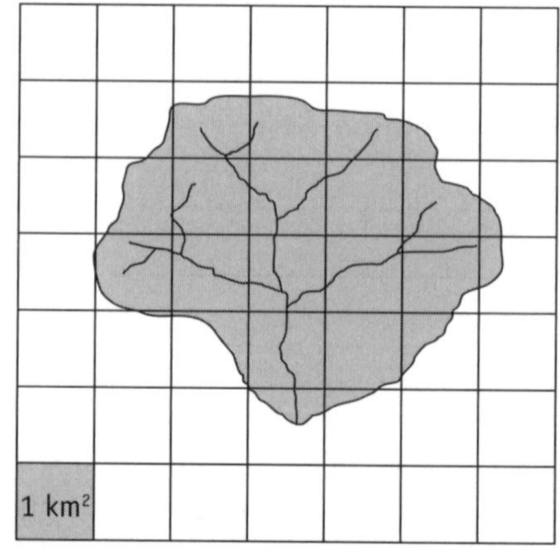

a Estimate the area of land covered by this catchment.

____

b Suppose the average rainfall over this catchment area is 75 mm. Calculate the volume of water, in megalitres, that would flow into the river system.

____

____

____

Measurement: Scale drawings

TOPIC TEST

SECTION I

Instructions
- This section consists of 5 multiple-choice questions.
- Each question is worth 1 mark.
- Fill in only ONE CIRCLE for each question.

Time allowed: 7 minutes **Total marks: 5**

1 Triangles *ABC* and *PQR* are similar.

Which expression could be used to find the value of x?

Ⓐ $y \times \frac{5}{17}$ Ⓑ $y \times \frac{9}{17}$

Ⓒ $y \times \frac{5}{9}$ Ⓓ $y \times \frac{17}{9}$

Not to scale

2 In a maternity ward 20 babies are born in one week. Twelve of the babies are boys.

What is the ratio of male babies to female babies?

Ⓐ 3:2

Ⓑ 2:3 Ⓒ 3:5 Ⓓ 2:5

3 The diagram shows a plastic model of the Eiffel Tower in Paris.
The actual Eiffel Tower is 300 m tall.
The scale used for the plastic model is 1:1500.
What is the height of the plastic tower?

Ⓐ 2 cm Ⓑ 12 cm Ⓒ 20 cm Ⓓ 50 cm

4 The distance from Jerry's Plains to Charlestown is 90 km.
A map is drawn showing the distance between the towns as 6 cm.
On the same map a straight section of railway track is 3.8 cm in length.
Which of these is closest to the actual length of the railway track?

Ⓐ 45 km Ⓑ 55 km Ⓒ 65 km Ⓓ 75 km

5 The table shows the number of votes for four candidates in an election.

What is the ratio of the total votes for Albert and Cook to the total votes for Bridge and Davison?

Ⓐ 3:1 Ⓑ 2:3

Ⓒ 3:4 Ⓓ 2:1

Results of an election	
Candidate	**Number of votes**
Albert	60
Bridge	20
Cook	80
Davison	50

TOPIC TEST

SECTION II

Instructions
- This section consists of 10 questions.
- Show all working.

Time allowed: 53 minutes **Total marks: 35**

6 Shannon used similar triangles to estimate the width of a river.
All dimensions are in metres.
What is the width (w) of the river? **2 marks**

Not to scale

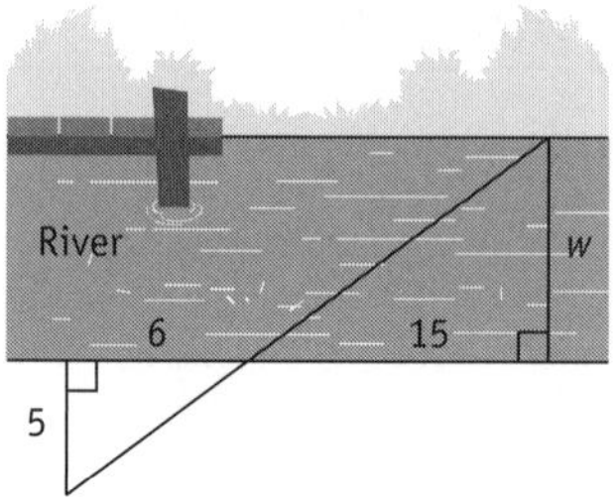

River
6
15
5
w

7 Tom used grid paper to represent a paddock $PQRS$ drawn to scale.
The distance from Q to R is 48 m.

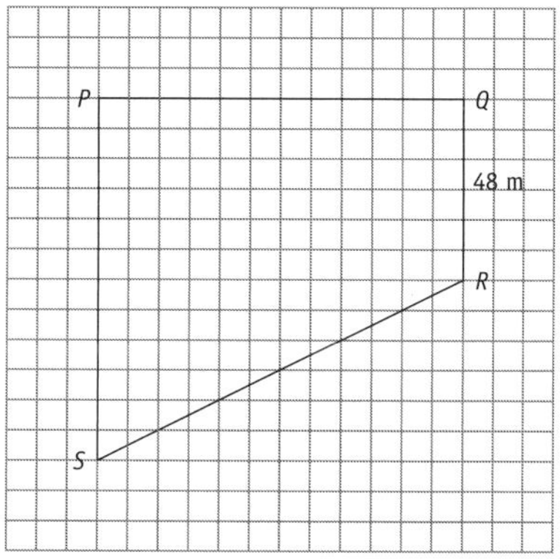

a What scale did Tom use? **1 mark**

b Find the area of the paddock in square metres. **2 marks**

P
Q
48 m
R
S

c The paddock is to be fenced using materials that cost \$11.45/m. Find the total cost. **3 marks**

8 The ratio of the ages of three sisters today is 3 : 4 : 5. The total of the girls' ages is 60.
What will be the ratio of the girls' ages in 10 years time? **3 marks**

9 A rectangular yard has dimensions 16 m by 10 m. It contains a paved area and a garden area.

The ratio of the area of the paved section to the garden is 3:2.

What is the perimeter of the garden? **3 marks**

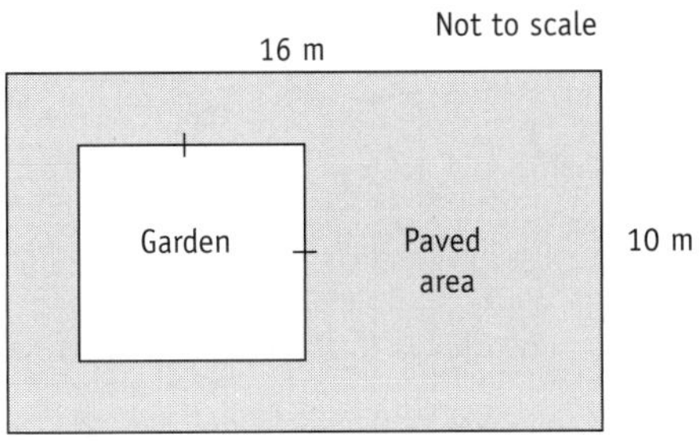

Not to scale

16 m

Garden

Paved area

10 m

10 At a particular time during the day, a tree of height 14.6 m casts a shadow. At the same time, a person who is 1.7 m tall casts a shadow 4 m long.

What is the length of the shadow cast by the tree at that time, correct to two decimal places? **2 marks**

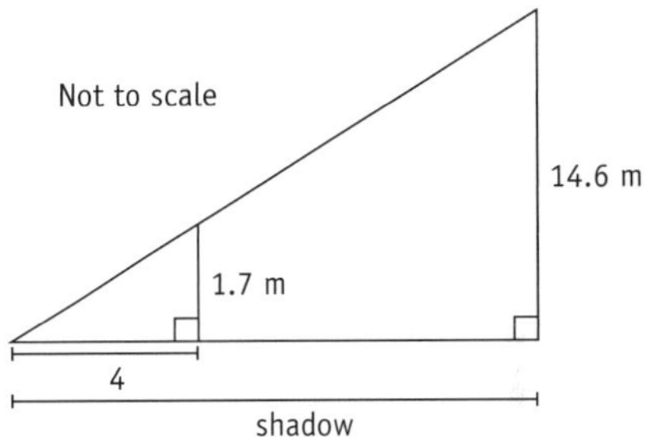

Not to scale

14.6 m

1.7 m

4

shadow

11 A map has a scale of 1:300 000.

a Two towns are 5.6 cm apart on the map.

What is the actual distance between the two towns, in kilometres? **2 marks**

b Two mountains are 84 km apart.

How far apart are the two mountains on the map, in centimetres? **2 marks**

12 The floor plan of a house is shown. The plan is drawn to scale.

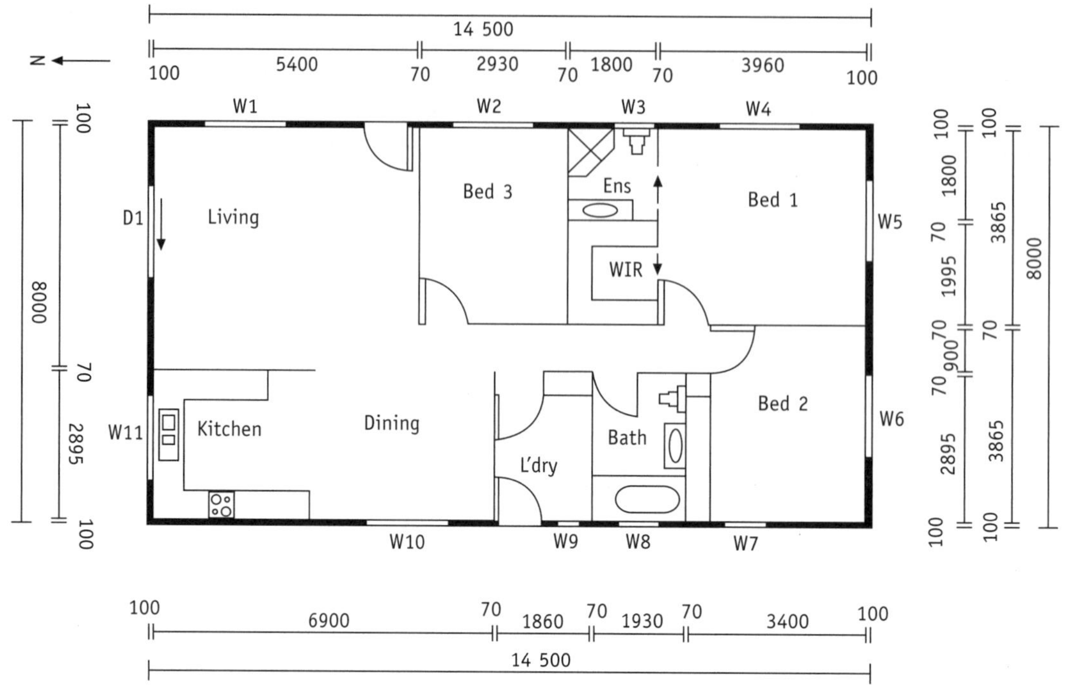

a What is the width of the internal walls, in millimetres? **1 mark**

---

b What is the total area of the three bedrooms, to the nearest square metre? **2 marks**

---

c To cool the house, the entire house is to be surrounded by a 2.4-m-wide verandah, as shown in the diagram below.

What is the area of the verandah? **2 marks**

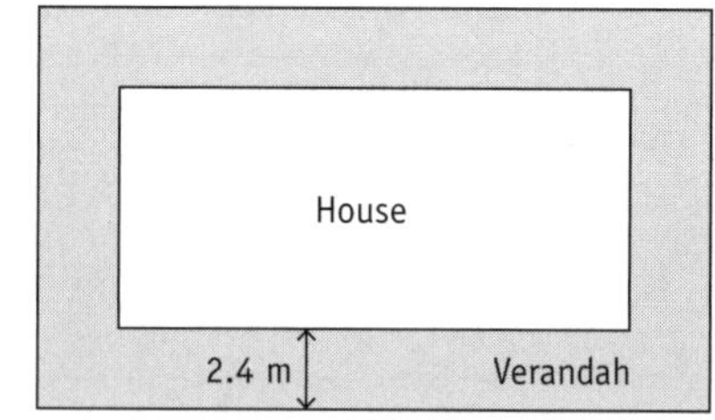

---

13 A point P lies between a wall, 3 m high, and a tree.

P is 4 m away from the base of the wall and 24 m away from the base of the tree.

From P, the angles of elevation to the top of the wall and to the top of the tree are equal.

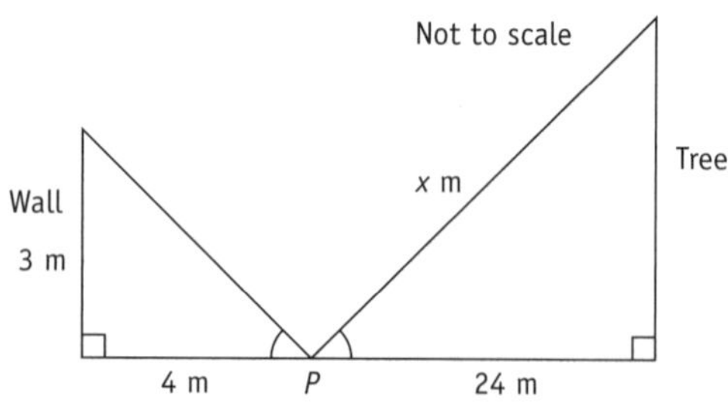

a Explain why the distance from P to the top of the wall is 5 m. **1 mark**

---

b Find the distance from P to the top of the tree. **1 mark**

____

____

14 Triangles ABC and PQR are similar.

a Find the scale factor from triangle ABC to triangle PQR. **1 mark**

____

____

____

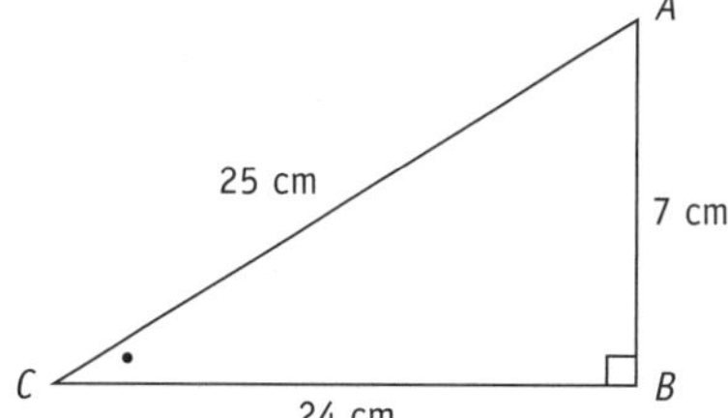

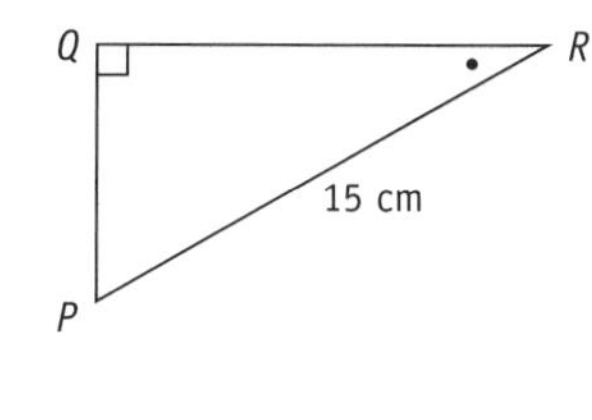

b Explain why the length of PQ is 4.2 cm. **1 mark**

____

____

c What is the area of triangle PQR? **3 marks**

____

____

____

____

15 The diagram shows a shelf ST attached to a wall.

A light on the wall at A shines on the shelf and casts a shadow BC.

Triangles AST and ABC are similar.

a Find the scale factor from triangle AST and triangle ABC. **1 mark**

____

____

____

____

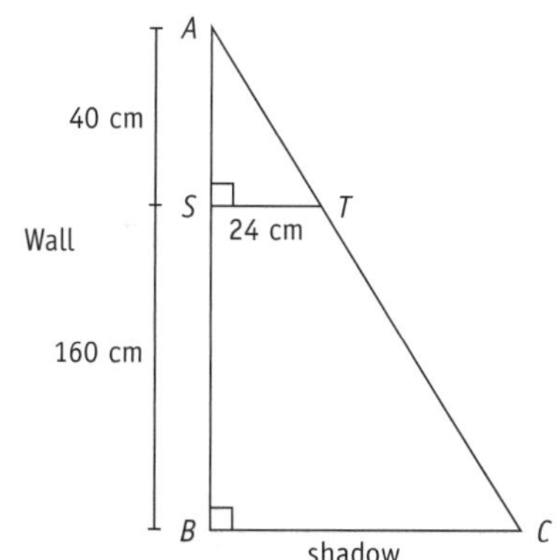

b Find the length of the shadow BC, in metres. **2 marks**

____

____

____

Excel MATHEMATICS STANDARD 1
Ch. 7, pp. 136–137

Questionnaires

QUESTION **1** This question appeared in a questionnaire: 'The workers believe it is the board of directors that should be dismissed not the workers themselves. Do you think they should be sacked?' Why is this not a good question?

QUESTION **2** This question appeared in a survey: 'Obviously it would be far better to take action immediately rather than risk further problems. Do you agree?' What is wrong with this question?

QUESTION **3**

a A question allows just two responses (yes or no). Why might this be?

b Another question allows five possible responses (definitely, probably, perhaps, probably not, definitely not). Why might this be preferable to either a yes or no response?

QUESTION **4** List some of the qualities of a good questionnaire.

QUESTION **5** List some of the things that should be avoided in a good questionnaire.

Suitability, strengths and weaknesses of displays

QUESTION 1 Choose the most appropriate display from histogram, line graph or bar graph to represent the data:

a a breakdown of how your income is spent ______________________

b the exam results of your class ______________________

c the temperature of a hospital patient over a day ______________________

QUESTION 2 Choose from sector graph, radar chart or dot plot, the most appropriate display for:

a the average maximum monthly temperatures ______________________

b the numbers of pupils scoring different marks out of ten in a spelling test ______________________

c a class's favourite sport ______________________

QUESTION 3 A survey has been conducted of the different ways students travel to school. The results are shown in the table.

Method	Walk	Cycle	Bus	Train	Car
Number	32	15	49	6	23

What type of graph would you choose to display this data? Briefly justify your answer.

____

____

____

____

____

____

QUESTION 4 The sector graph has been prepared to show the results of a survey of hair colour.

a What are the strengths of this display?

____

____

____

____

b What are the weaknesses?

____

____

____

____

Misrepresentation of displays

QUESTION 1 The graph below appeared in a magazine.

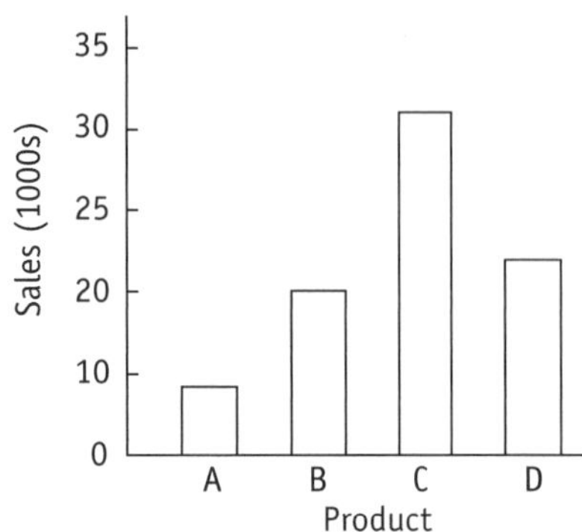

a What is wrong with the graph?

b How does this misrepresent the data?

c Beside the graph, draw it as it should be.

QUESTION 2 'While products A, C and D had similar results on the test, product B clearly performed much better.'

Briefly comment on this statement, explaining how the graph is misleading.

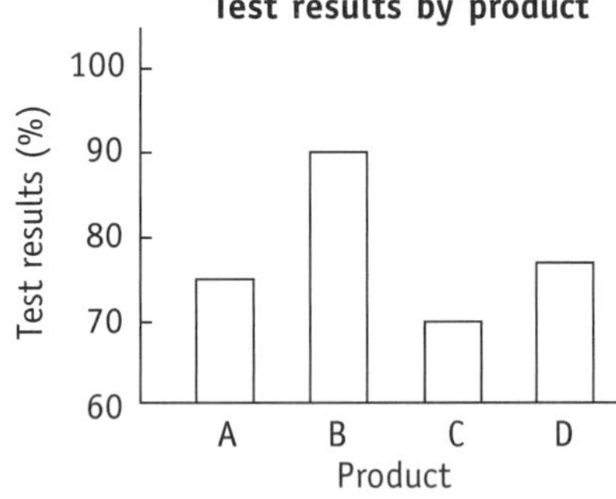

QUESTION 3 This graph appeared in a newspaper. What is wrong with it?

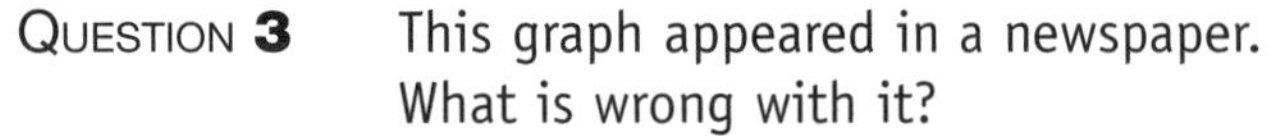

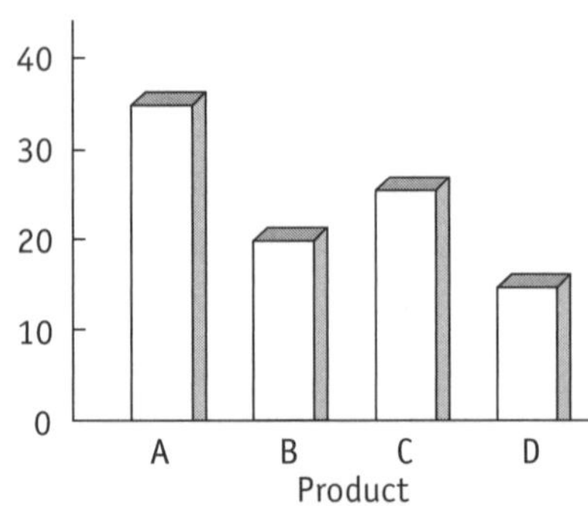

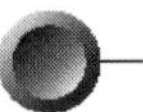

Measures of central tendency

Question 1 A stem-and-leaf plot is shown on the right. Find the:

20	4
21	789
22	3446
23	25

a mode ______

b median ______

c mean ______

Question 2 A dot plot is shown on the right. Find the:

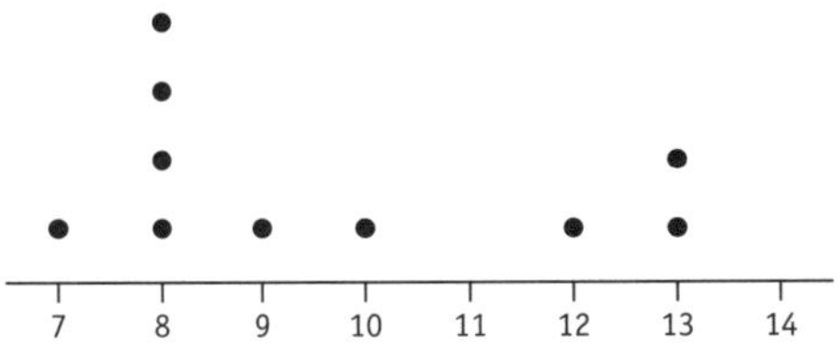

(Dot plot, axis 7–14: 7: 1 dot; 8: 4 dots; 9: 1 dot; 10: 1 dot; 12: 1 dot; 13: 2 dots)

a mode ______

b median ______

c mean ______

Question 3 Here are the results of 10 students in a weekly maths quiz.

8 7 6 9 9 5 7 8 6 7

Find the:

a mode ______ **b** median ______ **c** mean ______

Question 4 A social club played golf and recorded the results of the players for the 18th hole. The results are listed in the table below.

Score	3	4	5	6	7	8
Number of players	2	4	8	5	0	1

Find the:

a mode ______ **b** median ______ **c** mean ______

Question 5 The table shows the heights of 40 students.

Heights (cm)	Class centre (cm)	Number of students
143–149		2
150–156		5
157–163		7
164–170		13
171–177		10
178–184		3

a Complete the table by finding the class centres.

b Find the modal class.

______

c Use the class centres to calculate the mean.

______

Question 6 Create a set of data with the following characteristics: mode = 5, mean = 6, number of scores = 4.

______

Question 7 The mean age of a group of four young people is 12. When another person is included in the group the mean age increases by 2. What is the age of this fifth person?

______

Question 8 A teacher asked her students for the estimates of the current outside temperature. The results are displayed in the graph on the right. What is the mean of the students' guesses?

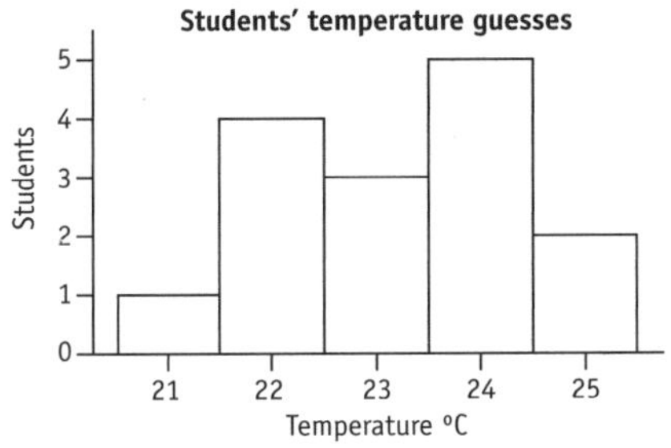

______

______

______

Students' temperature guesses

Temperature °C	21	22	23	24	25
Students	1	4	3	5	2

Measures of spread

QUESTION 1 Find the range of the following sets of scores:

a 12, 4, 19, 31, 8, 19, 7

b 1.8, 0.3, 15, 2.08, 20.1

c −1.3, 4.9, −2.5, −0.2, 3

QUESTION 2 What is the interquartile range of the following scores?

a 12, 7, 9, 19, 5, 10, 32

b 19, 76, 26, 54, 30, 16

c 6, 17, 11, 18, 40, 24, 8, 10

QUESTION 3 Use the following to determine the range of data:

a

12 13 14 15 16 17

b

Stem	Leaf
6	477
7	3689
8	04
9	3

c

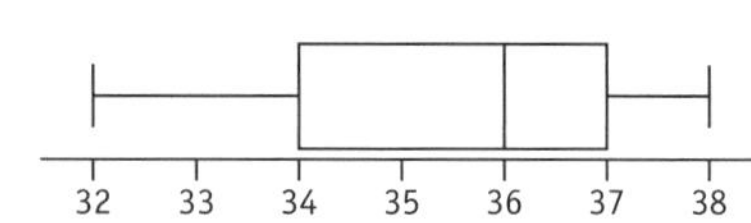

QUESTION 4 For each of the datasets in Question 4, find the interquartile range.

a

b

c

QUESTION 5 Each of these sets of scores is written in ascending order. The range of each set is 12. Find the value of the pronumeral:

a 3, 4, 8, 10, x

b y, 12, 13, 13, 16, 18, 20, 21

c −4, −3, −1, 5, z

QUESTION 6 Determine whether the number shown in **bold** is an outlier. Explain your answer with calculations.

a **2**, 18, 20, 24, 25, 28, 29

b 9, 10, 20, 28, 30, 34, **63**

Making conclusions

QUESTION 1 Here are two datasets of scores.

Set A: 9, 3, 8, 2, 8, 12 **Set B:** 12, 7, 8, 3, 6, 3, 10

Compare the datasets and complete the following statements.

a The range of scores in Set A (______) is __________ than the range of scores in Set B (______).

b The mode of scores in Set A (______) is __________ than the mode of scores in Set B (______).

c The median of scores in Set A (______) is __________ than the median of scores in Set B (______).

d The mean of scores in Set A (______) is __________ the mean of scores in Set B (______).

QUESTION 2 The stem-and-leaf plot shows the heights (cm) of students in a class. Compare the data for the height of the girls and boys and complete the following statements.

Heights of students (cm)

Girls		Boys
7	14	
742	15	67
8400	16	3689
3	17	44
	18	0

a The range of the heights of the girls (______ cm) is __________ than the range of heights of the boys (______ cm).

b The mode of the heights of the girls (______ cm) is __________ than the mode of heights of the boys (______ cm).

c The median of the heights of the girls (______ cm) is __________ than the median of heights of the boys (______ cm).

QUESTION 3 Students in a Biology class sat a test each Friday for two weeks. The results of both tests (Test A and Test B) are shown in the graph below.
Compare the data for the results of Test A and Test B and complete the following statements.

Results of Tests A and B
Test A
Test B
Students
0 1 2 3 4 5 6
5 6 7 8 9 10
Marks

a The number of students who sat Test A (______) was __________ than the number of students who sat Test B (______) .

b The range of marks in Test A (______) is __________ than the range of marks in Test B (______).

c The mode of marks in Test A (______) is __________ than the mode of marks in Test B (______).

QUESTION 4 The ages of players at a miniature golf course on a particular day were recorded and the results shown in the box plots below.

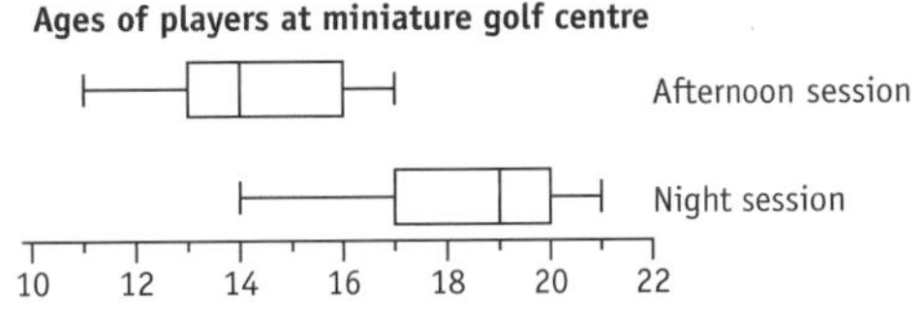

Compare the data for both sessions and complete the following statements.

a The range of ages of players in the afternoon session (______) is __________ than the range of ages of players in the night session (______).

b The IQR of players' ages in the afternoon session (______) is __________ than the IQR of players' ages in the night session (______).

c The median age of players in the afternoon session (______) is __________ than the median age of players in the night session (______).

Statistical analysis: Statistical investigation process for a survey

TOPIC TEST

SECTION I

Instructions
- This section consists of 5 multiple-choice questions.
- Each question is worth 1 mark.
- Fill in only ONE CIRCLE for each question.

Time allowed: 7 minutes **Total marks: 5**

1 Which is **not** a valid step in the process of statistical inquiry?

Ⓐ fabricating data Ⓑ organising data

Ⓒ summarising data Ⓓ analysing data

2 The range of the scores 3, 5, 12, 7, 13, 9, 2, 7, 10 is:

Ⓐ 7 Ⓑ 9

Ⓒ 11 Ⓓ 12

3 Which type of questions should be included in an effective questionnaire?

Ⓐ biased questions Ⓑ ambiguous questions

Ⓒ long-winded complicated questions Ⓓ clear and concise questions

4 The median of the scores 8, 3, 6, 7, 4, 7, 9, 2, 5 is:

Ⓐ 4 Ⓑ 6

Ⓒ 6.5 Ⓓ 7

5 A dataset has the five-number summary shown on the right.

If the interquartile range (IQR) is 19, what is the value of the first quartile?

Ⓐ 16 Ⓑ 18

Ⓒ 19 Ⓓ 20

Minimum value	11
First quartile	?
Median	21
Third quartile	37
Maximum value	43

TOPIC TEST

SECTION II

Instructions
- This section consists of 10 questions.
- Show all working.

Time allowed: 53 minutes **Total marks: 35**

6 Survey questions are listed below.
In the space provided, give a reason why the survey question may be considered to be poorly designed. **1 mark each**

a What is your favourite rugby league team?
Tick the box beside your favourite team.
☐ Broncos ☐ Roosters ☐ Raiders

b Why do you enjoy drinking cola-flavoured drinks?

c How happy are you about the size and location of the planned shopping centre development?

d Should caring parents vaccinate their children?
Tick the box.
☐ Yes ☐ No

7 Patrick conducted a survey of students' opinions of an excursion. He chose to survey the first five people to get off each of the buses as they arrived back at school.

a Explain why this is not a random sample. **1 mark**

b Why might the results of this survey be biased? **1 mark**

8 After the last census, government officials began making plans to construct a new school at Kurraglen even though there were very few pupils of school age living in the area. Do you think this is likely to be a government bungle? Justify your answer. **1 mark**

9 The number of gold medals won by Australian athletes in each of the Olympic games held from the 1956 games in Melbourne to the 2016 games in Rio is given in the table.

Year	1956	1960	1964	1968	1972	1976	1980	1984	1988	1992	1996	2000	2004	2008	2012	2016
Medals	13	8	6	5	8	0	2	4	3	7	9	16	17	14	7	8

a What is the mean number of gold medals won? ____________ **1 mark**

b What is the mode? ____________ **1 mark**

c What is the range? ____________ **1 mark**

d What is the median? ____________ **1 mark**

10 The number of gold medals won by Australian athletes in each of the first 12 Olympic games held is given in the table.

Year	1896	1900	1904	1908	1912	1920	1924	1928	1932	1936	1948	1952
Medals	2	3	0	1	2	0	3	1	3	0	2	6

a What is the mean number of gold medals won? **1 mark**

____

b What is the mode? ____________ **1 mark**

c What is the range? ____________ **1 mark**

d What is the median? **1 mark**

____

e Comment briefly on the similarities and differences between these results and those from Question 9. **2 marks**

____

____

____

____

____

____

11 A dataset is listed below.

12, 19, 17, 23, 29, 18, 25, 11, 16, 20

Find the:

a mean ____________ **1 mark**

____

b median ____________ **1 mark**

____

c interquartile range ____________ **1 mark**

____

12 A hockey referee recorded the number of goals scored in seven games of hockey.

1, 3, 4, 6, 6, 7, ☐

There is one missing number.

a If the range for the data is 8, explain why the missing number is 9. **1 mark**

____

b What is the interquartile range? **2 marks**

____

____

____

c Complete a box plot. **2 marks**

Goals scored in nine games

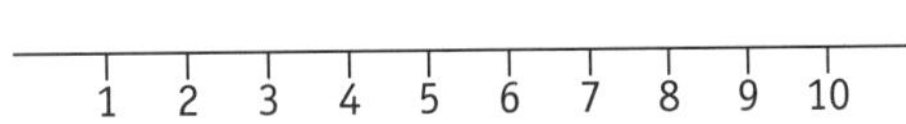

13 Students in a business studies class were given a test out of 50.

The results are shown in the stem-and-leaf plot below.

Results in Business Studies test

Males		**Females**
8	1	69
8740	2	2489
5543	3	77
1	4	269
0	5	

Compare the data for the results of the male and female students and complete the following statements. **4 marks**

a The number of males (______) was __________ than the number of females (______).

b The range of the males' results (______) was __________ than the range of the females' results (______).

c The mode of the males' results (______) was __________ than the mode of the females' results (______).

d The median of the males' results (______) was __________ than the median of the females' results (______).

14 Before students completed a new biology topic their teacher had them sit a pre-test.

At the end of the topic the students sat a post-test to evaluate their understanding of the topic.

Both tests were out of 10 and the results are shown in the dot plots below.

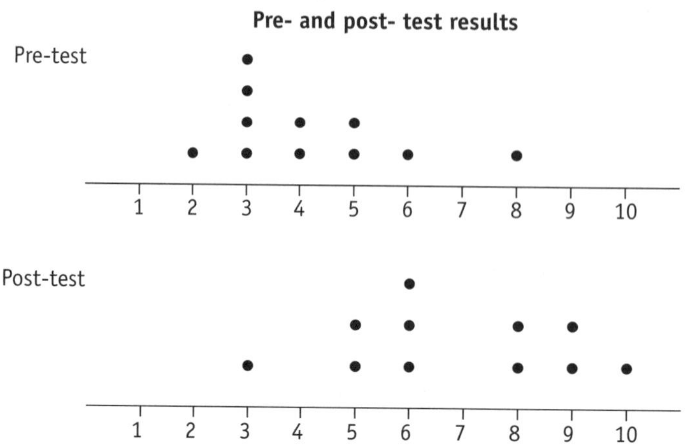

Compare the data for the results of the pre-test and the post-test by commenting on the differences between range, median and interquartile range. **3 marks**

____

____

____

____

____

15 A survey was conducted to find the average time spent exercising by a group of young people and the results were summarised in the box plots below.

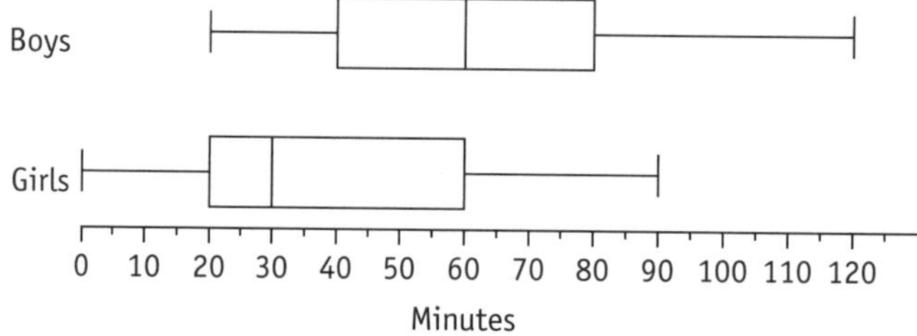

Compare the data for the boys and girls by commenting on the differences between range, median and interquartile range. **3 marks**

____

____

____

____

____

Constructing a scatterplot

QUESTION 1 The table shows the number of weeks spent on a diet and the weight loss over that time for a group of six people.

Time (weeks)	4	5	7	10	12	15
Weight loss (kg)	1	3	2	3	4	3

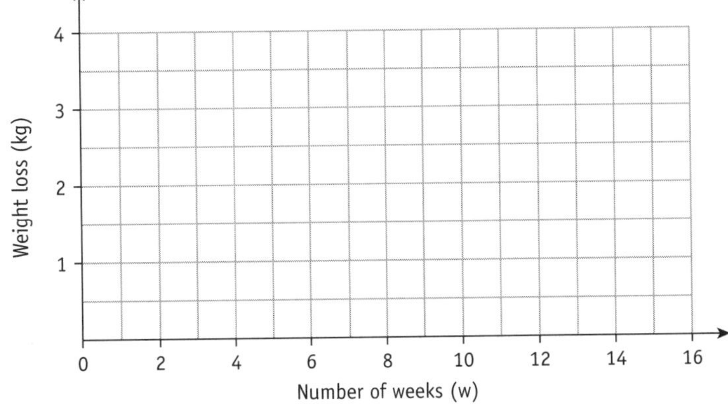

a How many people:

i were on a diet for more than 8 weeks?

____

ii lost more than 2 kg?

____

b Use the data to draw a scatterplot on the graph.

QUESTION 2 The table below shows the maximum temperature across 10 days and the number of drinks purchased using a vending machine in a hospital.

Temperature (°C)	26	32	35	31	26	23	25	28	31	36
Number of drinks	24	34	48	40	28	18	24	29	36	44

a On how many days:

i did the temperature exceed 30°?

____

ii were more than 25 drinks sold?

____

b Draw a scatterplot of the data in the table.

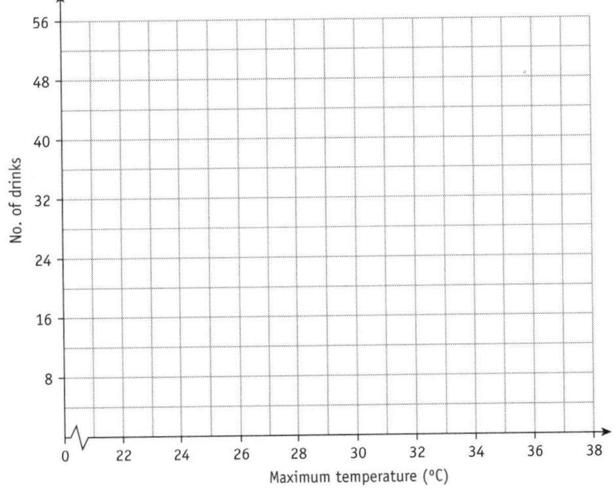

QUESTION 3 The table below shows the height and handspan of eight female students.

Height (cm)	140	170	165	170	155	145	135	150
Handspan (cm)	16	18	21	22	18	17	15	19

Draw a scatterplot of the data in the table.

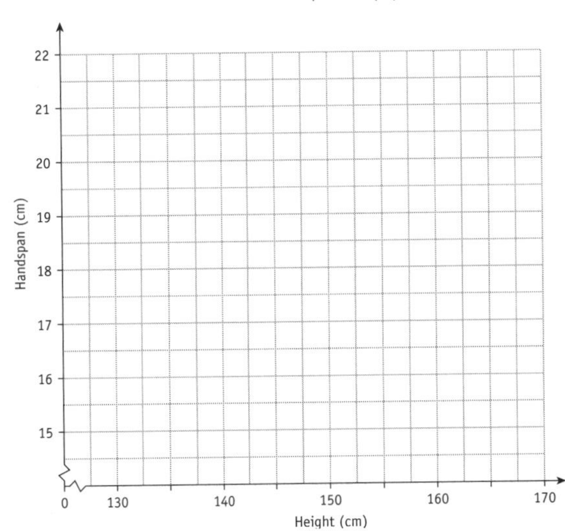

Describing scatterplots

Question 1 The following six scatterplots show associations between two variables, x and y.

i
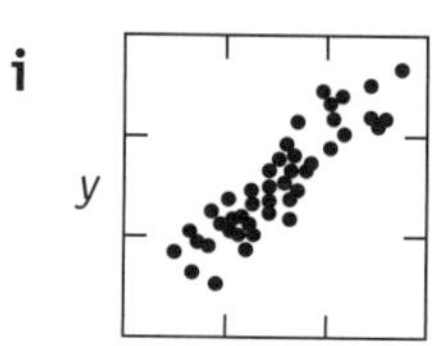

ii
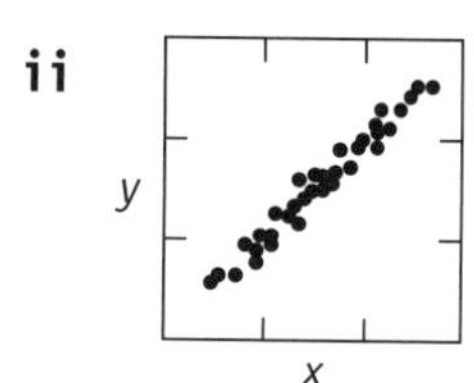

iii
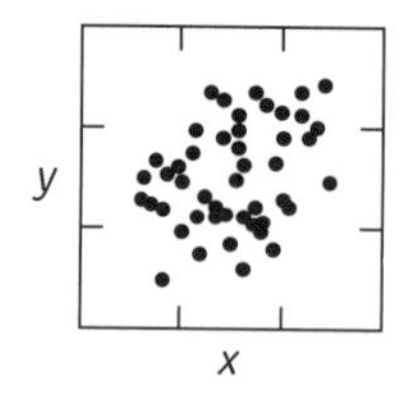

iv
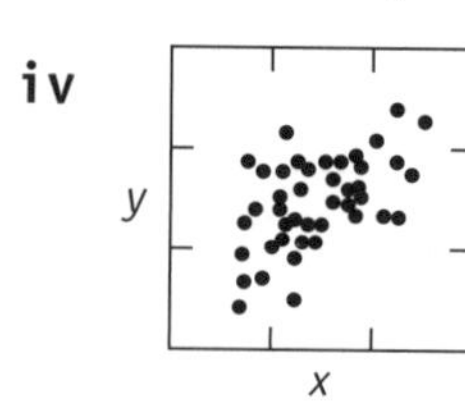

v
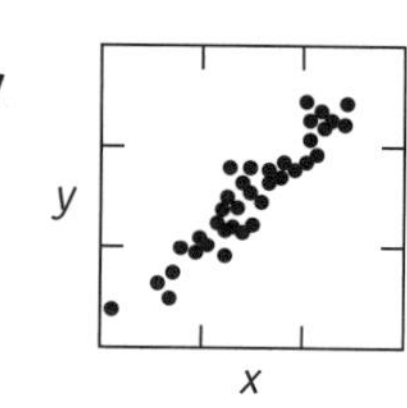

Order these scatterplots in increasing strength of association; that is, with increasing correlation coefficients.

____

Question 2 The following are scatterplots of some associations.

i
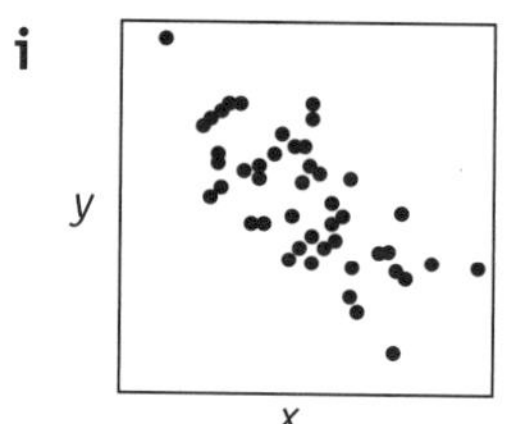

ii
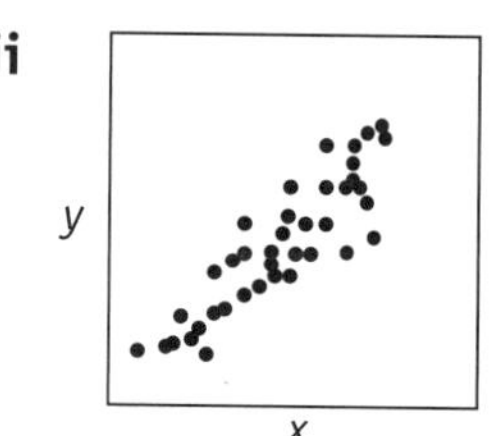

iii
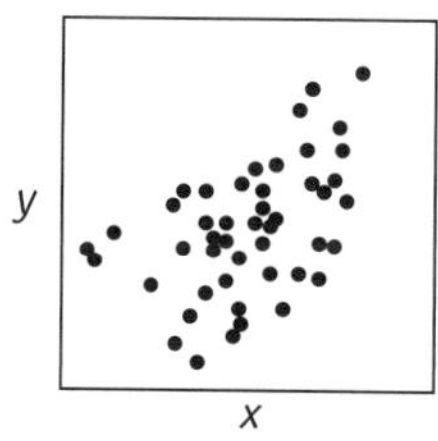

iv
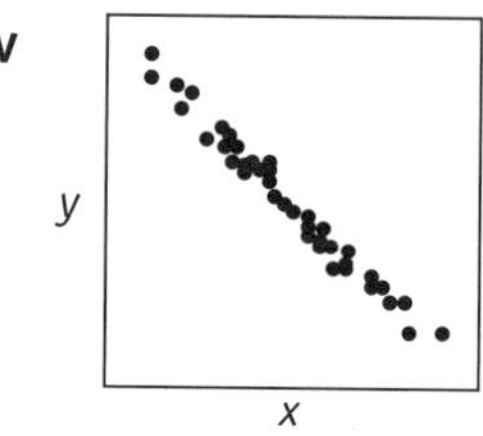

v
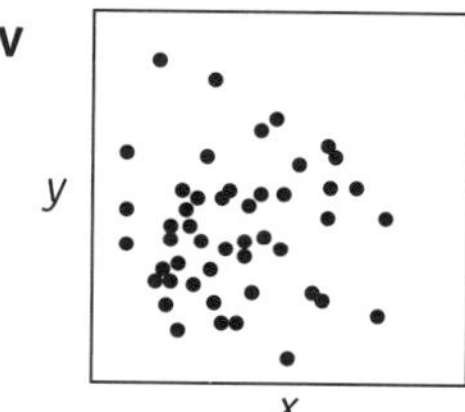

Identify the association showing:

a no correlation ____________

b a weak positive correlation ____________

c a weak negative correlation ____________

d a strong positive correlation ____________

e a strong negative correlation ____________

Question 3 Describe the association of the two variables in the following scatterplots as either **linear** or **non-linear**.

a
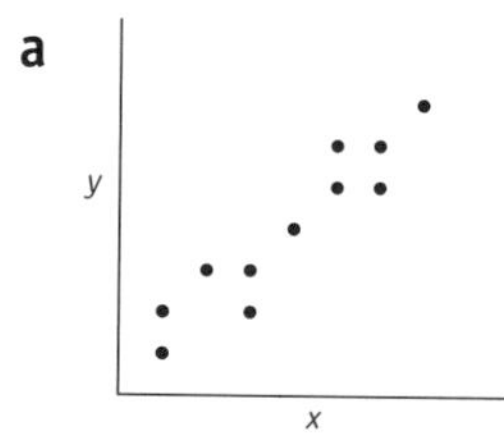
____

b
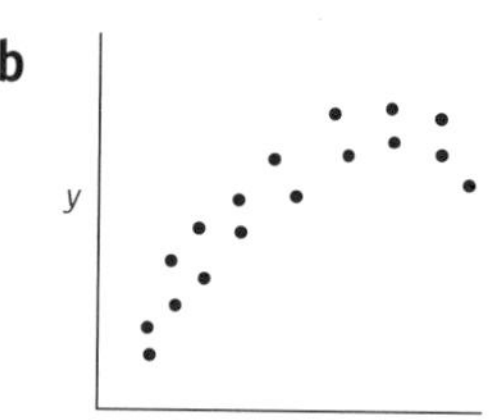
____

c
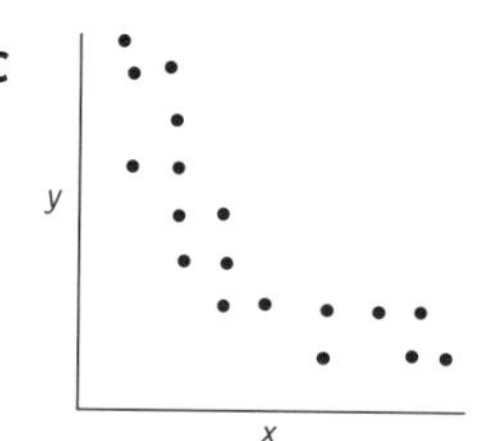
____

d
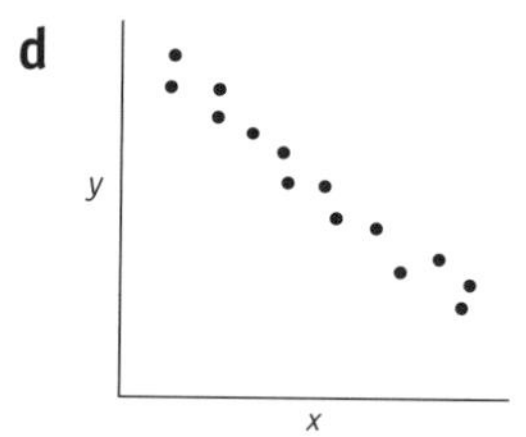
____

e
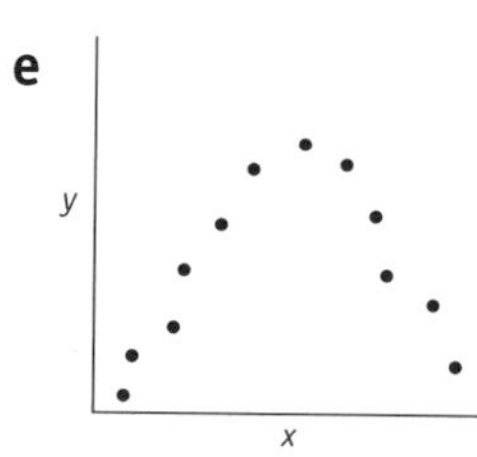
____

Question 4 Use the diagrams below to draw a scatterplot, with 10 points, representing the following associations:

a independent variable p and dependent variable q with a strong negative linear association

b independent variable a and dependent variable b with a moderate non-linear association

Statistical analysis: Exploring and describing data from two variables

Excel MATHEMATICS STANDARD 1
Ch. 7, pp. 141–142

Drawing a line of best fit by eye

QUESTION **1** Use your ruler to draw a line of best fit by eye:

a

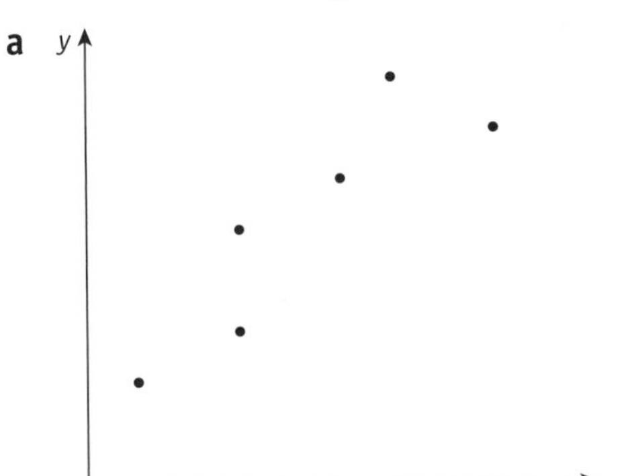

b

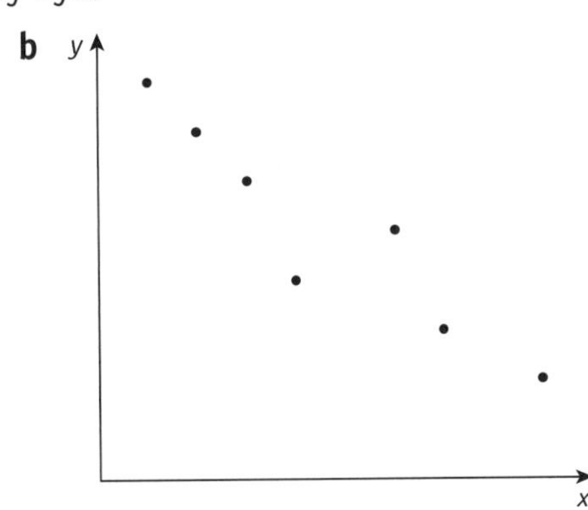

c

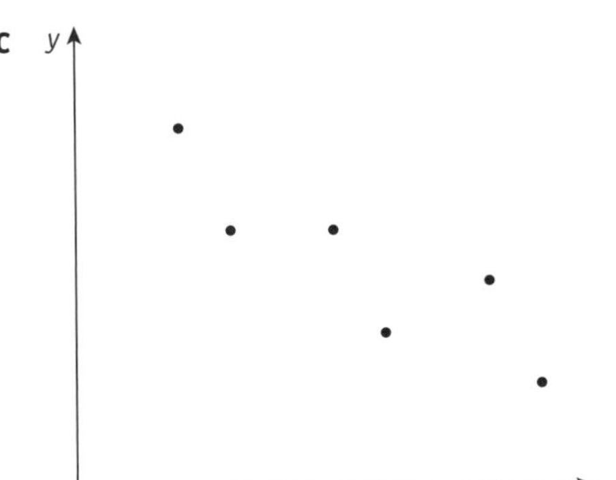

d

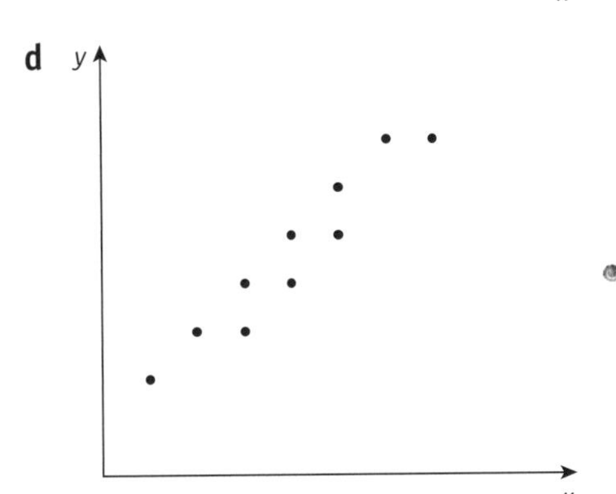

e

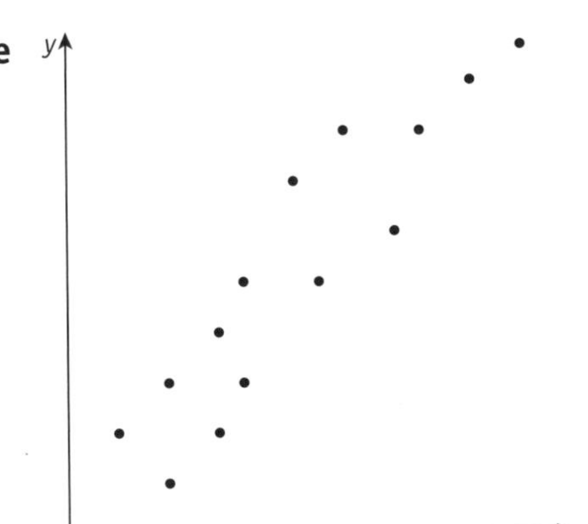

f

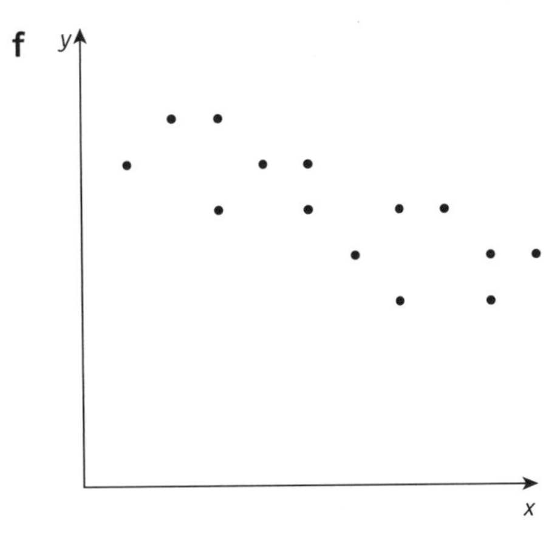

g

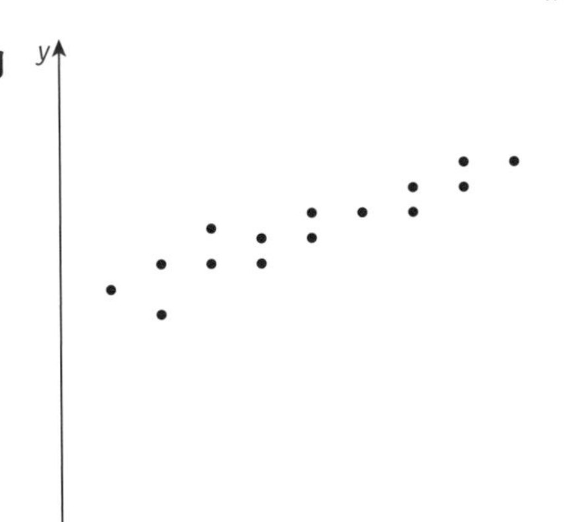

h

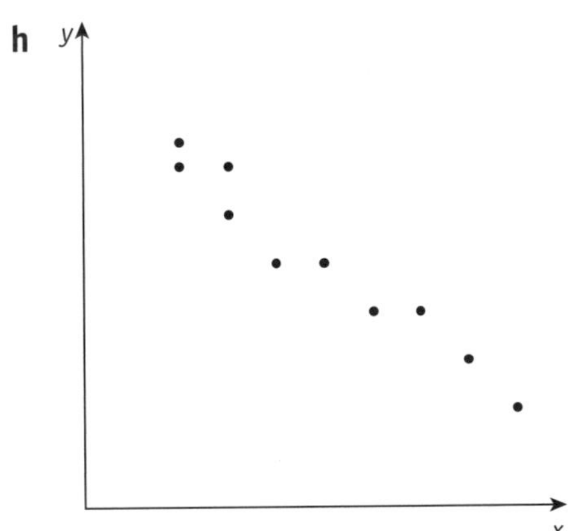

Equation of the line of best fit

Question 1 Find the equation of the line in the form $y = mx + c$ where:

a $m = 6$ and $c = 12$ **b** $m = 0.45$ and $c = -23$ **c** $m = \frac{4}{5}$ and $c = 8$

Question 2 Find the value of c in the line $y = 5x + c$ which passes through the point:

a (8, 63) **b** (21, 54) **c** (12.8, 3.6)

Question 3 Find the gradient of the line of best fit which passes through the points:

a (12, 18) and (22, 68) **b** (8, 23) and (16, 25) **c** (26, 63) and (39, 11)

Question 4 Find the equation of the line of best fit linking x and y, which passes through:

a (26, 37) and (51, 72) **b** (12, 48) and (20, 16) **c** (19, 101) and (51, 265)

Question 5 A line of best fit drawn on a scatterplot links basketballers' heights (h) to the average number of points (p) they scored in a match. The line passes through the points (187, 7) and (192, 15).

a Explain why the line has a gradient of 1.6.

b By substituting the point (187, 7) into the equation $p = 1.6h + c$, find the value of c.

Question 6 The number of skiers (s) each weekend at a ski resort depends on the depth of snow (d) in metres. A scatterplot records data over a series of days using d as the independent variable and s as the dependent variable. A line of best fit is drawn which passes through the points (1.6, 4300) and (2.4, 6400).

a Determine the equation of the line.

b In this context, what does the value of the constant c represent?

Statistical analysis: Exploring and describing data from two variables

Using the line of best fit to make predictions

QUESTION **1** Andrew investigated the relationship between the speed of his exercise bike and his pulse. After riding his bike for one minute at different speeds (km/h) he measured his pulse (beats/min). He recorded the data on a scatterplot and drew a line of best fit by eye.

a Andrew determined the equation of the line of best fit is $p = 3.8s + 65$.

What was his resting pulse, before he started any exercise?

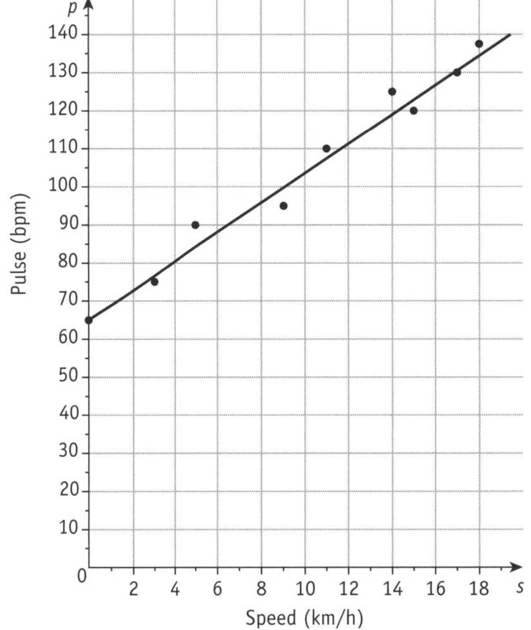

b Estimate his pulse when he is riding at 12 km/h.

____

c Determine the speed he is riding if his pulse is 120 beats per minute.

____

d How reliable do you think estimates outside this data would be? Justify your answer.

____

____

QUESTION **2** The heights (h) in centimetres and length of armspan (a) in centimetres of 12 students are recorded in the table below.

Student	*A*	*B*	*C*	*D*	*E*	*F*	*G*	*H*	*I*	*J*	*K*	*L*
height (h)	180	174	168	172	169	173	175	168	164	167	176	178
armspan (a)	176	175	171	173	168	172	171	172	165	169	172	175

a Use the data to draw a scatterplot using height as the independent variable and armspan as the dependent variable.

b Label the data for students J and L. Draw a line passing through J and L representing a line of best fit for the data for all 12 students.

c Use your line to estimate the:

i armspan of a student with a height of 176 cm

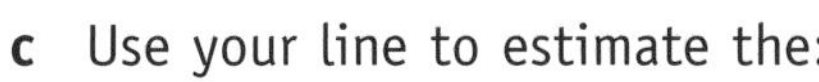

____

____

ii height of a student with an armspan of 170 cm

____

____

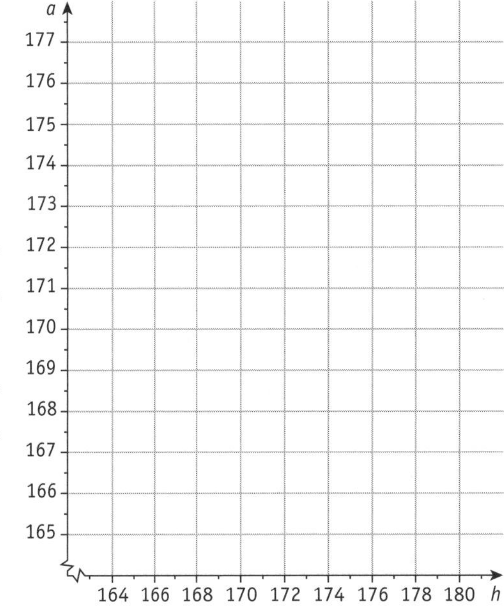

Statistical analysis: Exploring and describing data from two variables

TOPIC TEST — SECTION I

Instructions
- This section consists of 5 multiple-choice questions.
- Each question is worth 1 mark.
- Fill in only ONE CIRCLE for each question.

Time allowed: 7 minutes **Total marks: 5**

1 The diagram shows four scatterplots.
Which scatterplot shows an association which is moderate, linear and negative?

Ⓐ Ⓑ Ⓒ Ⓓ

2 The equation of a line of best fit drawn on a scatterplot is $C = 2.5n + 46$.
Theon used the equation to estimate the value of C when $n = 16$.
Which of these will be the estimate for C?

Ⓐ 12 Ⓑ 64.5 Ⓒ 86 Ⓓ 130

3 Four different students used a scatterplot to draw a line of best fit by eye.

Which of these shows the best line?

Ⓐ
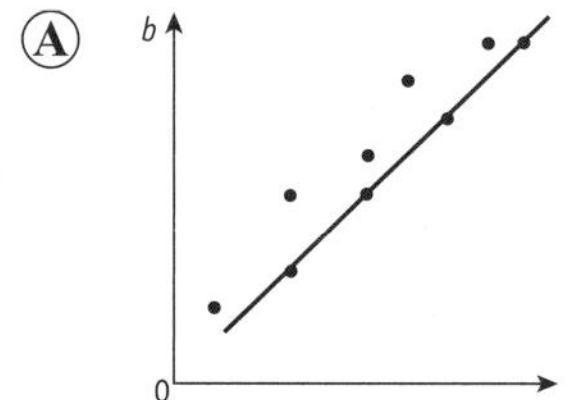

Ⓑ

Ⓒ
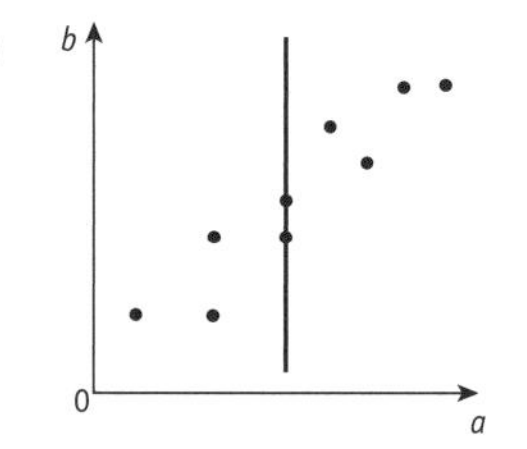

Ⓓ
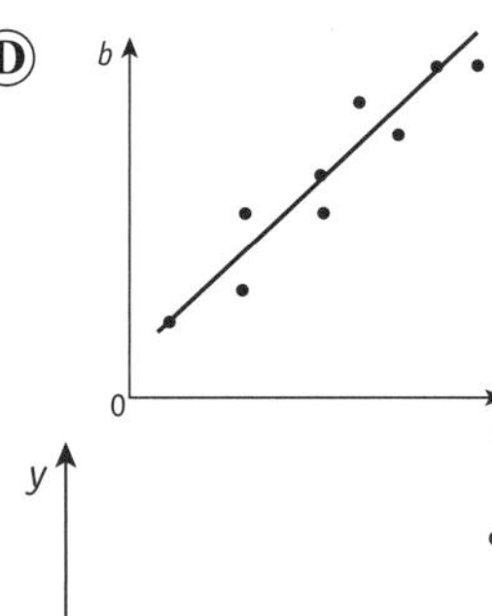

4 A scatterplot is shown on the right.
Which of these best describes the association between x and y?

Ⓐ strong, negative, linear
Ⓑ weak, positive, linear
Ⓒ strong, positive, non-linear
Ⓓ strong, positive, linear

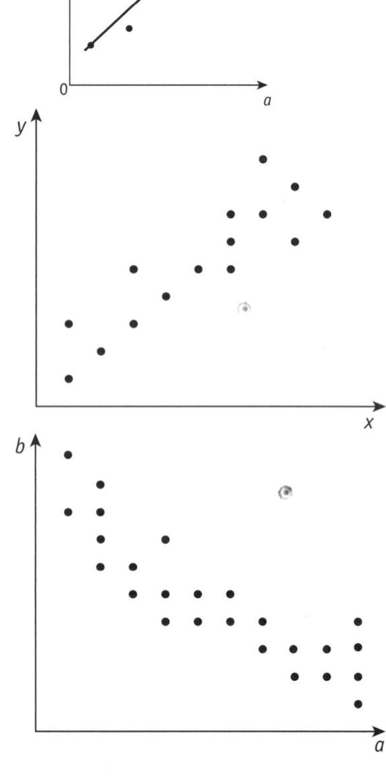

5 The scatterplot shows the association between a and b.

Which of these is a reasonable observation?

Ⓐ As the value of a increases, the value of b increases.
Ⓑ As the value of a increases, the value of b decreases.
Ⓒ As the value of a increases, the value of b remains the same.
Ⓓ As the value of a remains the same, the value of b decreases.

TOPIC TEST

SECTION II

Instructions
- This section consists of 10 questions.
- Show all working.

Time allowed: 53 minutes **Total marks: 35**

6 A group of people were involved in a 15-week fitness program to lose weight.
During the program the participants' Body Mass Index (BMI) was calculated and the results recorded on a scatterplot.

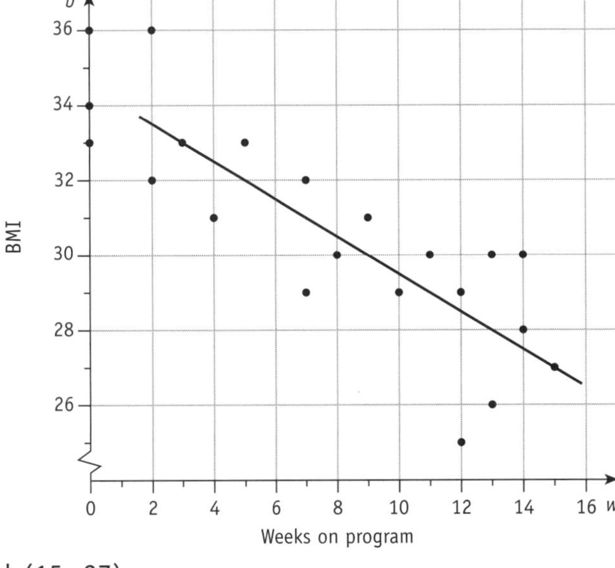

a Describe the association between the number of weeks participants were on the program (w) and their BMI (b) using terms such as weak/strong, positive/negative and linear/non-linear. **1 mark**

____

____

____

b A line of best fit by eye has been drawn on the scatterplot.
The line passes through the two points (3, 33) and (15, 27).
Using a calculation, explain why the gradient of the line is −0.5. **1 mark**

c The equation of the line is $b = -0.5w + c$. Find the value of c. **2 marks**

____

____

____

d Based on the line drawn, what BMI could be expected after 10 weeks on the program? **2 marks**

____

____

____

e Could the model be used to find the number of weeks until the BMI has dropped to 15? Justify your answer. **2 marks**

____

____

____

7 A gym surveyed eight members to find a link between the number of hours per week (n) they attended and their resting heart rate (R). The results are recorded in the table below.

Weekly exercise in hours (h)	1	3	4	4	5	5	6	6	8
Resting heart rate (R)	75	70	75	70	65	50	55	40	40

a Record the data on the scatterplot below. **2 marks**

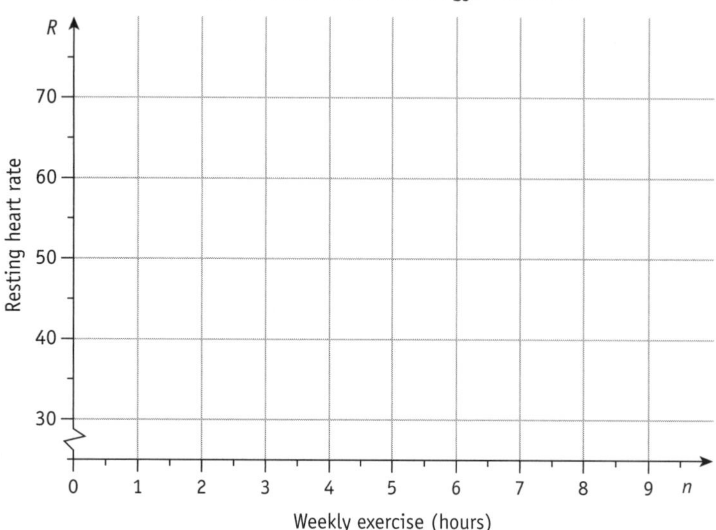

b Using the bivariate data, draw a line of best fit on the scatterplot. **1 mark**

c Michael uses the gym for 7 hours each week.

Use your line to estimate his resting heart rate. **1 mark**

---

8 On each of the diagrams below, plot eight points that illustrate a set of data with the following association:

a strong positive and linear association **1 mark**

b weak, negative and linear association **1 mark**

c no association **1 mark**

9 Mr Henderson measured the heights and handspans, in centimetres, of 10 students in his Year 12 mathematics class. The results are recorded in the table below.

Height (x)	164	178	175	173	179	181	177	180	165	169
Handspan (y)	16	18	17	18	20	20	20	21	17	16

a Represent this data on the scatterplot below. **2 marks**

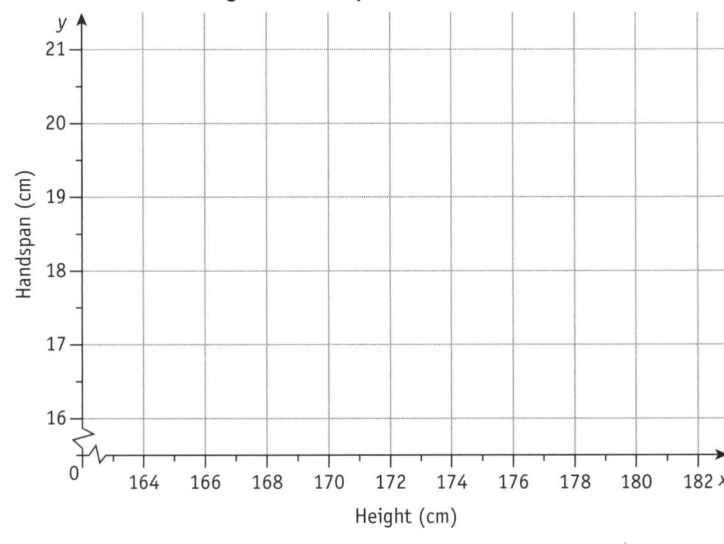

b Mr Henderson draws a line of best fit passing through the points (173, 18) and (181, 20). Represent this line on the scatterplot above. **1 mark**

c Show that the gradient of the line is 0.25. **1 mark**

____

____

d The equation of the line is $y = 0.25x - 25.25$.

Laura was absent from the class.

If Laura is 172 cm tall, use the equation to estimate the length of her handspan. **2 marks**

____

____

____

10 The table relates the percentage of lecture hours attended by 10 students and their results in an end-of-semester examination.

Lecture hours attended (%)	85	60	45	50	95	90	100	85	75	90
Exam result (%)	80	50	35	60	75	90	85	70	55	80

By recording these results on the scatterplot below, comment on the association of lecture hours attended and the result in the end-of-semester examination. **3 marks**

____

____

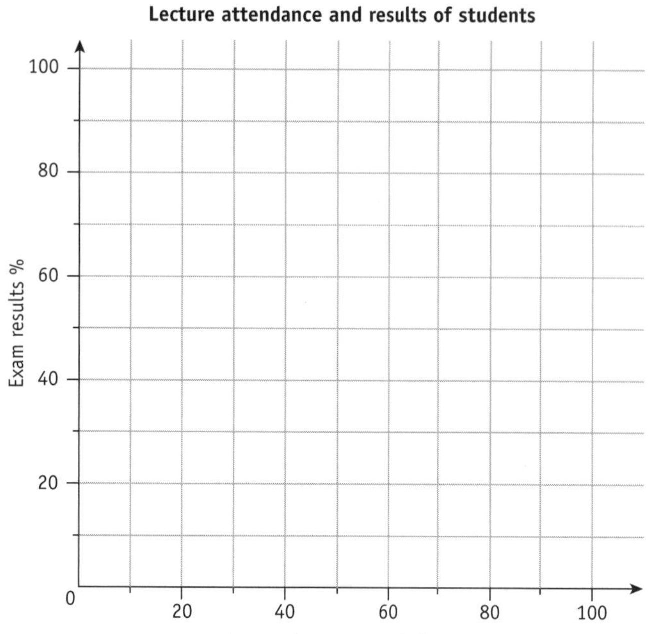

11 A scientist collected data on the length of the femur (thighbone) and the height of a skeleton.

The data is shown in the table below.

Length of femur (cm)	48	42	45	47	43	40	48	47	42	44
Height of skeleton (cm)	178	166	176	182	168	154	184	180	168	174

a Represent the data on the scatterplot below. **2 marks**

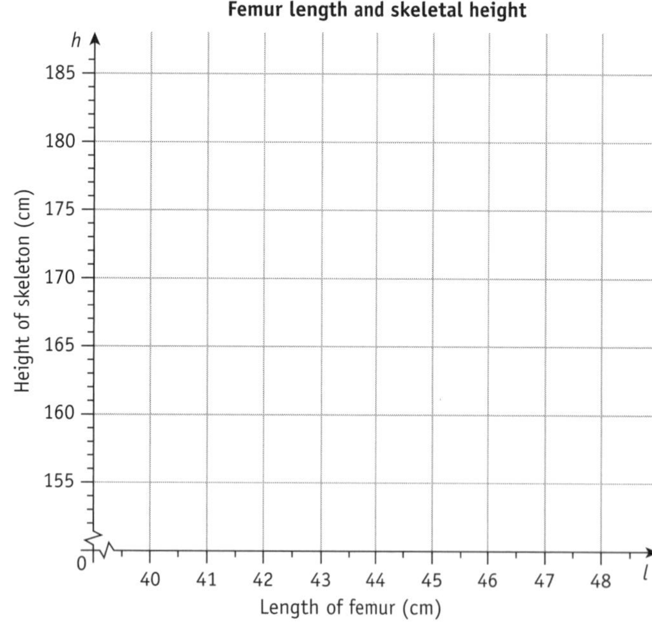

b Using the scatterplot, comment on the association of the length of the femur and the height of the skeleton. **1 mark**

____

____

c The scientist draws a line of best fit passing through the points (43, 168) and (47, 180).

Represent this line on the scatterplot above. **1 mark**

d Explain why the gradient of this line is 3. **1 mark**

____

____

e Let l = length of the femur and h = height of the skeleton.

By using the gradient and the point (43, 168), explain why the equation of the line is $h = 3l + 39$. **2 marks**

____

____

____

f The scientist measures the height of another skeleton as 177 cm. Using the equation of the line, estimate the length of the femur. **2 marks**

____

____

____

g Another skeleton is investigated.

The length of the femur is 20 cm.

Use the equation to estimate the height and comment on the accuracy of this model. **2 marks**

____

____

____

CHAPTER 10
Networks: Introduction to networks

Excel MATHEMATICS STANDARD 1
Ch. 8, pp. 158–162

Identifying network terminology

QUESTION **1** For the networks shown, state:

a

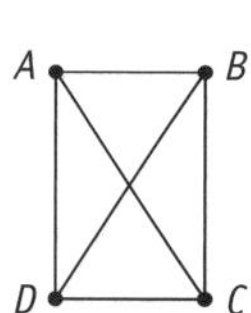

b

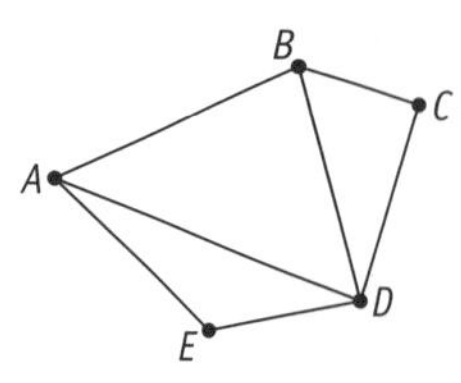

c

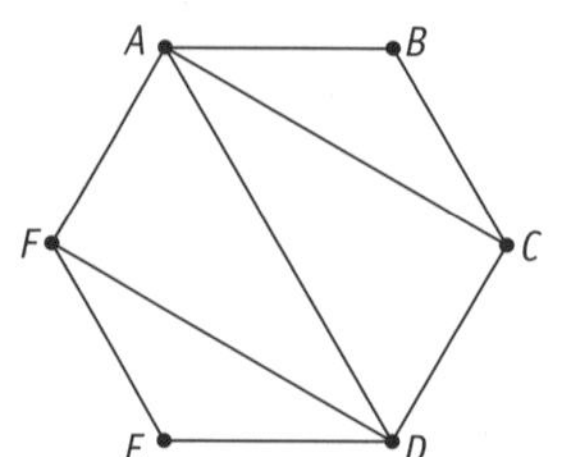

d

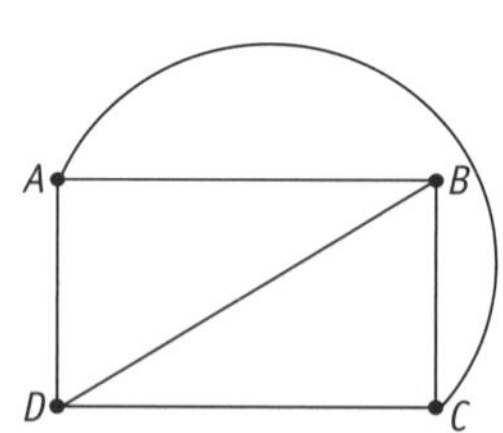

i the number of vertices

ii the number of edges (arcs)

iii the degree of vertex C

iv vertices adjacent to B

QUESTION **2** Draw a network diagram with these features:

a vertices A, B and C, where each vertex has degree 2

b vertices P, Q, R and S, where each vertex has degree 2

c vertices W, X, Y and Z, where each vertex has degree 3

QUESTION **3** For the directed networks shown, state the number of vertices, edges (arcs) and the degree of vertex C:

a

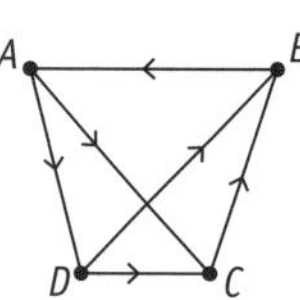

b

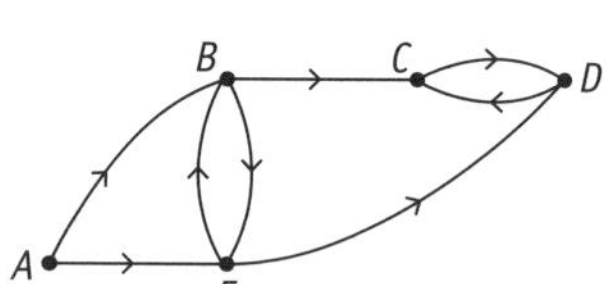

c

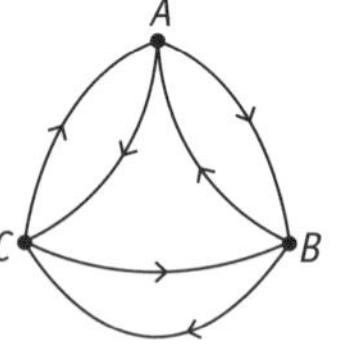

QUESTION **4**

a Name the isolated vertex.

b What is the weight of edge AB?

c What is the degree of B?

d What is the weight of path BCD?

e What is the weight of circuit $BCDB$?

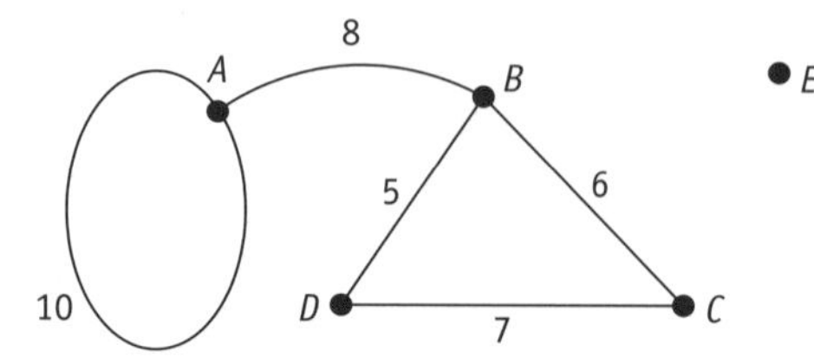

Networks and tables 1

QUESTION 1 Use the tables to add weights to the edges of the network diagrams:

a

	A	*B*	*C*	*D*
A	–	6	10	7
B	6	–	5	–
C	10	5	–	8
D	7	–	8	–

b

	G	*H*	*J*	*K*	*L*
G	–	5	–	9	6
H	5	–	7	–	4
J	–	7	–	8	5
K	9	–	8	–	7
L	6	4	5	7	–

c

	P	*Q*	*R*	*S*	*T*	*U*	*V*
P	–	3	–	–	–	–	11
Q	3	–	4	–	–	–	6
R	–	4	–	7	8	–	9
S	–	–	7	–	5	–	–
T	–	–	8	5	–	8	–
U	–	–	–	–	8	–	3
V	11	6	9	–	–	3	–

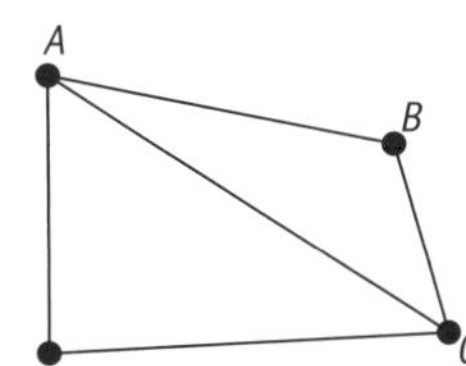

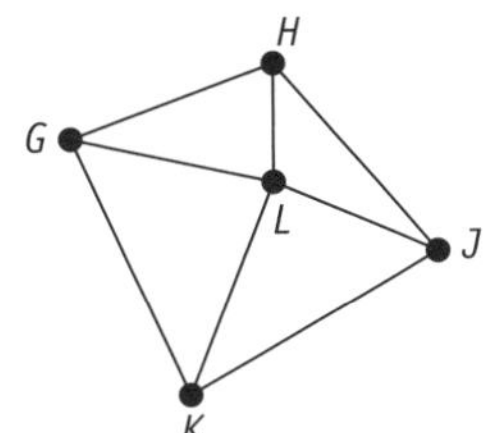

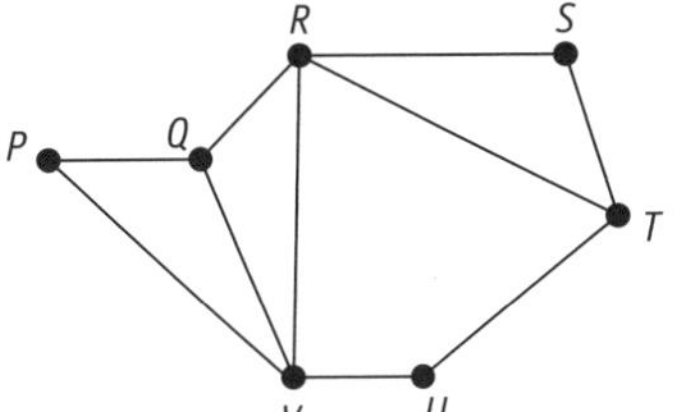

QUESTION 2 Use the network diagrams to complete the tables:

a

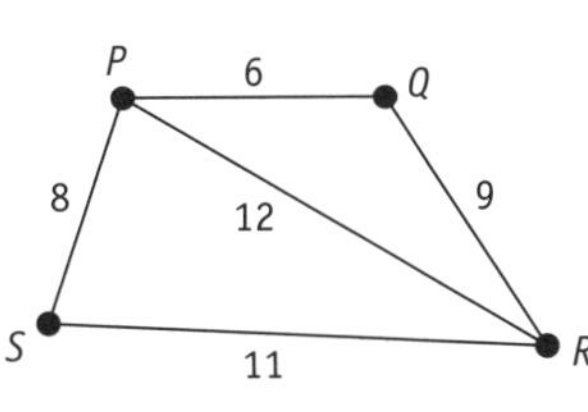

	P	*Q*	*R*	*S*
P				
Q				
R				
S				

b

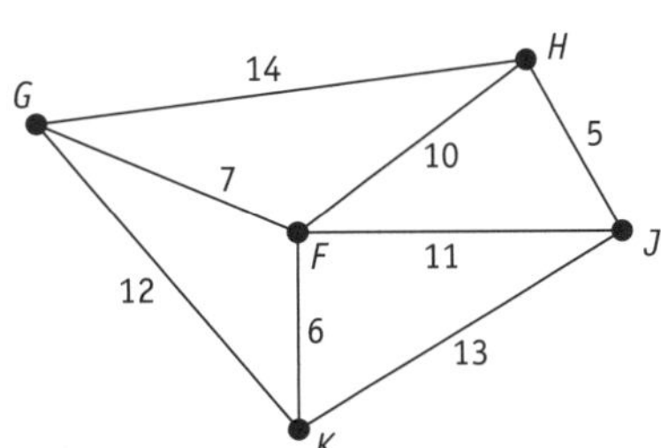

	F	*G*	*H*	*J*	*K*
F					
G					
H					
J					
K					

c

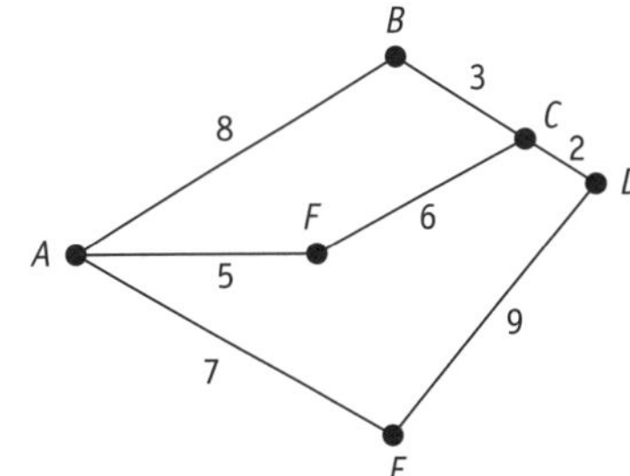

	A	*B*	*C*	*D*	*E*	*F*
A						
B						
C						
D						
E						
F						

QUESTION 3 The table shows the distances by air between the Hunter Valley airports: Cessnock (*C*), Maitland (*M*), Newcastle (*N*) and Singleton (*S*). Complete the network diagram to represent the data in the table.

Flight distances (km)				
	Cessnock	**Maitland**	**Newcastle**	**Singleton**
Cessnock	–	24	54	34
Maitland	24	–	32	39
Newcastle	54	32	–	68
Singleton	34	39	68	–

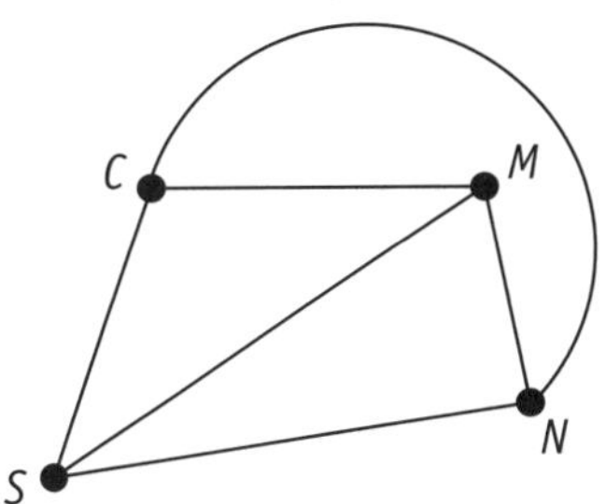

Networks and tables 2

QUESTION 1 Use the network diagrams to complete the tables:

a

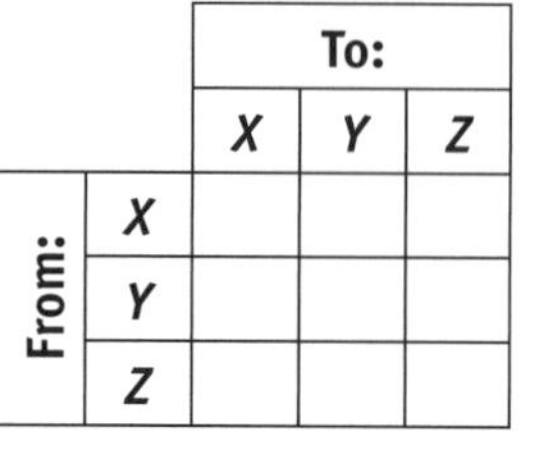

		To:		
		X	Y	Z
From:	X			
	Y			
	Z			

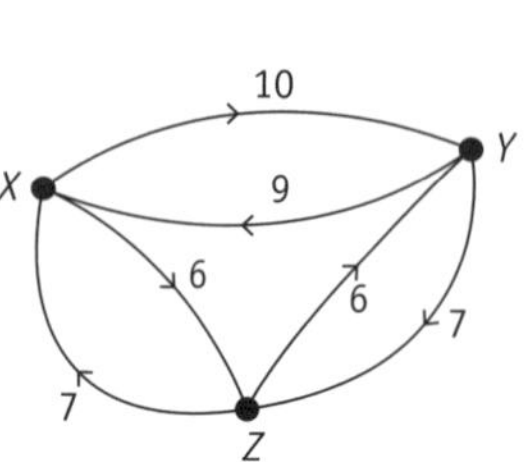

b

		To:			
		A	B	C	D
From:	A				
	B				
	C				
	D				

c

		To:					
		P	Q	R	S	T	U
From:	P						
	Q						
	R						
	S						
	T						
	U						

QUESTION 2 Use the tables to add weights to the edges of the network diagrams:

a

		To:			
		P	Q	R	S
From:	P	–	4	–	–
	Q	5	–	–	7
	R	–	6	–	–
	S	–	8	–	–

b

		To:				
		A	B	C	D	E
From:	A	–	2	4	–	
	B	2	–	–	–	–
	C	–	–	–	–	3
	D	–	5	4	–	–
	E	–	–	–	3	–

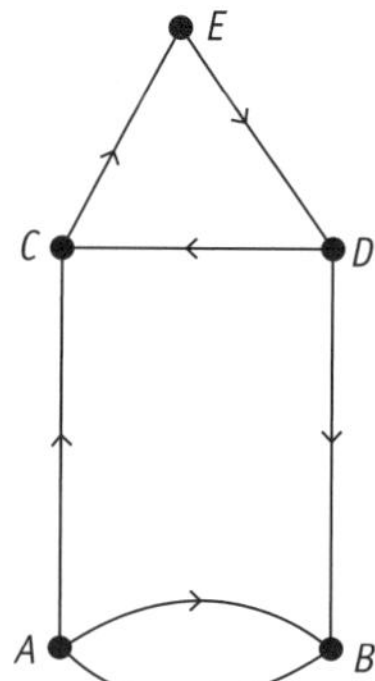

c

		To:				
		V	W	X	Y	Z
From:	V	–	3	–	–	5
	W	4	–	5	–	–
	X	–	–	–	6	–
	Y	–	–	7	–	8
	Z	4	–	–	–	–

QUESTION 3 Use the table to complete the network diagram by using arrows and writing the weights on each edge.

		To:						
		A	B	C	D	E	F	G
From:	A	–	12	–	–	6	5	–
	B	–	–	15	–	–	–	–
	C	–	–	–	6	–	–	–
	D	–	–	7	–	–	–	8
	E	7	–	–	8	–	–	–
	F	–	–	–	–	5	–	4
	G	–	–	6	–	–	3	–

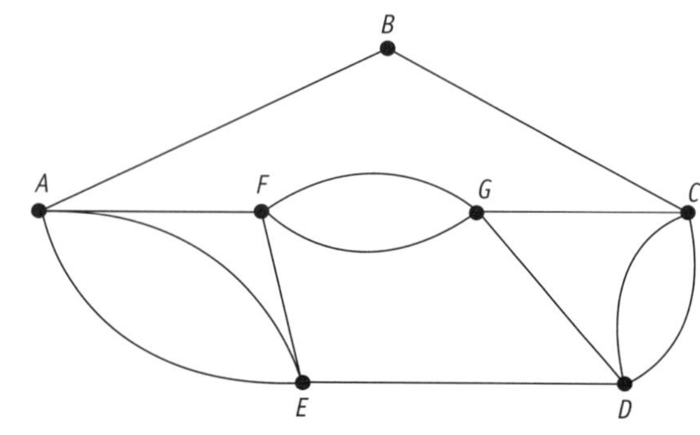

Networks: Introduction to networks

Excel MATHEMATICS STANDARD 1
Ch. 8, pp. 162–165

Maps and networks

QUESTION 1 The sketches show islands in the middle of a river joined to the riverbanks by bridges. For example, in part **a** there are two bridges from *A* to *B*. Complete the network diagrams:

a

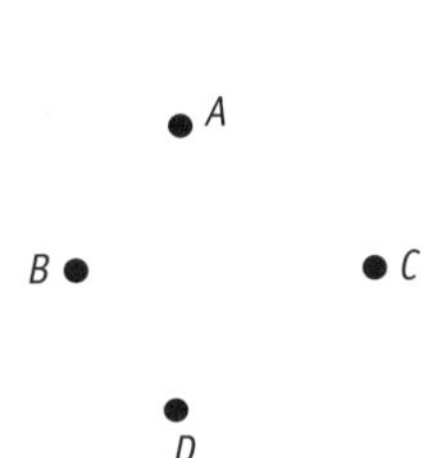

b

QUESTION 2 Use the maps to complete the network diagrams:

a

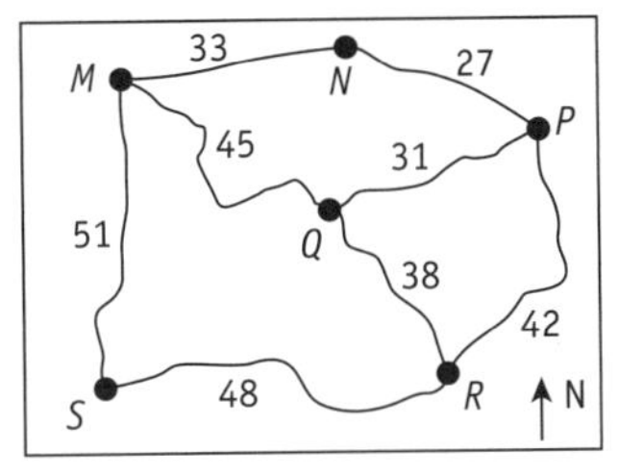

b

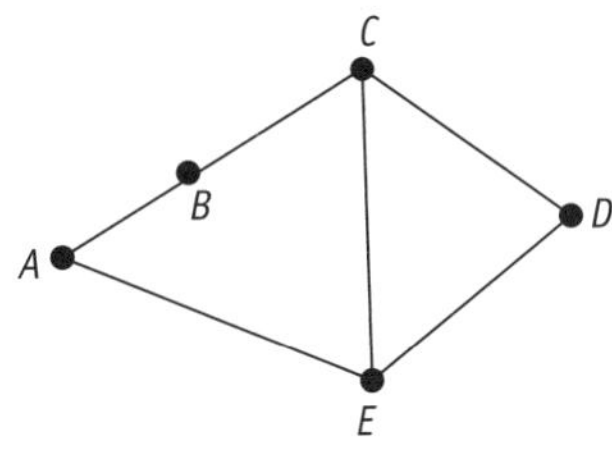

QUESTION 3 Draw a network diagram to represent the roads between the towns *A*, *B*, *C*, *D* and *E*, if:

- Town *A* is 32 km from town *B* and 43 km from town *E*
- Town *B* is 54 km from town *D* and 29 km from town *C*
- Town *D* is 41 km from town *C* and 34 km from town *E*

QUESTION 4 A 'round-robin competition' is where each team plays the other teams in their 'pool' once. Draw network diagrams to show the games played for a pool of:

a four teams: *A*, *B*, *C* and *D*

b six teams: *P*, *Q*, *R*, *S*, *T* and *U*

QUESTION 5 Three airlines are named according to their base city and can only fly from their base airport to and from two other airports. Airline *A* flies to and from *B* and *E*, *C* flies to and from *B* and *E*, and *D* flies to and from *C* and *E*. The cost of a ticket for each airline is the same for any of its routes. Airline *A* charges $80, *C* charges $70 and *D* charges $90. Draw a weighted network diagram to represent the flight routes and costs.

Networks: Introduction to networks

TOPIC TEST

SECTION I

Instructions
- This section consists of 5 multiple-choice questions.
- Each question is worth 1 mark.
- Fill in only ONE CIRCLE for each question.

Time allowed: 7 minutes **Total marks: 5**

1 In the network diagram, how many vertices have an odd degree?

(A) 3

(B) 4

(C) 5

(D) 9

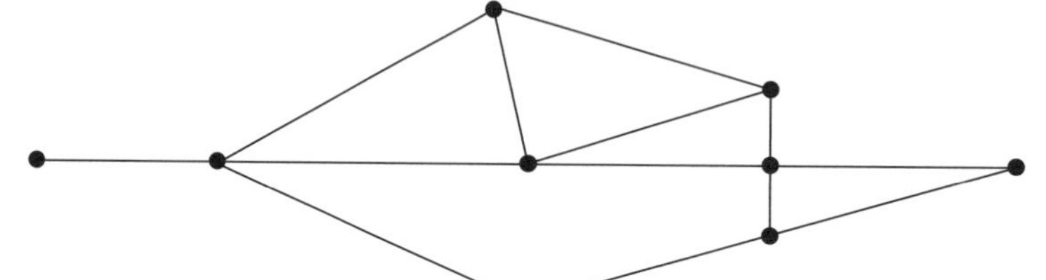

2 The table shows the friendships between four girls: Abbie (*A*), Belinda (*B*), Charlie (*C*) and Didi (*D*).

	A	*B*	*C*	*D*
A		✓		✓
B	✓		✓	✓
C		✓		✓
D	✓	✓	✓	

A network diagram is to be drawn where each vertex represents a girl and an edge joining vertices represents a friendship. Which of these is the correct network diagram?

(A)
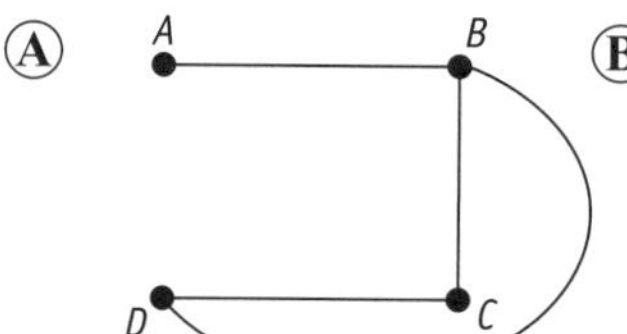

(B)
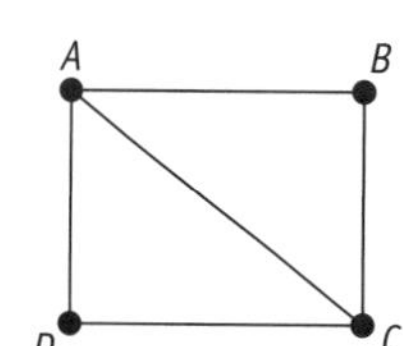

(C)
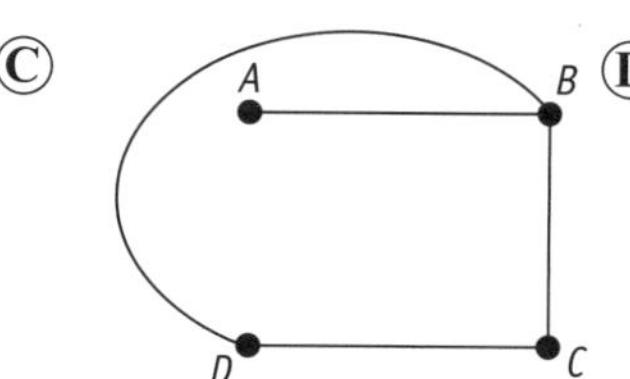

(D)
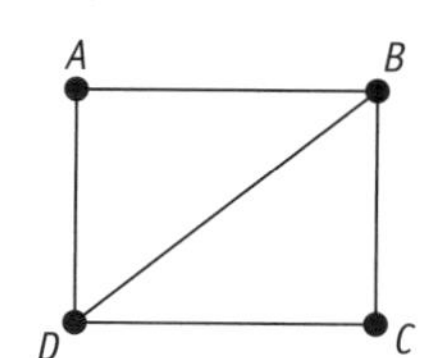

3 A network diagram is given. What is the degree of vertex *D*?

(A) 1 (B) 2

(C) 3 (D) 4

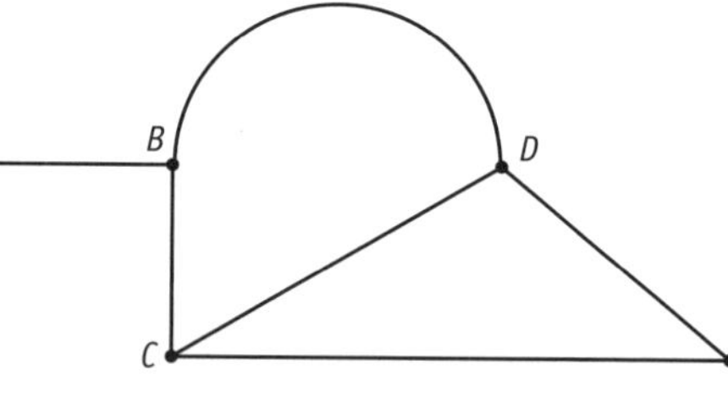

4 The table has been used to draw the network diagram below. Some of the weights have been shown.

What is the value of *k*?

(A) 1 (B) 2

(C) 4 (D) 5

	A	*B*	*C*	*D*	*E*
A	–	5	3	7	–
B	5	–	–	–	2
C	3	–	–	1	5
D	7	–	1	–	4
E	–	2	5	4	–

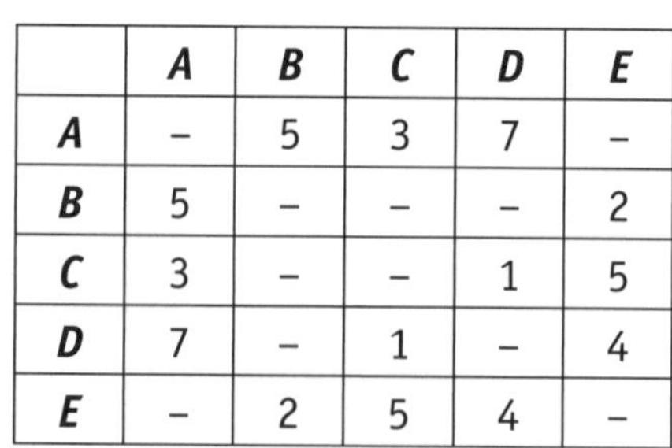

5 The diagram shows a network.

What is the sum of the degrees of the vertices in this network?

(A) 5 (B) 10

(C) 12 (D) 15

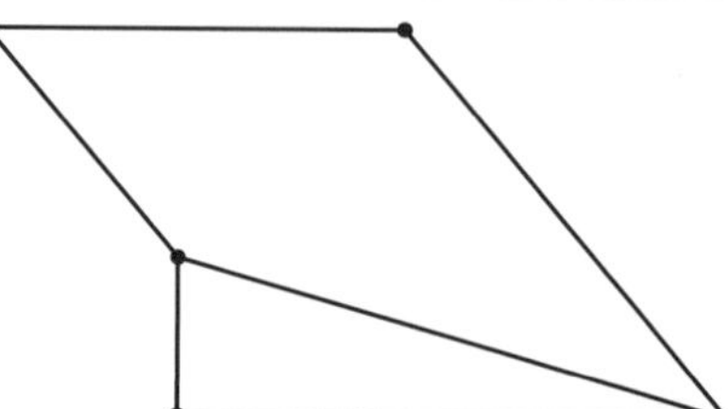

TOPIC TEST

SECTION II

Instructions
- This section consists of 12 questions.
- Show all working.

Time allowed: 53 minutes **Total marks: 35**

6 For the network, state:

a the number of vertices ______________________ **1 mark**

b the degree of vertex C ______________________ **1 mark**

c the name of the vertices adjacent to E ______________________ **1 mark**

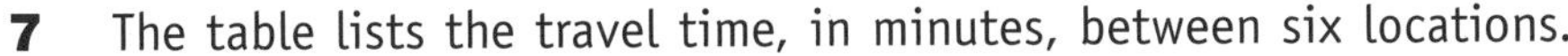

7 The table lists the travel time, in minutes, between six locations.

		To:					
		A	*B*	*C*	*D*	*E*	*F*
From:	*A*	–	8	–	–	–	5
	B	–	–	7	–	–	–
	C	–	–	–	5	6	–
	D	–	–	9	–	8	–
	E	6	9	8	–	–	–
	F	–	–	–	–	8	–

Use the table to complete the network diagram by using arrows and writing the weights on each edge. **3 marks**

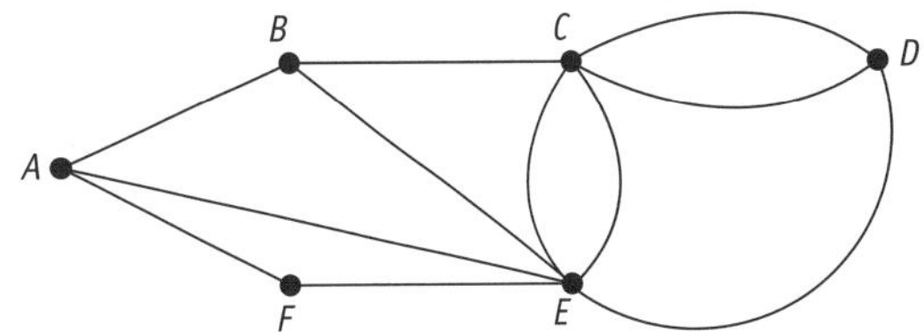

8 Draw a network diagram to represent the roads between the eight towns *A, B, C, D, E, F, G and H* if:
- Town B is 12 km from town A, 15 km from town C, 8 km from town D
- Town E is 8 km from town C, 7 km from town D and 9 km from town G
- Town H is 14 km from town G and 10 km from town C

2 marks

9 Lara investigated the time it takes to fly between various cities.
She recorded the information in the table below.

	Adelaide	Broken Hill	Mildura	Dubbo	Sydney
Adelaide	–	1 h 20 min	35 min	–	1 h 45 min
Broken Hill	1 h 20 min	–	1 h 15 min	2 h 5 min	2 h 50 min
Mildura	35 min	1 h 15 min	–	–	–
Dubbo	–	2 h 5 min	–	–	1 h 10 min
Sydney	1 h 45 min	2 h 50 min	–	1 h 10 min	–

Complete the network diagram to show how towns are connected, with weights on the edges showing flight times. **3 marks**

Mildura

Adelaide

Broken Hill

Dubbo

Sydney

10 For the network diagram below, find:

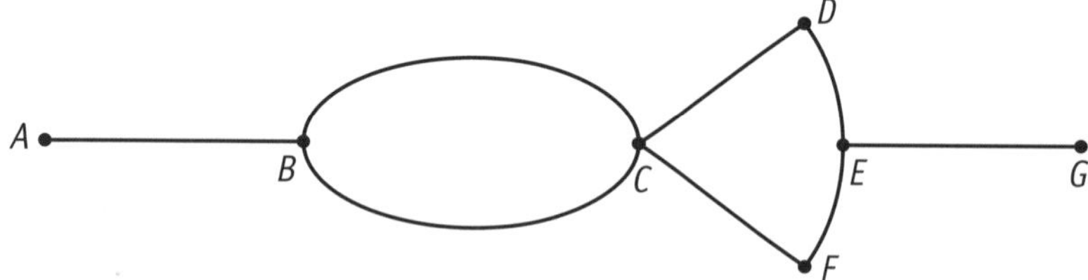

a the number of vertices **1 mark**

____

b the number of vertices with an odd degree. **1 mark**

____

11 Complete the network diagram below to illustrate the following table. **3 marks**

	A	*B*	*C*	*D*	*E*	*F*
A	–	12	9	11	13	–
B	12	–	10	–	–	14
C	9	10	–	8	–	5
D	11	–	8	–	–	6
E	13	–	–	–	–	8
F	–	14	5	6	8	–

• *B*

A • • *C* • *F*

• *D*

• *E*

12 A network diagram is shown below. Complete the table that summarises the information. **2 marks**

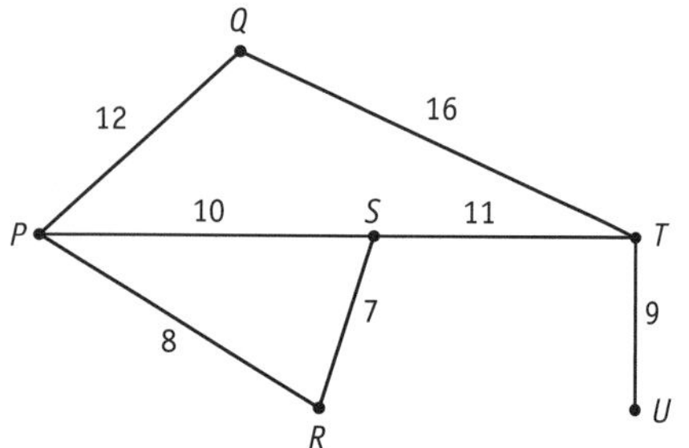

	P	*Q*	*R*	*S*	*T*	*U*
P						
Q						
R						
S						
T						
U						

13 Use the table to draw a directed network. **3 marks**

		To:						
		P	***Q***	***R***	***S***	***T***	***U***	***V***
From:	***P***	–	8	–	–	–	–	–
	Q	–	–	12	14	11	–	–
	R	–	–	–	–	–	16	–
	S	–	–	7	–	–	9	–
	T	–	–	–	–	–	10	–
	U	–	–	–	–	–	–	13
	V	–	–	–	–	–	–	–

14 The table represents a network.

Find:

a the degree of vertex *B* **1 mark**

____

b the number of vertices with even degrees **1 mark**

____

	A	***B***	***C***	***D***	***E***
A	–	6	–	3	5
B	6	–	4	–	7
C	–	4	–	5	–
D	3	–	5	–	2
E	5	7	–	2	–

15 The diagram below shows a directed network.

Use the network to complete the table. **3 marks**

		To:						
		A	***B***	***C***	***D***	***E***	***F***	***G***
From:	***A***							
	B							
	C							
	D							
	E							
	F							
	G							

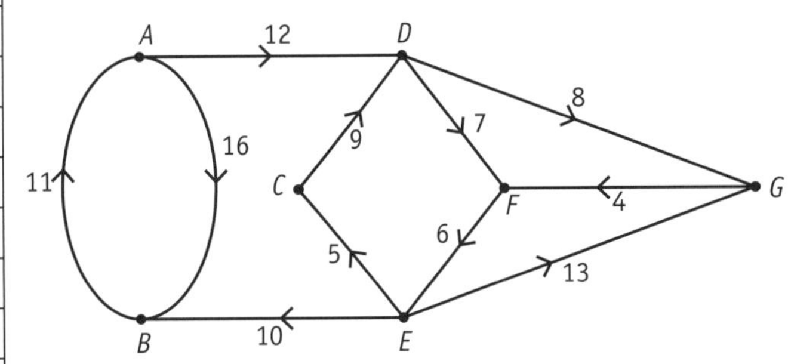

16 Here is a table representing distances on one-way roads between six locations.

		To:					
		A	***B***	***C***	***D***	***E***	***F***
From:	***A***	–	13	15	–	–	24
	B	–	–	–	–	–	19
	C	–	10	–	–	7	–
	D	16	–	–	–	–	20
	E	–	18	–	11	–	–
	F	–	–	–	–	12	–

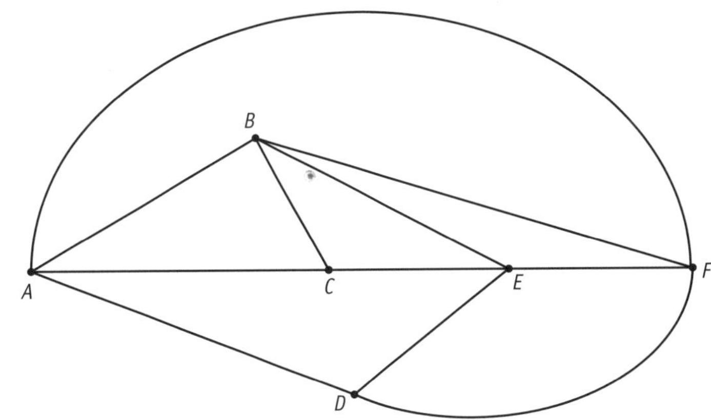

The network is drawn below.

Complete the diagram by recording weights and arrows on each edge. **3 marks**

17 A network of water pipes joins locations P, Q, R, S and T.
Pipes connect P to each of the other locations.
Also, pipes join S to Q and S to T, while a pipe joins R to T.
Each pipe connected to P is 16 m long while the other pipes are 18 m in length.
The cost of laying the water pipe is $16.20 per metre.

a Complete the network diagram and include the weights on each edge. **2 marks**

R Q

P

T S

b What is the total cost? **2 marks**

____

____

____

c At a later time another pipe is used to connect R to S. The cost of laying this new pipe is $420. If the cost per metre has increased by $1.30, what is the length of this new pipe? **2 marks**

____

____

____

Chapter 11
Networks: Shortest paths

Excel MATHEMATICS STANDARD 1
Ch. 8, pp. 168–170

Paths

Question 1 The following network diagrams are to be walked so that every edge is to be travelled once, and only once.

a

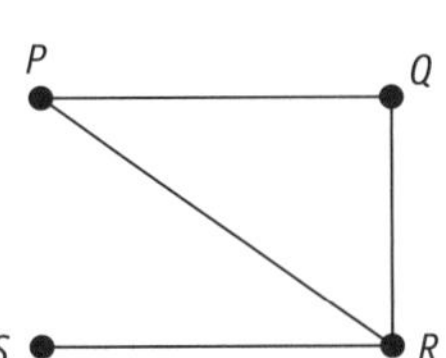

Write the possible start/finish vertex. __________

b

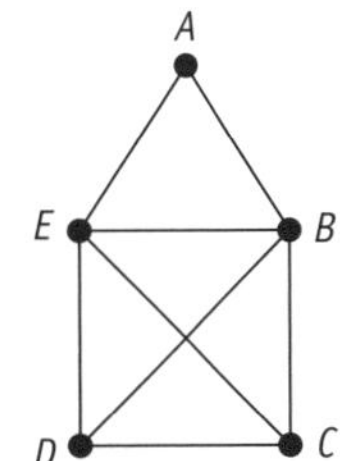

Write the start and finish vertices. __________

Question 2 In the diagrams, name the different paths that are possible to travel from *A* to *E*.

a

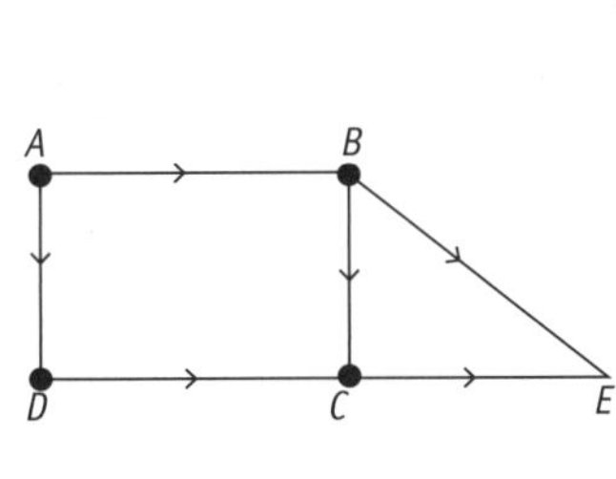

____

____

b

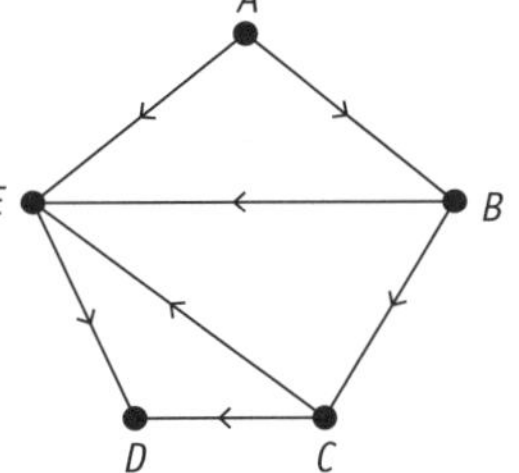

____

____

c

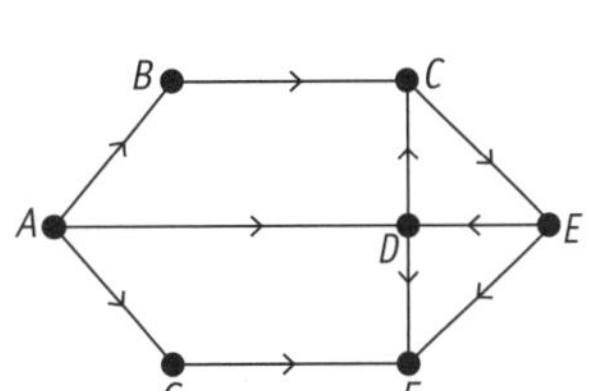

____

____

d

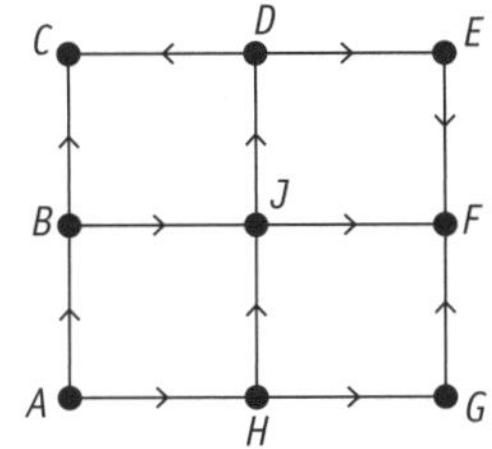

____

____

Question 3 Here are two network diagrams. Simon needs to travel from *A* to *B*. He is not allowed to backtrack—in this case, Simon always moves towards the right or towards the top of the diagram. How many paths are possible?

a

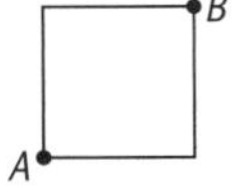

____

b

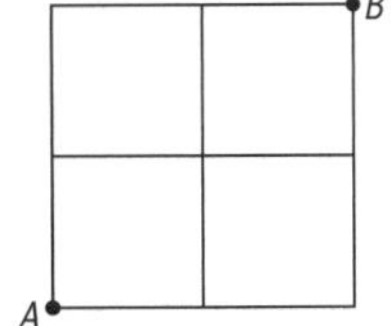

____

Question 4 How many different paths are possible between *A* and *C*, if backtracking is not allowed?

a

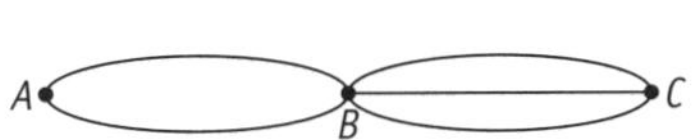

____

b

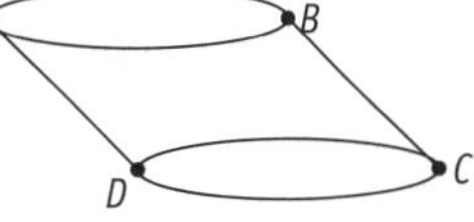

____

c

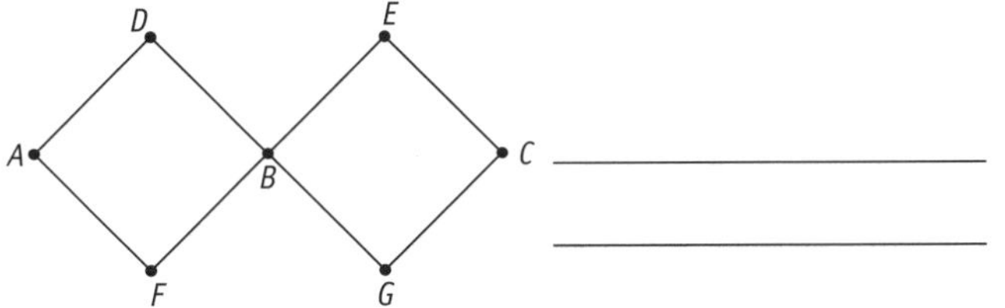

____

____

____

d

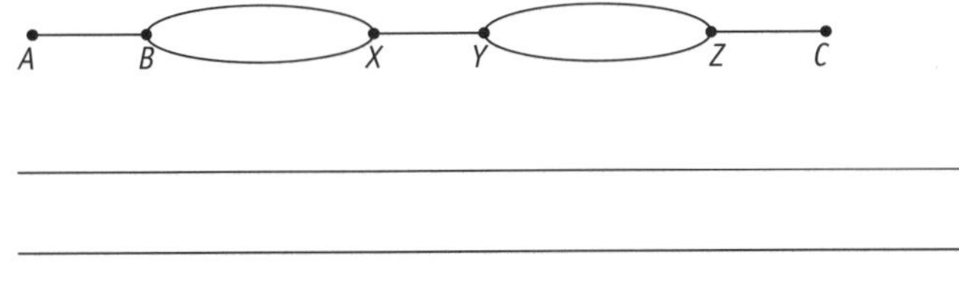

____

____

____

Networks: Shortest paths

Excel MATHEMATICS STANDARD 1
Ch. 8, pp. 168–170

Shortest paths

QUESTION 1 If all dimensions on these networks are in metres find, by calculation, the shortest path and write the length of this path:

a from *J* to *N*

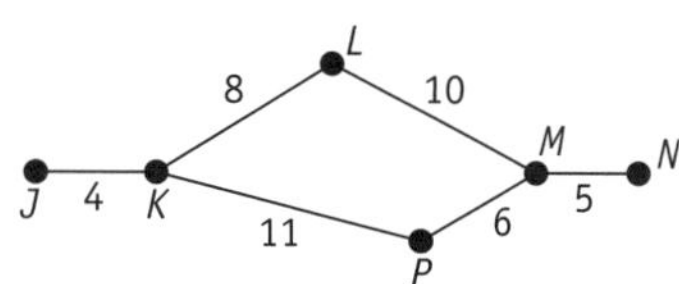

b from *A* to *E*

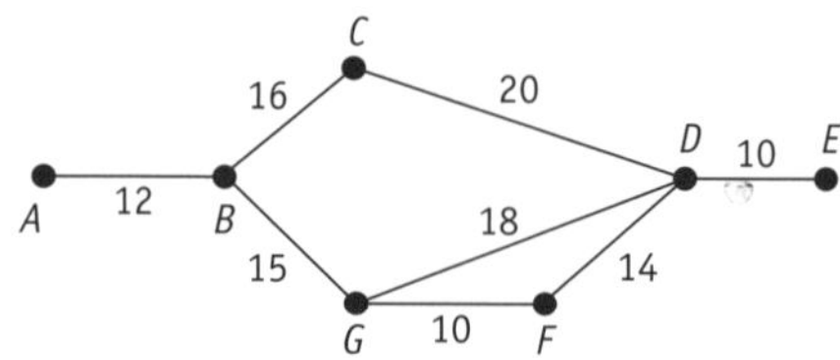

c from *Q* to *T*

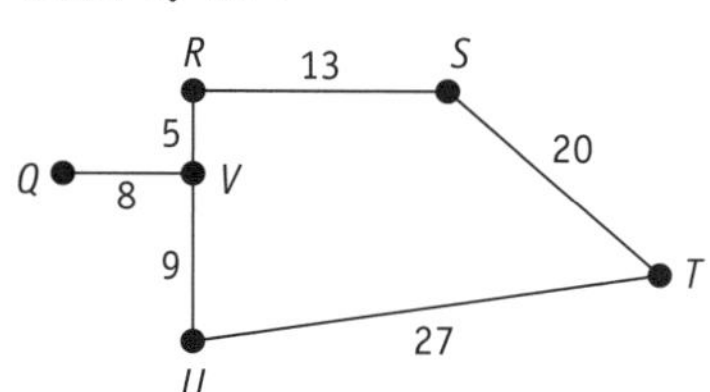

d from *A* to *D*

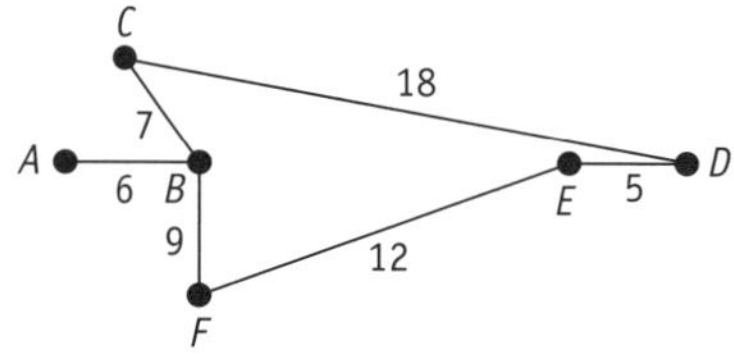

e from *M* to *R*

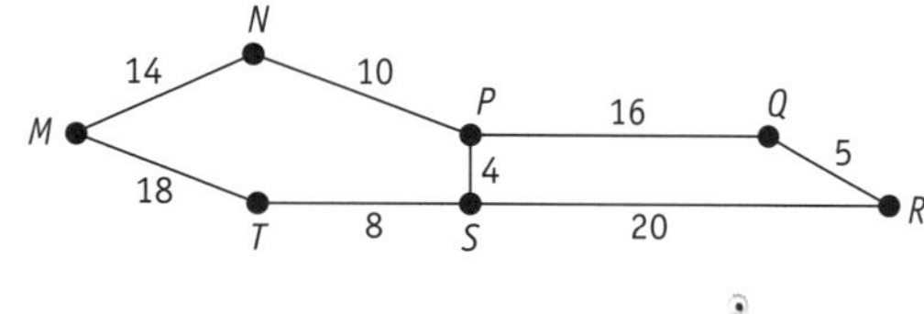

f from *T* to *X*

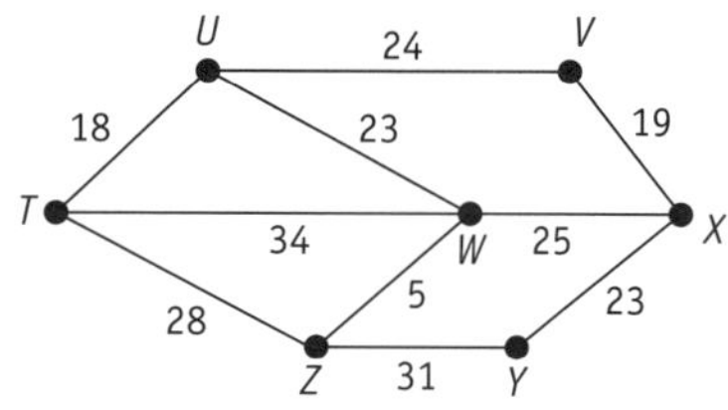

QUESTION 2 The times (in minutes) to travel between locations are shown on the networks below. Find the shortest time to travel between:

a *A* to *D*

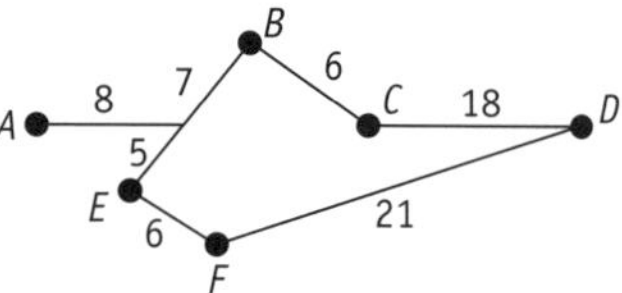

b *P* to *S*

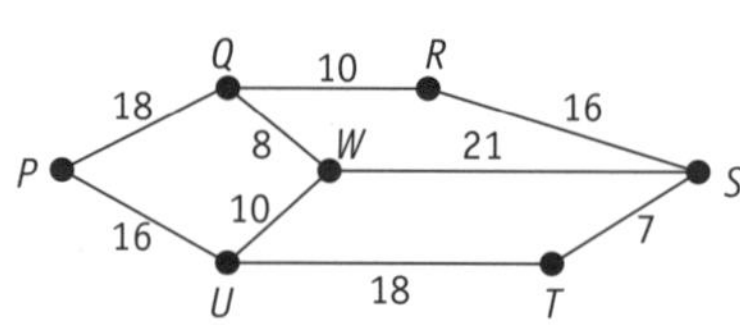

c *W* to *Y*

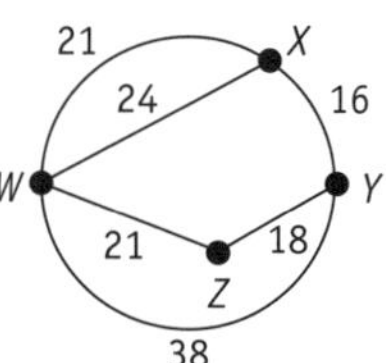

d *M* to *R*

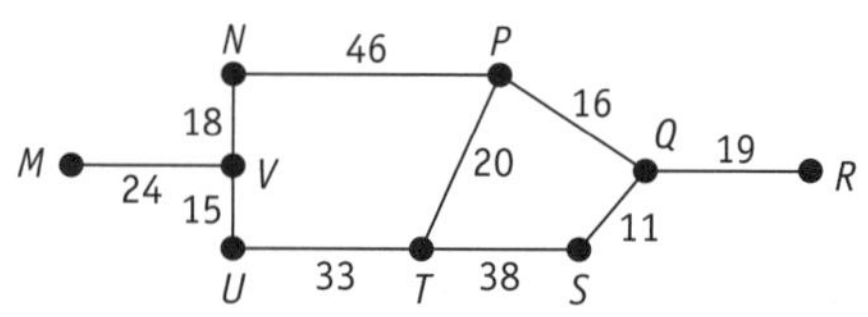

Networks: Shortest paths

Spanning trees

QUESTION 1 For each network below, by removing edges, draw two possible spanning trees.

a

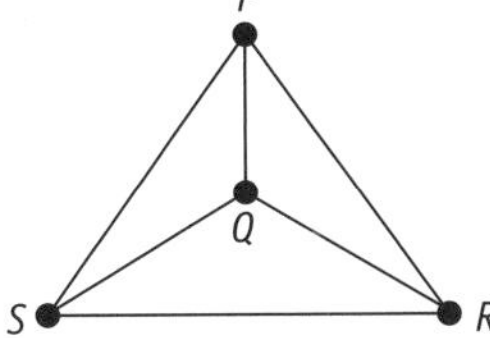

b

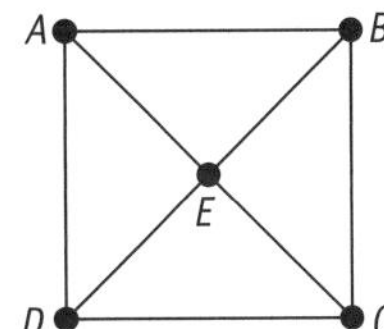

QUESTION 2 Using the weighted network on the right, find the weight of each of these spanning trees:

a

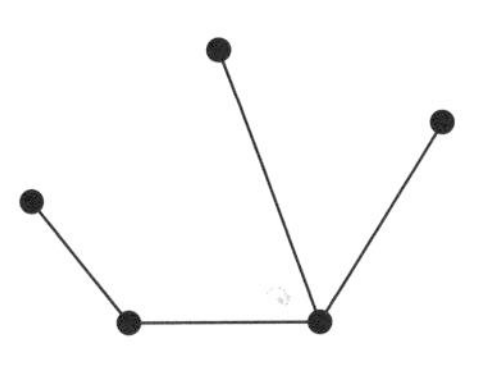

b

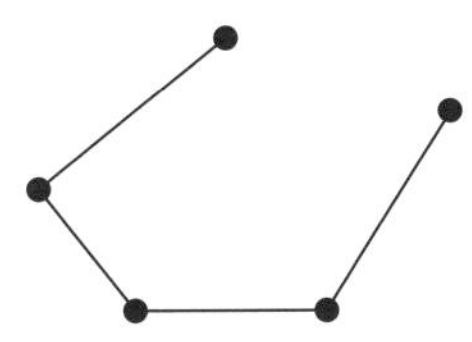

c

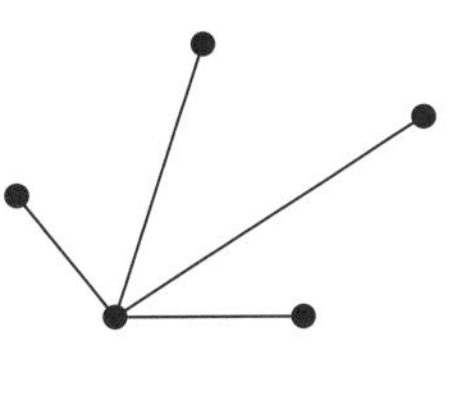

QUESTION 3 Use the weighted network to the right to find the weight of each of these spanning trees:

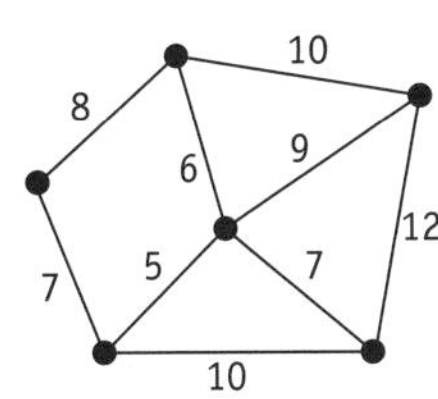

a

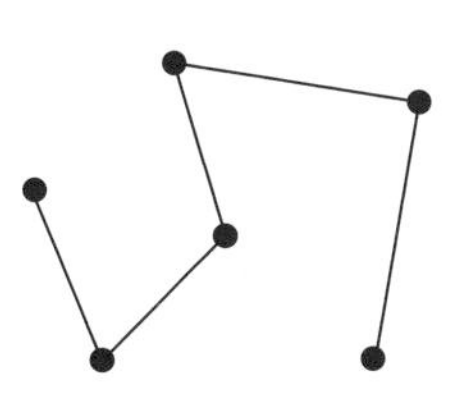

b

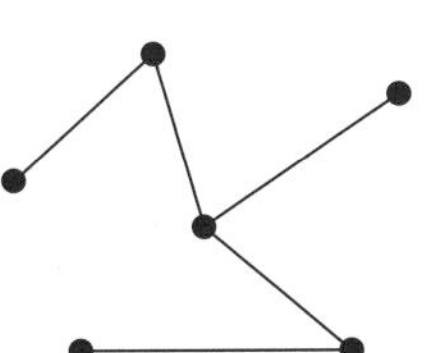

c

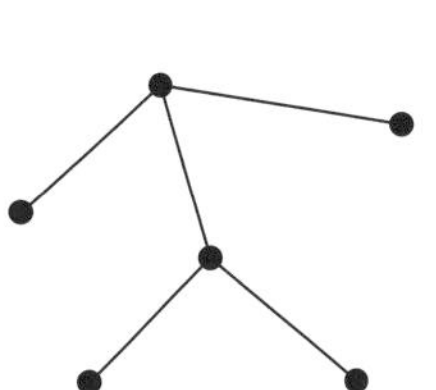

QUESTION 4 Use the weighted network to the right to find the weight of each of these spanning trees:

a

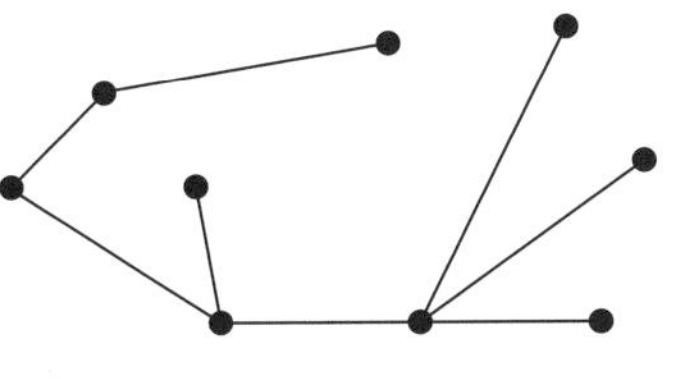

b

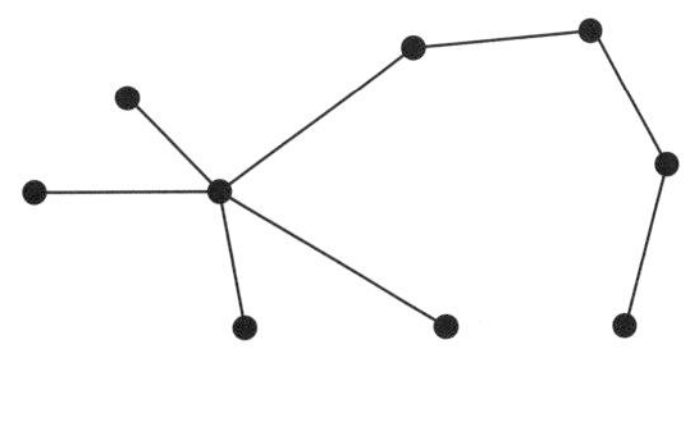

Networks: Shortest paths

Minimum spanning tree

QUESTION 1 Paisley used Kruskal's algorithm to calculate the minimum spanning tree for this network.

a She used a table to help her. Complete Paisley's table:

Edge	Weight
PT	1
QS	2
QT	
QR	

c Draw the minimum spanning tree.

b What is the weight of the minimum spanning tree?

____

QUESTION 2 Nathan used Prim's algorithm to find the minimum spanning tree for the following network.

a Nathan used a table to help him. Complete his table:

Action	Weight	Vertex visited
Start at *G*	–	*G*
Use *GA*		*A*
Use *AF*		*F*
Use *AB*		*B*
	4	
		D
Use *DE*		*E*

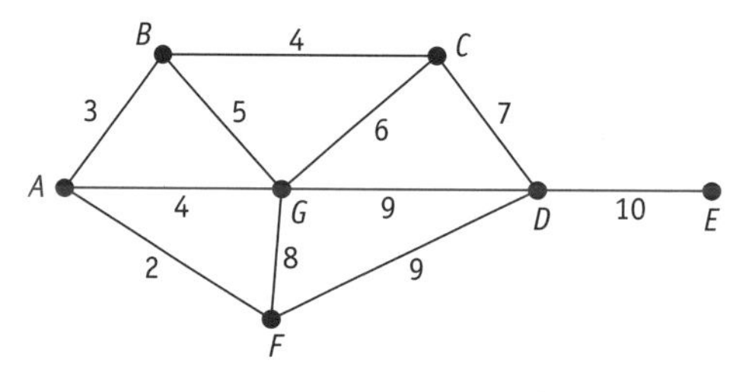

b What is the weight of the minimum spanning tree?

____

____

____

____

c Draw the minimum spanning tree.

QUESTION 3 For the networks below, find the weight of the minimum spanning tree:

a ______________________________

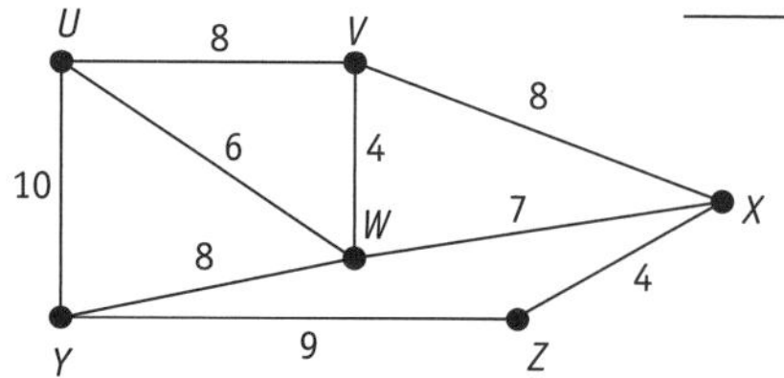

b ______________________________

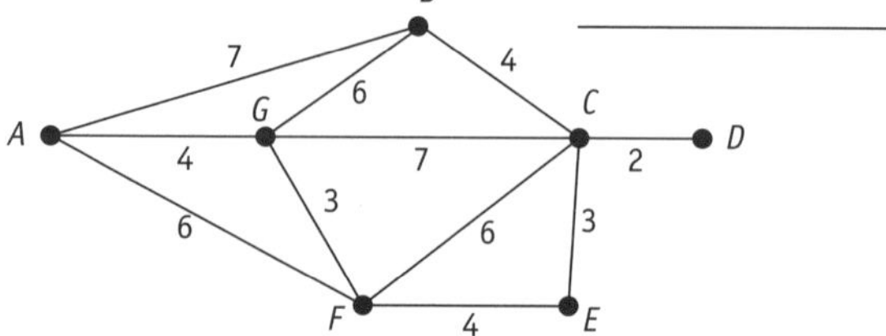

c ______________________________

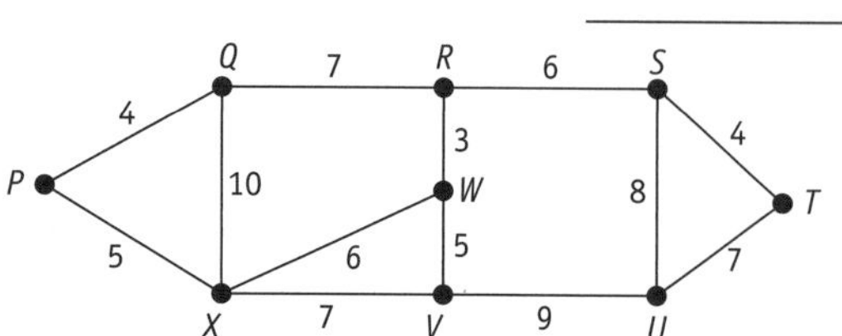

d ______________________________

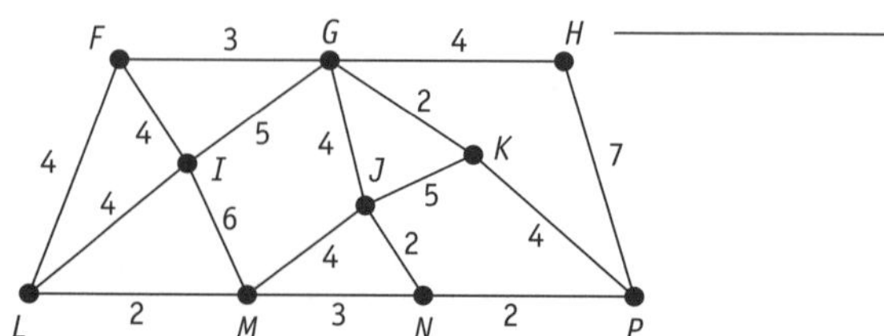

Networks: Shortest paths

Solving problems involving minimum spanning trees

QUESTION 1 Five towns are to be joined to the city of Alexander with fibre-optic cable. Use the network diagram to determine the minimum length of cable required.

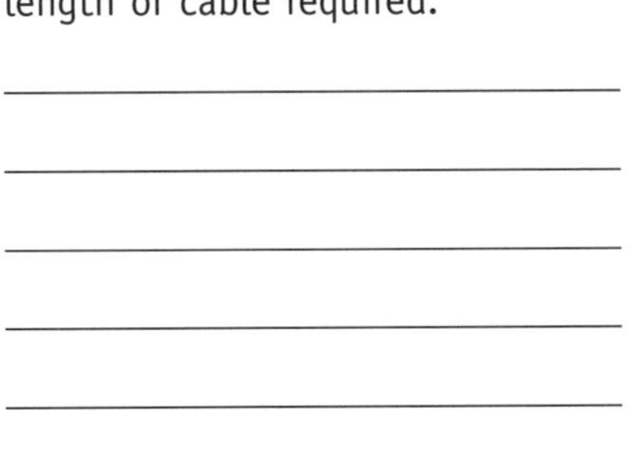

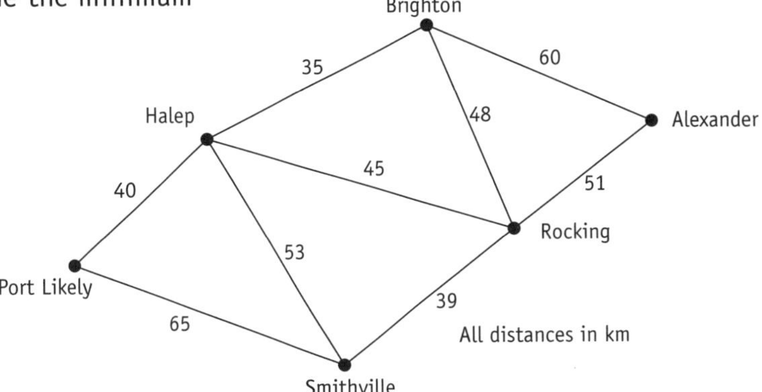

QUESTION 2 A village has unsealed roads which become very muddy after rain. The mayor has decided to pave some roads. What will be the minimum length of roads she needs to pave to ensure that every person living and working in the village can walk to and from their home without walking through mud?

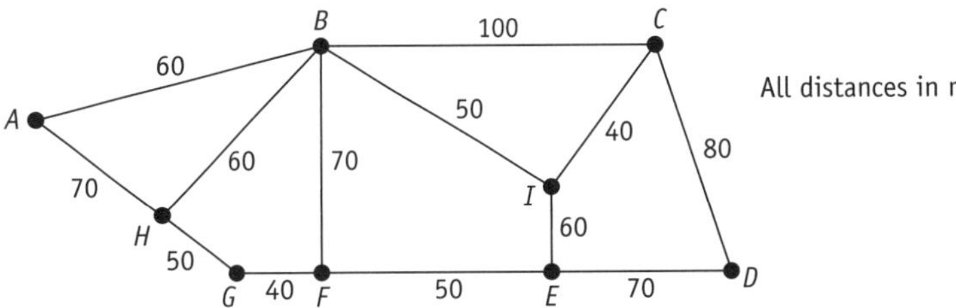

QUESTION 3 A security system is to be installed in a department store.

a What is the length of the minimum spanning tree for the network?

b The cost of the installation is \$85/metre. What is the minimum cost of the installation?

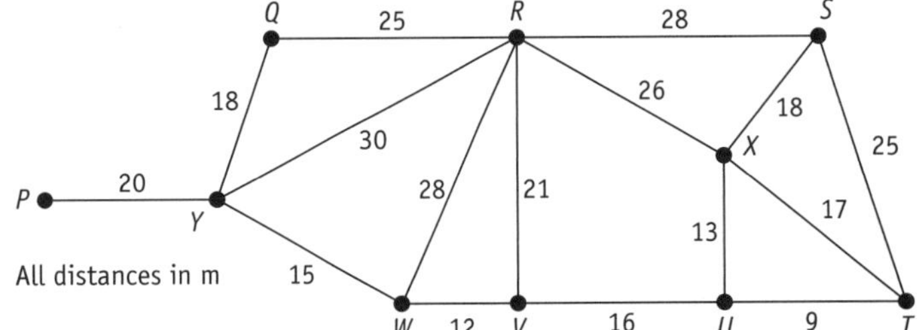

QUESTION 4 A newly opened botanical gardens has pathways between seven significant points of interest (POIs). The committee in charge of the gardens wishes to install water bubblers at these POIs by laying water pipes along pathways. The cost of laying the pipes is \$45/metre and each bubbler costs \$580.

a What is the minimum cost of laying the pipes and providing the seven bubblers?

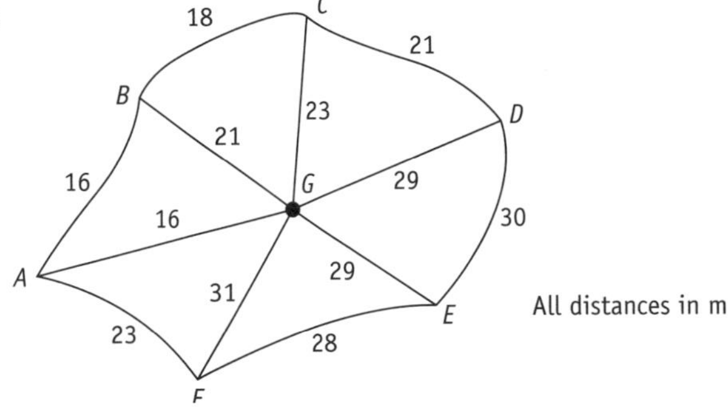

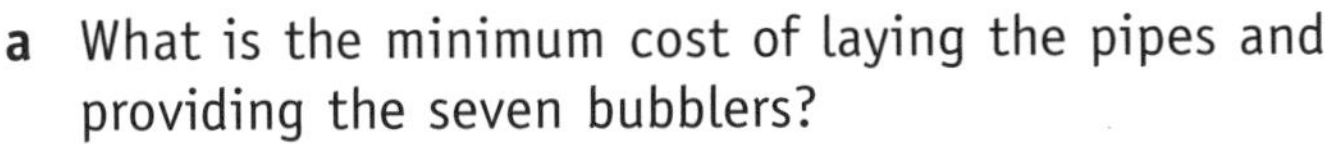

b Suppose the committee decides to close the feature at *B* and block all paths leading to it. This means there will only be six bubblers at the other POIs. How much money is saved by the decision?

Networks: Shortest paths

TOPIC TEST

SECTION I

Instructions
- This section consists of 5 multiple-choice questions.
- Each question is worth 1 mark.
- Fill in only ONE CIRCLE for each question.

Time allowed: 7 minutes **Total marks: 5**

1 The network diagram shows the distances in kilometres between locations A to H. The shortest distance between A and H is 23 km. Which of these is a possible value for k?

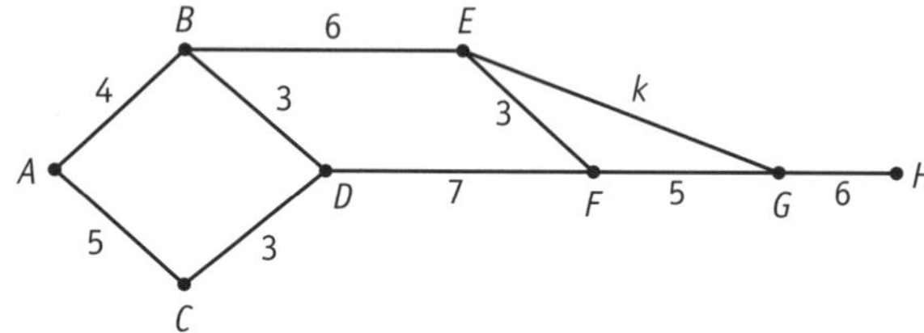

Ⓐ 7 Ⓑ 8

Ⓒ 9 Ⓓ 10

2 What is the weight of the minimum spanning tree of the network?

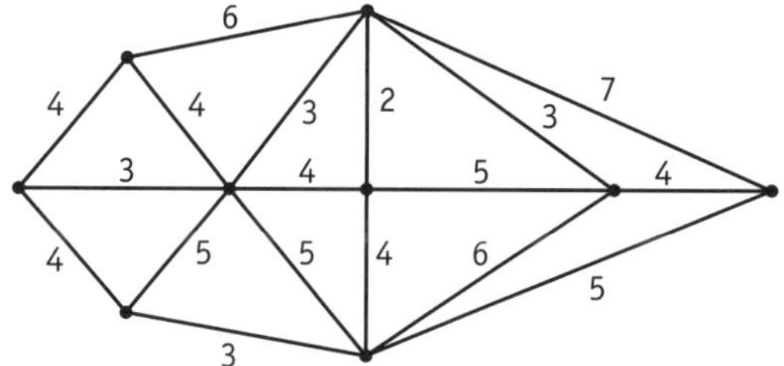

Ⓐ 24

Ⓑ 26

Ⓒ 28

Ⓓ 30

3 In the directed network on the right, starting at P, which of these vertices can be reached?

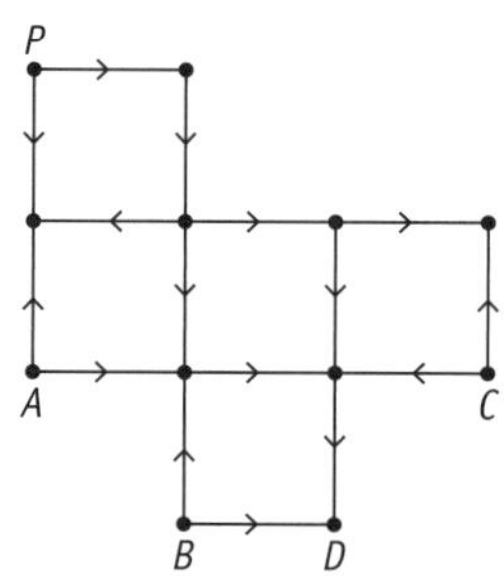

Ⓐ A Ⓑ B

Ⓒ C Ⓓ D

4 Which of these is the shortest path to travel from A to E?

Ⓐ $AGHFDE$

Ⓑ $ABFDE$

Ⓒ $AGHJE$

Ⓓ $AGHJDE$

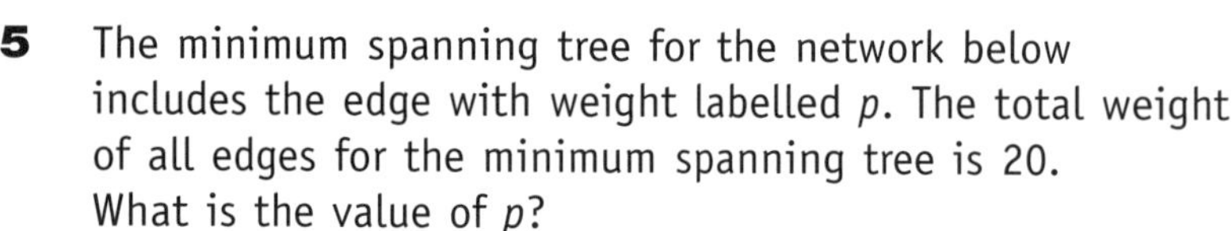
5 The minimum spanning tree for the network below includes the edge with weight labelled p. The total weight of all edges for the minimum spanning tree is 20. What is the value of p?

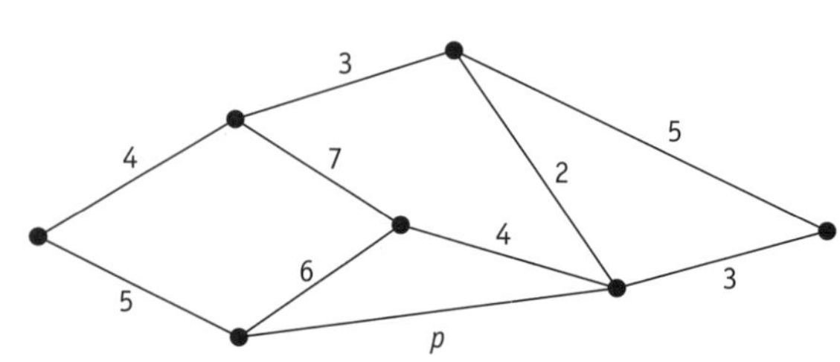

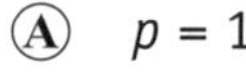
Ⓐ $p = 1$ Ⓑ $p = 2$

Ⓒ $p = 3$ 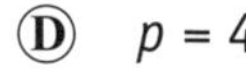 Ⓓ $p = 4$

TOPIC TEST

SECTION II

Instructions
- This section consists of 7 questions.
- Show all working.

Time allowed: 53 minutes **Total marks: 35**

6 The table shows the distances, in kilometres, to travel between six locations.

		To:					
		A	***B***	***C***	***D***	***E***	***F***
From:	***A***	–	6	–	5	2	–
	B	–	–	3	5	–	7
	C	–	4	–	–	–	–
	D	–	–	6	–	4	–
	E	5	–	–	–	–	10
	F	–	–	6	–	–	–

a Use the table to complete the network diagram below by recording arrows and weights on the edges. **2 marks**

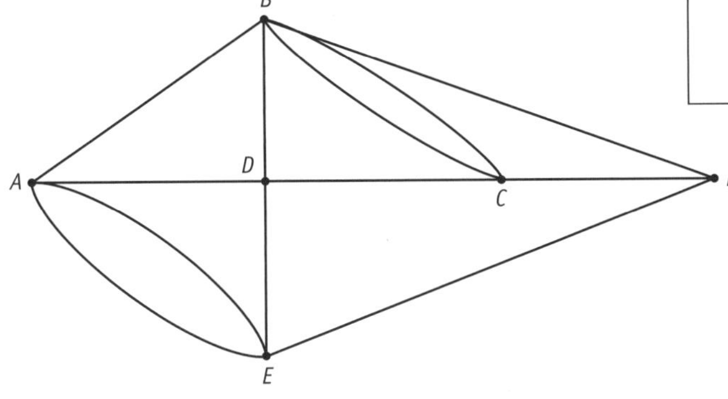

b What is the shortest distance to travel from *A* to *F*? **1 mark**

____

c Rushton travels from *A* to *F* but must pass through *C*. What is the length of the shortest path? **1 mark**

____

____

7 The following table shows the travelling times, in minutes, between five towns. The dash indicates there is no road between these two towns.

a Complete the network diagram showing the information in the table. **3 marks**

	P	***Q***	***R***	***S***	***T***
P	0	70	20	65	–
Q	70	0	40	–	20
R	20	40	0	–	65
S	65	–	–	0	30
T	–	20	65	30	0

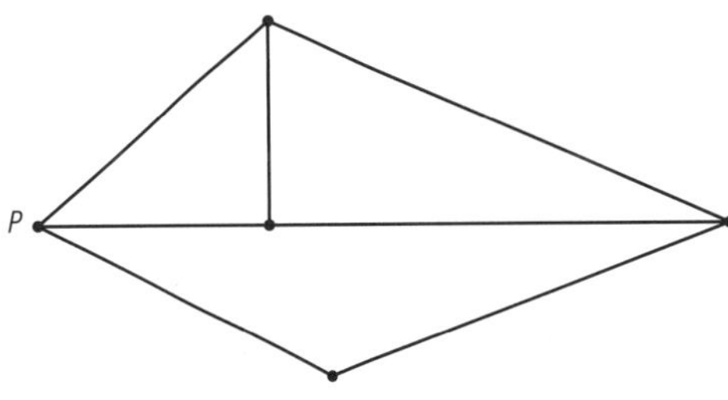

b Find the shortest travelling time between *P* and *T*. **1 mark**

____

____

____

8 The manager of Country Living Campsite is developing the facility by installing an upgraded water pipe to each of eight cabins named A to H.

The network diagram below shows the location of the cabins and the distances, in metres, between each of the cabins.

The manager's residence, shown as location X, presently has the upgraded water pipe.

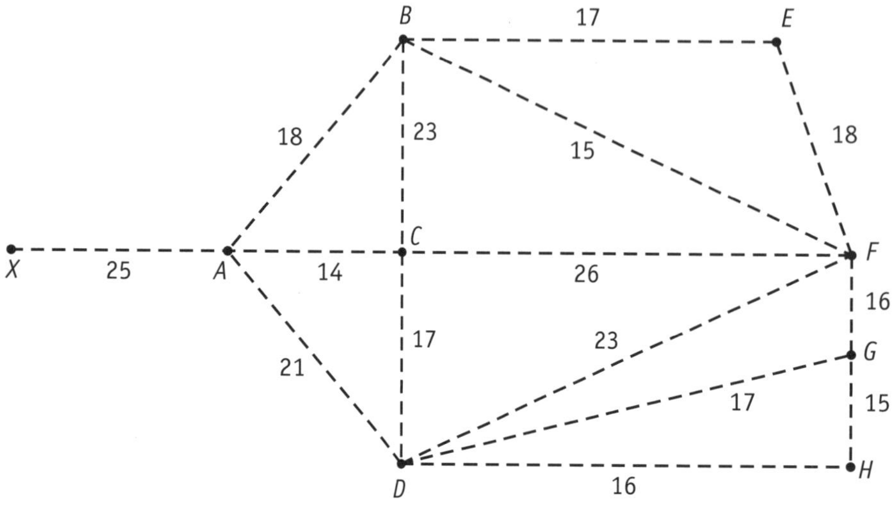

a Use the diagram to draw a minimum spanning tree joining the manager's residence and the eight cabins. **1 mark**

b The cost of laying the water pipe is \$12.60 per metre.

What is the minimum cost to connect the residence and the eight cabins? **2 marks**

____

____

____

9 These networks represent streets. If a delivery van needs to drive every street once only, identify a possible route that needs to be taken by the driver. **1 mark each**

a

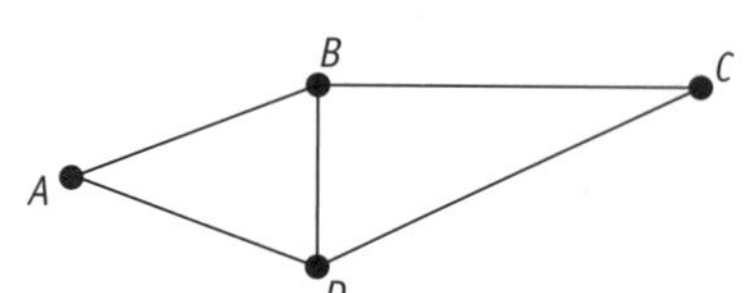

____

b

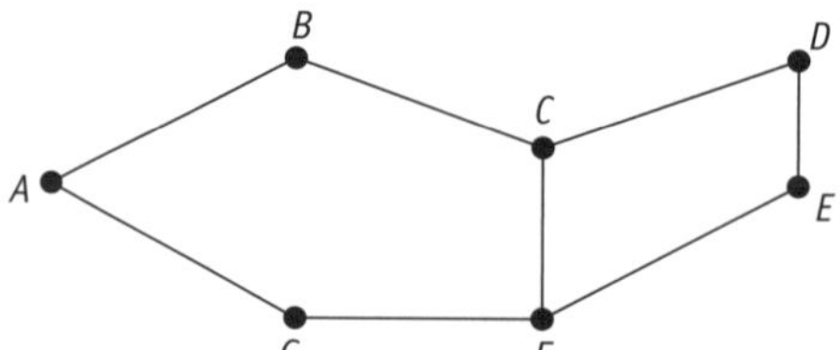

____

c

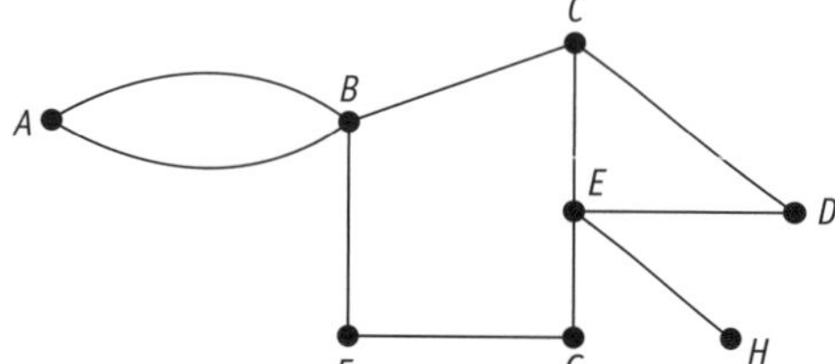

____

10 Identify by drawing a minimum spanning tree (MST) on the following networks and state the weight of the MST: **3 marks each**

a

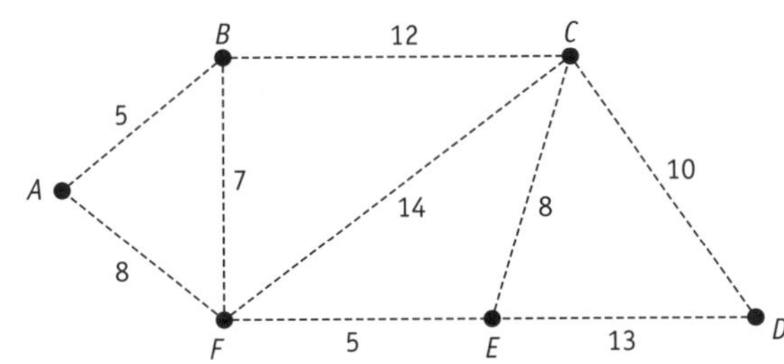

____

b

____

c

____

11 A university requires seven computer servers to communicate with each other through a connected network of cables across the campus. The servers are named *A*, *B*, *C*, *D*, *E*, *F* and *G* and are shown on the network diagram on the right. The edges on the graph represent proposed cables that would connect the computer servers. The numbers on the edges shows the cost, in dollars, of installing each cable.

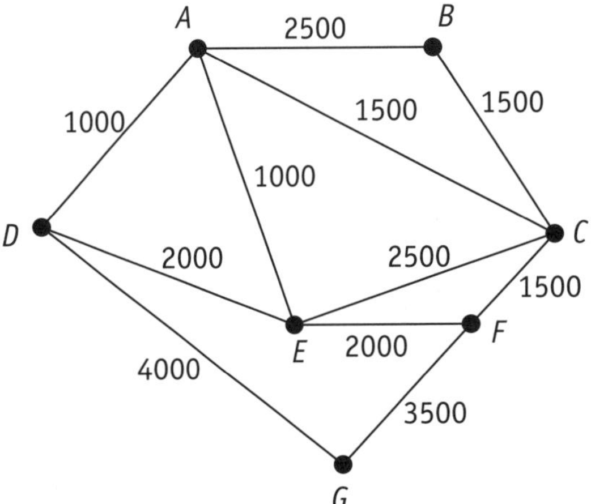

a What is the cheapest cost, in dollars, of installing cables between server *D* and server *C*? **1 mark**

____

b The proposed network needs to be checked by walking along its entire length. To avoid walking along a section more than once, at which server should the inspection start and where should it finish? **1 mark**

____

c The cheapest installation ensuring that each server connects to another server is by using a minimum spanning tree. Draw the minimum spanning tree on the diagram and write down the length. **2 marks**

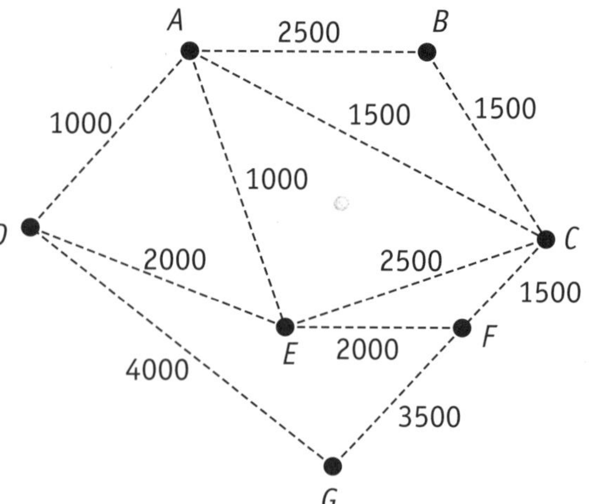

____

d The university's IT manager has decided to reduce the number of servers to six. What is the change in installation costs if the manager removes server *A* from the minimum spanning tree? **2 marks**

____

12 A new theme park is being designed with nine rides. On the network diagram on the right, the vertices represent the locations of the rides. The numbers on the edges represent the distances, in metres, between the rides.

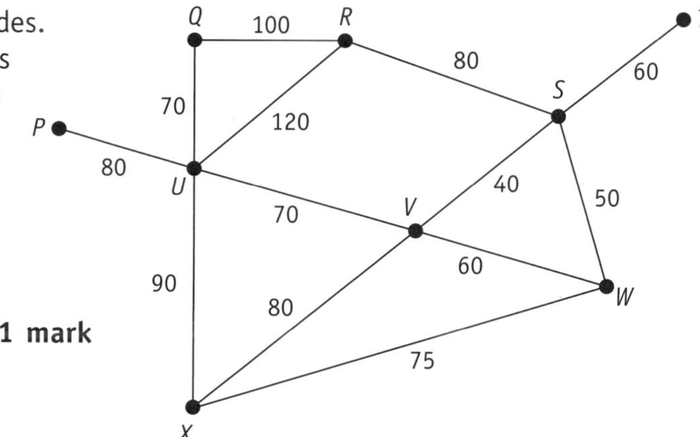

a Electrical cables are required to power the rides and these cables will be located using the network diagram. What is the minimum length of cabling required to join *S* to *U*? **1 mark**

____

b Cables will form a connected graph and the shortest length of cable will be used.

What term is given to the network used where the length of cable is minimised? **1 mark**

____

c Use the diagram on the right to show the cabling used and find the length. **2 marks**

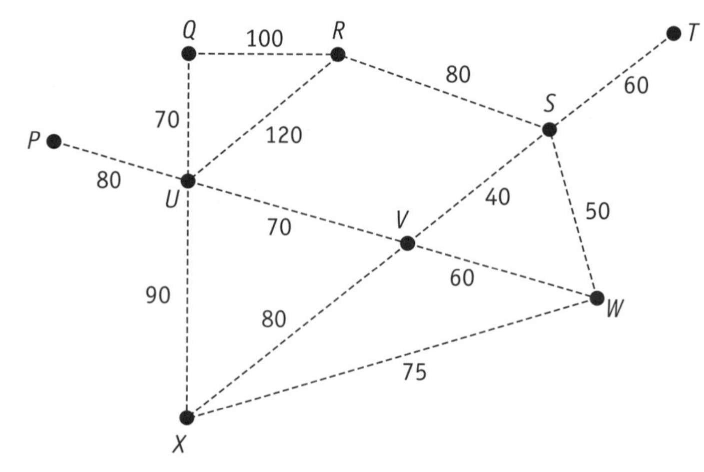

____

d The theme park company has budgeted \$48 000 for the cabling. Aztech Services has quoted a labour cost of \$12 600 plus \$65/m to lay the cable. Is the quote more or less than the company's budget? Show calculations to support your answer. **2 marks**

____

____

____

CHAPTER 12

Sample HSC Examination 1

Total time: 2 hours **Total marks: 80**

SECTION I

Marks: 10

Instructions

- Attempt Questions 1 to 10.
- Allow about 15 minutes for this section.
- Each question is worth 1 mark.
- Fill in only ONE CIRCLE for each question.

1 A network diagram is shown below consisting of five vertices.
What is the degree of vertex R?

Ⓐ 1 Ⓑ 2

Ⓒ 3 Ⓓ 4

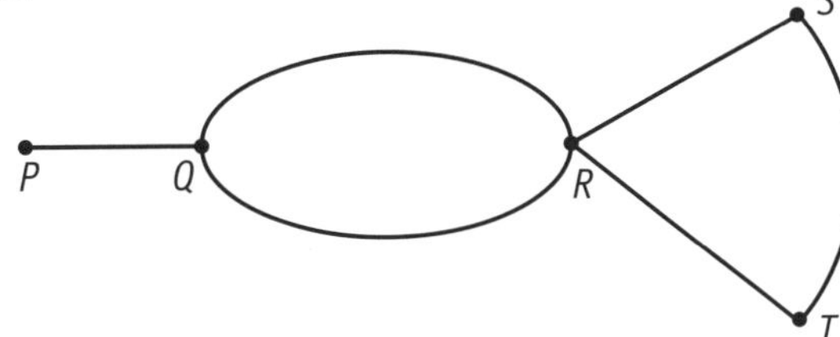

2 Honey is sold in four different-sized containers.
Which is the best buy?

Ⓐ 300 g for $4.80 Ⓑ 400 g for $6.20 Ⓒ 500 g for $7.60 Ⓓ 600 g for $9.90

3 The container shown is initially empty.
Water is being added to the container at a constant rate.
Which graph best shows the depth of water in the container as time varies?

Ⓐ

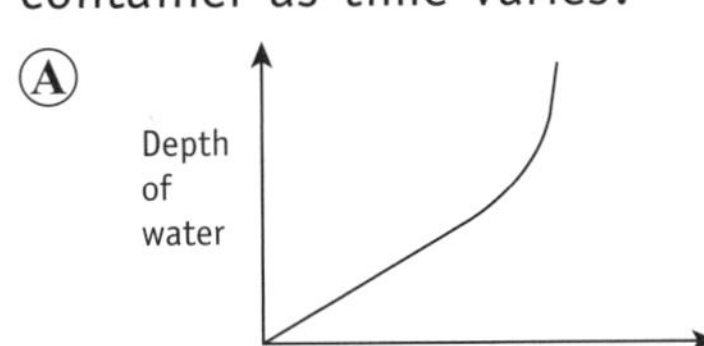

Ⓑ

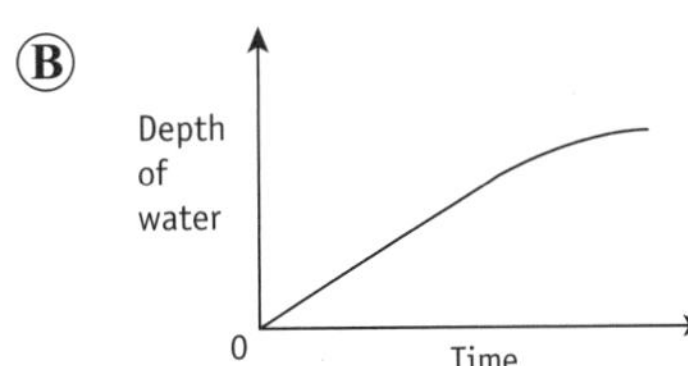

Ⓒ

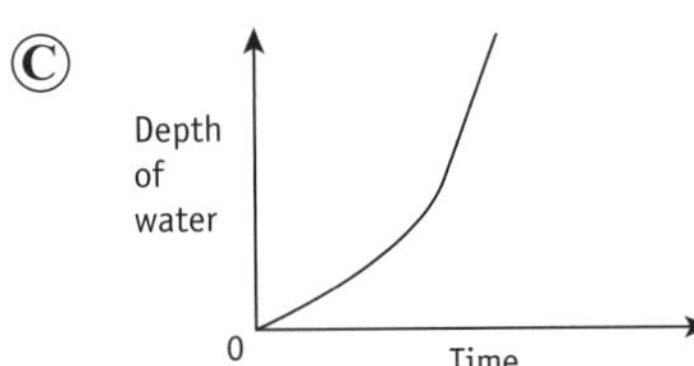

Ⓓ

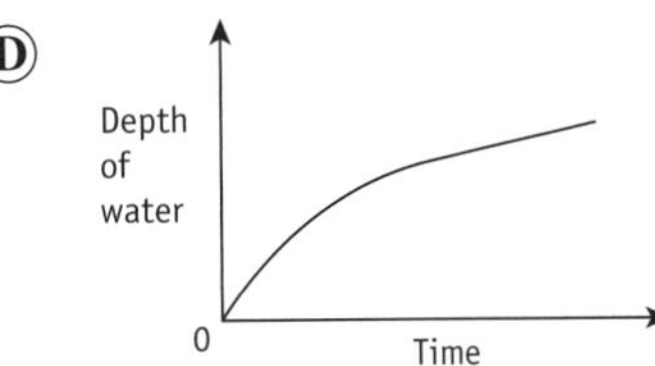

4 In a golf club, the ratio of men to women members is 5:2.
There are 105 men who are members of the golf club.
What is the total membership of the golf club?

Ⓐ 107 Ⓑ 147 Ⓒ 165 Ⓓ 175

5 Which of these relationships shows a strong positive linear association?

Ⓐ 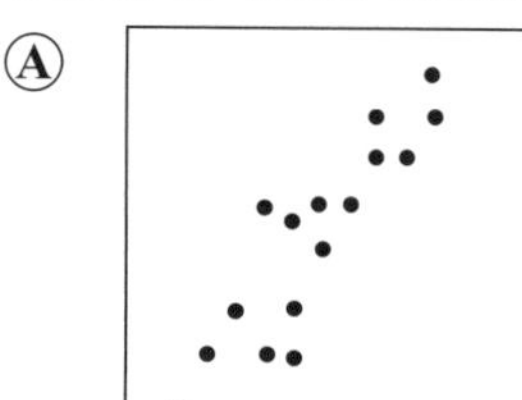 Ⓑ Ⓒ Ⓓ

6 The cost of electricity is $0.21/kWh. What is the total cost of running a 2400-watt heater for 6 hours?

Ⓐ $0.03 Ⓑ $0.30 Ⓒ $3.02 Ⓓ $30.24

7 The figure shows a right triangle and a semicircle. The hypotenuse of the triangle is the diameter of the semicircle.
What is the area of the semicircle?

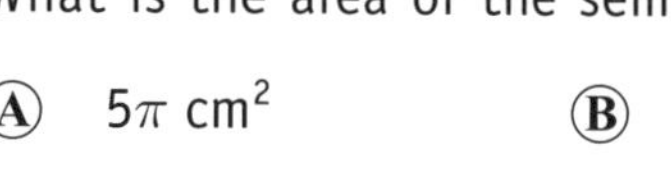

Ⓐ 5π cm^2 Ⓑ $\frac{25\pi}{2}$ cm^2

Ⓒ 25π cm^2 Ⓓ 100π cm^2

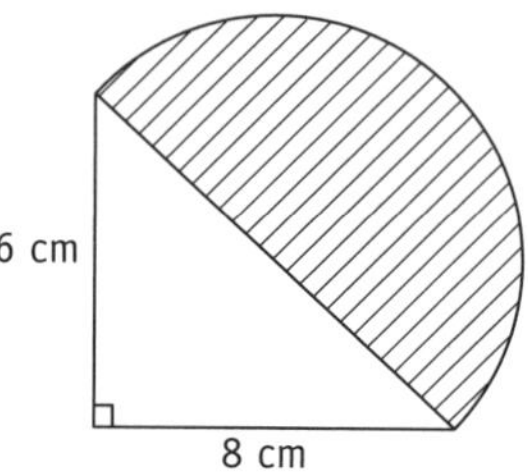

8 A car that was purchased for $39 900 is worth $20 400 after 5 years.
The annual amount of depreciation, using the straight-line method, is:

Ⓐ $3900 Ⓑ $4080 Ⓒ $7980 Ⓓ $12 060

9 Susan invested $7000. She earns interest at 3% per annum, compounded monthly. What is the future value of the investment after 4 years?

Ⓐ $7878.56 Ⓑ $7891.30 Ⓒ $8893.60 Ⓓ $9980.21

10 The minimum spanning tree for the network below includes the edge with weight labelled p.
The total weight of all edges for the minimum spanning tree is 20.
What is the value of p?

Ⓐ $p = 1$ Ⓑ $p = 2$

Ⓒ $p = 3$ Ⓓ $p = 4$

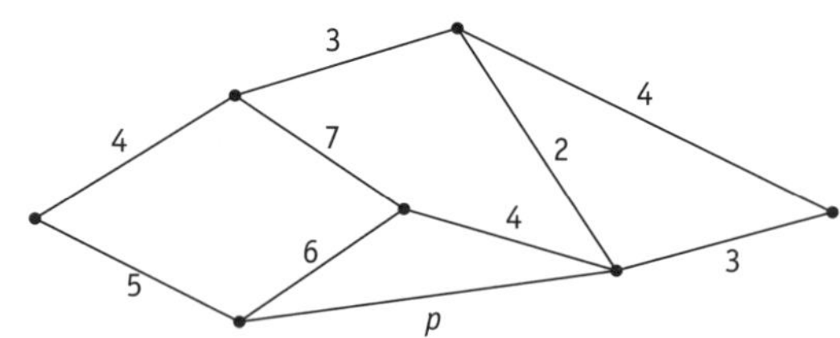

Sample HSC Examination 1

SECTION II

Marks: 70

Instructions

- Attempt Questions 11 to 30.
- Allow about 1 hour and 45 minutes for this section.
- Answer the questions in the spaces provided. These spaces provide guidance for the expected length of response.
- Your responses should include relevant mathematical reasoning and/or calculations.

QUESTION **11** (2 marks)

Juliet works from 9 am to 2 pm each Sunday.
Her normal rate of pay is $28 per hour.
She is paid the normal rate of pay for morning work and double time for afternoon work.
How much does Juliet earn each Sunday? **2 marks**

QUESTION **12** (2 marks)

Calculate the simple interest on $480 at 4.5% p.a. for 3 years. **2 marks**

QUESTION **13** (2 marks)

Shane purchased the following items while shopping in his local supermarket:

- toothpaste* $6.80
- peaches $3.29
- 2-L milk $4.35
- soft drink* $2.99

Prices that include 10% GST are marked with *.
What is the total amount of GST paid by Shane? **2 marks**

QUESTION **14** (3 marks)

The diagram shows a shape made of a rectangle and three-quarters of a circle.

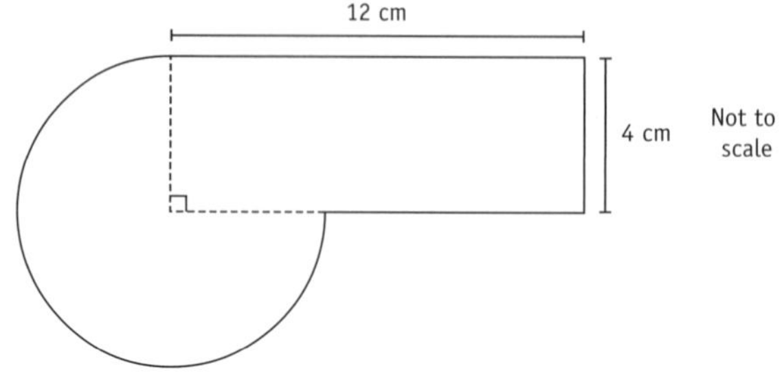

Find the area of the shape to the nearest square centimetre. **3 marks**

QUESTION **15** (4 marks)

The travel graph shows a journey taken by Emma. She left her home at 8 am to drive to her cousin's farm, arriving at noon.
Emma stayed at her cousin's farm for an hour before returning home, using the same roads.

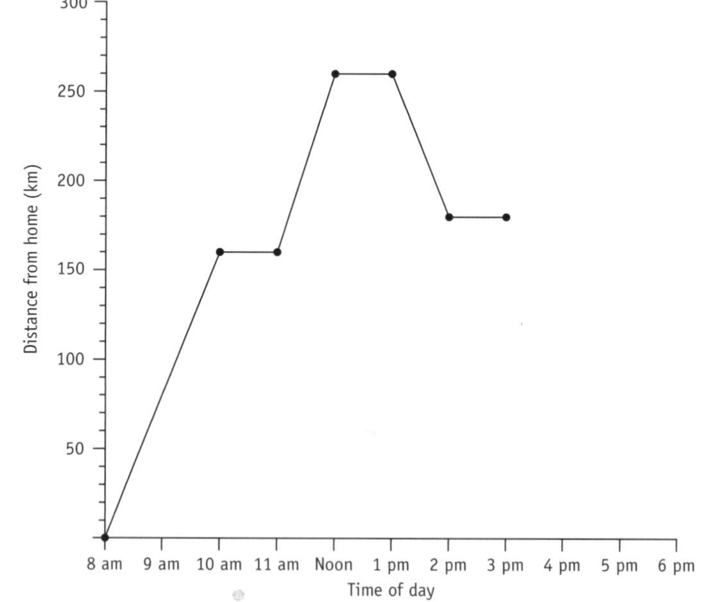

a What was Emma's average speed between 8 am and 10 am? **1 mark**

b After leaving her cousin's farm, how far did Emma drive before she had a one-hour rest break? **1 mark**

c After her rest break, Emma drove straight home, at an average speed of 90 km/h. Using this information, complete the travel graph. **1 mark**

d How long was Emma away from home? **1 mark**

Sample HSC Examination 1

QUESTION **16** (3 marks)

The ages of people at last year's family Christmas dinner were recorded.
1, 1, 20, 24, 26, 27, 27, 30, 31, 31, 32, 32, 34, 50, 52, 54, 58, 62, 86
Is the age of the oldest person at the Christmas dinner an outlier?
Justify your answer with calculations. **3 marks**

QUESTION **17** (3 marks)

A biased six-sided dice is labelled using the numbers 1, 2, 3, 4, 5 and 6.
The dice is rolled many times and a relative frequency table is completed based on the results.

Result	Relative frequency
1	10%
2	10%
3	
4	0%
5	20%
6	20%

a What is the relative frequency of rolling a 3? **1 mark**

b If the dice is rolled 400 times, how many times can you expect an even number to appear? **2 marks**

QUESTION **18** (3 marks)

Meghan investigated the cost of holding a wedding reception at a certain venue:

- facility hire $1800
- food/drink per person $120 per person
- PA/music $100 + $60/h
- photo booth $250.

What would be the total cost for 110 guests for a wedding reception from 6:00 pm to 10:30 pm? **3 marks**

Sample HSC Examination 1

QUESTION **19** (3 marks)

Given the formula $P = \frac{T(m-n)}{6}$, find the value of m when $P = 10$, $T = 8$ and $n = 12$. **3 marks**

QUESTION **20** (2 marks)

The table below shows a number of blood-pressure categories.

Blood pressure stages			
Blood pressure category	**Systolic (mm/Hg)**		**Diastolic (mm/Hg)**
Optimal	less than 120	and	less than 80
Normal	120–129	and/or	80–84
High to normal	130–139	and/or	85–89
Mild hypertension	140–159	and/or	90–99
Moderate hypertension	160–179	and/or	100–109
Severe hypertension	more than or equal to 180	and/or	more than or equal to 110

Determine the blood pressure categories for the following people:

a Jeremy with a blood pressure of 123/83 **1 mark**

b Sonia with a blood pressure of 161/104 **1 mark**

QUESTION **21** (4 marks)

Connie's Computers offers an in-home repair service.
Connie charges a call-out fee of $55 plus $12 per 15-minute block or part thereof.
Connie arrived at Juan's home at 11:50 am and charged $127 for her services.
What is the latest time Connie had completed her work? **4 marks**

Sample HSC Examination 1

Question **22** (4 marks)

The following table shows the travelling times in minutes between towns *A*, *B*, *C*, *D* and *E*.

A dash shown in the table means these towns are not connected directly.

a Use the table to draw a network diagram.

	A	**B**	**C**	**D**	**E**
A	0	35	70	–	25
B	35	0	45	–	–
C	70	45	0	20	40
D	–	–	20	0	15
E	25	–	40	15	0

2 marks

b Find the shortest travelling time between *A* and *C*, and write the path. **2 marks**

Question **23** (3 marks)

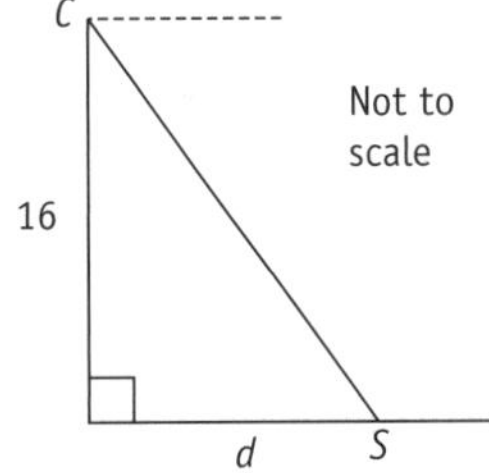

Cadence is standing on the top of a 16-metre cliff and looks down to a swimmer. The angle of depression from Cadence to the swimmer is 54°.

How far is the swimmer from the cliff, correct to two decimal places? **3 marks**

Question **24** (5 marks)

The diagram shows three towns *A*, *B* and *C*. Town *B* is 250 km due west of *C* and town *A* is due north of town *B*. The bearing of town *A* from town *C* is 300°.

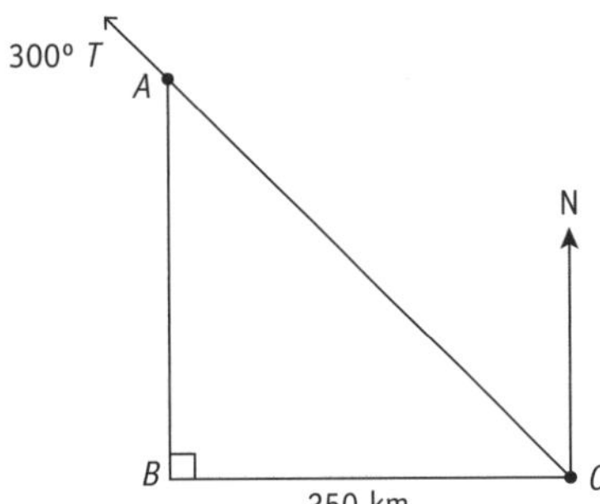

a Explain why angle *ACB* is 30°. **1 mark**

b How far south of town *A* is town *B*? Give your answer to the nearest kilometre. **2 marks**

c At 9:00 am a plane is flying directly overhead town *A* towards town *C*. If the plane flies at 400 km/h, at what time will it be directly overhead town *C*? **2 marks**

QUESTION **25** (4 marks)

Solve the following equations:

a $7x - 20 = 3x + 16$ **2 marks**

b $\sqrt{x + 3} = 8$ **2 marks**

QUESTION **26** (8 marks)

The table gives an average water use per person per day.

Use	Litres/day
Bath/shower	72
Sink	18
Toilet	35
Washing clothes	45
Washing dishes	29
Cooking	8
Miscellaneous	6
Total	

a Complete the table to insert the value for total. **1 mark**

b Some people will look at this table and argue that they don't use this much water. How can you respond? **1 mark**

c How much water is used in the sink over a year? **1 mark**

d How much water (in kilolitres) does the average person use in a year? **1 mark**

e One of these areas uses 21.1% of the daily water use. Which is it? **1 mark**

f What percentage of water is spent on washing dishes? **1 mark**

g A water-efficient toilet can save about 20% on a person's water use of that device.

i What is the daily saving in this area? **1 mark**

ii What is the annual saving in this area? **1 mark**

QUESTION **27** (4 marks)

Tom purchased a tractor for $80 000.

a Use the declining-balance formula to find the salvage value of the tractor after 10 years if it depreciates in value by 12% p.a. Leave your answer to the nearest dollar. **2 marks**

b Tom also calculated the salvage value of the tractor after 10 years using a straight-line method of depreciation of $6000 p.a. What is the difference in the amount of depreciation using both methods? **2 marks**

QUESTION **28** (3 marks)

A plane leaves Melbourne (UTC+10) at 8:30 am on Wednesday and arrives in Tokyo (UTC+8) at 5:00 pm that evening. If the flight distance is 8160 km, what is the average speed of the aeroplane, to the nearest kilometre per hour? **3 marks**

QUESTION **29** (3 marks)

The suggested maximum heart rate (MHR) is 220 beats per minutes minus your age.

a What is the age of a person if their suggested MHR is 165 beats per minute? **1 mark**

b An alternative method to determine a person's maximum heart rate (MHR), in beats per minute, is to use the formula MHR = 208 − 0.7a, where a is the person's age in years. Use this formula to determine the age of a person if their suggested MHR is 165 beats per minute, to the nearest year. **2 marks**

QUESTION **30** (5 marks)

A company recorded the age and measured the systolic blood pressure of 12 of its employees.

The age (a) and systolic blood pressure (p) of the employees are recorded in the table below.

Employee	*A*	*B*	*C*	*D*	*E*	*F*	*G*	*H*	*I*	*J*	*K*	*L*
Age (a)	48	39	31	25	55	63	44	47	40	29	53	49
Systolic blood pressure (p)	156	141	127	125	162	160	145	151	145	130	145	148

Record the data on the diagram below by drawing a scatterplot. Use the scatterplot to draw a line of best fit ensuring the line passes through the dots representing Employees *G* and *J*. Determine the equation of this line. **5 marks**

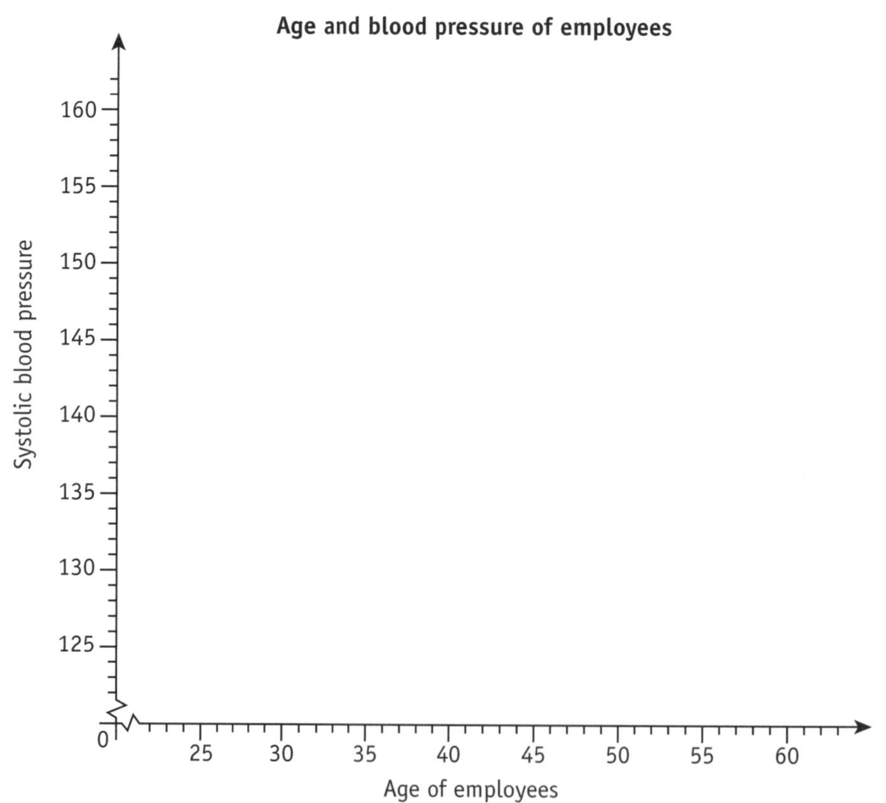

Sample HSC Examination 2

Total time: 2 hours **Total marks: 80**

SECTION I

Marks: 10

Instructions
- Attempt Questions 1 to 10.
- Allow about 15 minutes for this section.
- Each question is worth 1 mark.
- Fill in only ONE CIRCLE for each question.

1 The graph shows the typing speed for a student.

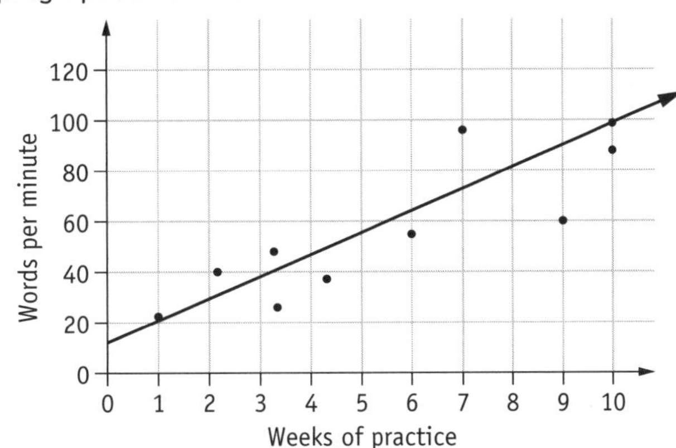

Based on the line of best fit, what would be the approximate typing speed, in words per minute, of a person who has practised for 8 weeks?

(A) 40 (B) 48 (C) 62 (D) 81

2 In the triangle drawn, the value of y is given by:

(A) $y = 9\cos 63°$ (B) $y = 9\sin 63°$

(C) $y = \dfrac{9}{\cos 63°}$ (D) $y = \dfrac{9}{\sin 63°}$

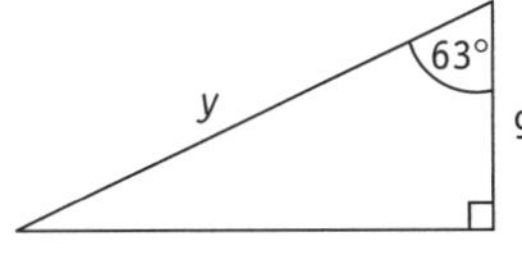

3 Matthew took out a loan of $9000 at the flat interest rate of 8% p.a. over a term of 36 months. How much will he have to repay?

(A) $6840 (B) $11 337 (C) $11 160 (D) $2160

4 The equation of this graph could be:

(A) $y = x^2 + 3$ (B) $y = x^3 + 3$

(C) $y = \dfrac{3}{x}$ (D) $y = 3^x$

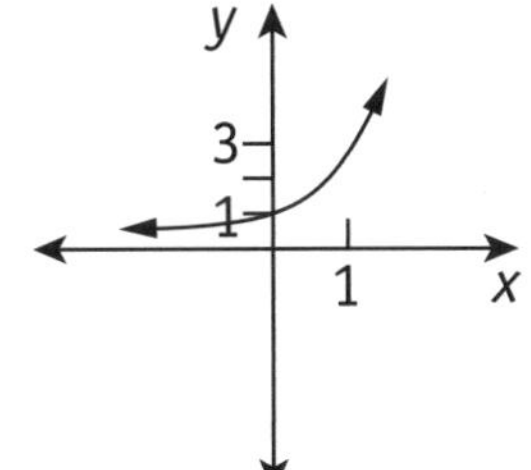

5 A map is drawn using a scale of 1 : 250 000. The towns of Marabin and Cosalletti are 75 km apart. What is the distance between the two towns on the map?

(A) 1.875 cm (B) 3 cm

(C) 18.75 cm (D) 30 cm

6 Xavier graphed the lines $y = 2x - 4$ and $y = 5 - x$ on a number plane. What is the point of intersection of the two lines?

Ⓐ (1, −2) Ⓑ (1, 4) Ⓒ (3, 1) Ⓓ (3, 2)

7 A network diagram is drawn by connecting edges to vertices. How many vertices have an odd degree?

Ⓐ 2 Ⓑ 3

Ⓒ 4 Ⓓ 6

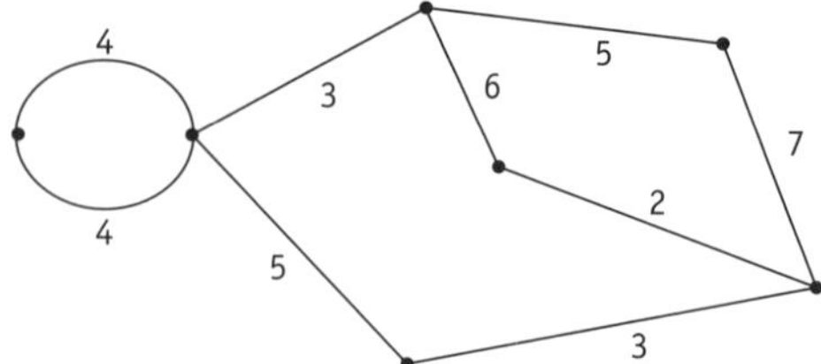

8 Which of these is closest to the future value of an investment of $12 000 at 8% p.a., compounded quarterly, for 3 years?

Ⓐ $14 880 Ⓑ $15 117 Ⓒ $15 219 Ⓓ $19 301

9 The table is used to add weights to the network diagram. What is the value of t?

		To:				
		A	*B*	*C*	*D*	*E*
From:	*A*	–	3	–	–	–
	B	2	–	4	3	–
	C	–	5	–	–	2
	D	–	2	–	–	–
	E	–	–	4	–	–

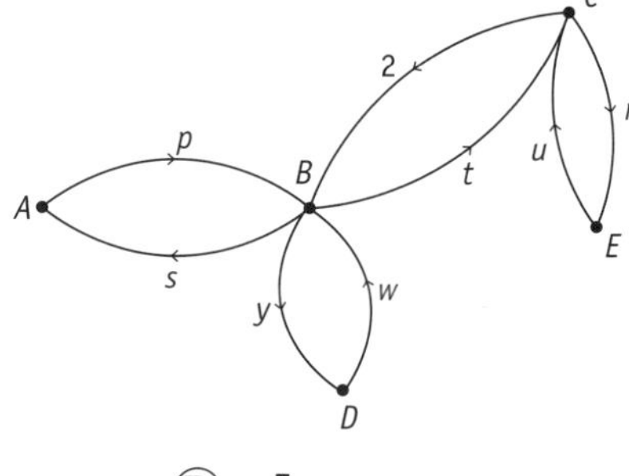

Ⓐ 2 Ⓑ 3 Ⓒ 4 Ⓓ 5

10 A concrete pipe has an inner diameter of 20 cm and is 3 cm thick. What volume of concrete is used to make a pipe section 15 m long?

Ⓐ 0.3 m^3 Ⓑ 0.43 m^3 Ⓒ 3251.55 m^3 Ⓓ 6078.98 m^3

Sample HSC Examination 2

SECTION II

Marks: 70

Instructions
- Attempt Questions 11 to 29.
- Allow about 1 hour and 45 minutes for this section.
- Answer the questions in the spaces provided. These spaces provide guidance for the expected length of response.
- Your responses should include relevant mathematical reasoning and/or calculations.

QUESTION **11** (3 marks)

Amelia imports and screen-prints T-shirts.

Technology has been used to draw straight-line graphs to represent Amelia's cost (C) of producing the T-shirts and the revenue (R) from selling them.

The x-axis displays the number of T-shirts and the y-axis displays the cost/revenue in dollars.

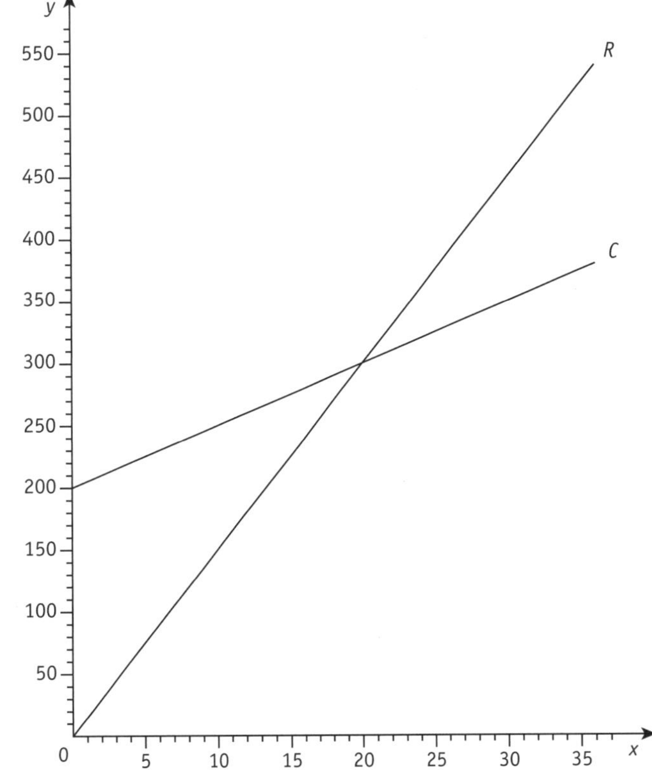

a How many T-shirts does Amelia need to sell to break even? **1 mark**

____

b How much profit does Amelia make if she sells 28 T-shirts? **2 marks**

____

____

____

QUESTION **12** (3 marks)

The diagram shows two right triangles ABC and ABD. Calculate the size of angle θ, to the nearest degree. **3 marks**

Not to scale

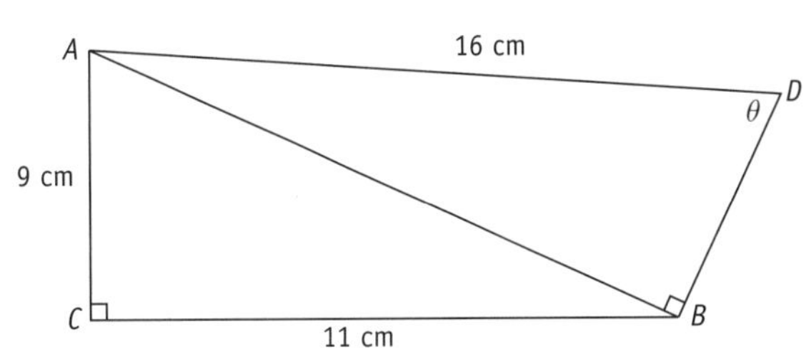

QUESTION **13** (3 marks)

Anna drives her car for 620 km on 50 L of petrol.

a What is the fuel consumption expressed in litres per 100 km? **1 mark**

____

____

____

b If petrol cost 169.9 c/L, what is the cost of petrol for Anna to drive 470 km? Give your answer to the nearest dollar. **2 marks**

QUESTION **14** (2 marks)

The diagram shows two buildings which are 36 m apart.

The angle of elevation of the top of Building *A* from the top of Building *B* is 28°.

A length of wire is to be strung from the edge of the top of Building *A* to the edge of the top of Building *B*.

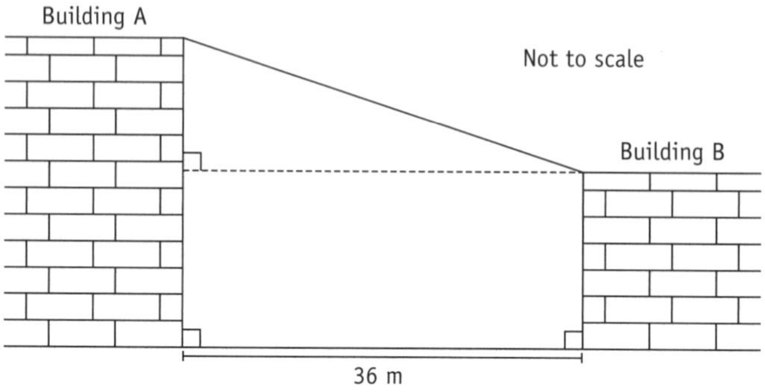

Building A

Not to scale

Building B

36 m

What is the length of the wire, correct to two decimal places? **2 marks**

QUESTION **15** (2 marks)

A house was purchased in 1990 for $285 000.
Inflation has increased the value of the house at an average annual rate of 3.4%.
What was the value of the house in 2020, to the nearest thousand dollars? **2 marks**

QUESTION **16** (2 marks)

A ship left *B* and travelled on a bearing of 220° for 84 km to *A*.
How far west of *B* is *A*?
Give your answer to the nearest kilometre. **2 marks**

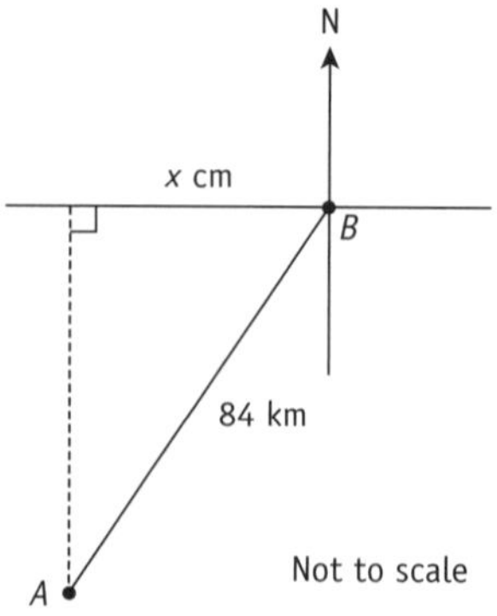

N

x cm

B

84 km

A

Not to scale

Sample HSC Examination 2

QUESTION **17** (3 marks)

A tree and a flagpole both cast shadows on the school quadrangle.
The flagpole is 3.6 m high and cast a 4-m-long shadow.
At that moment the difference in the length of the two shadows was 8 m.
What is the height of the tree? **3 marks**

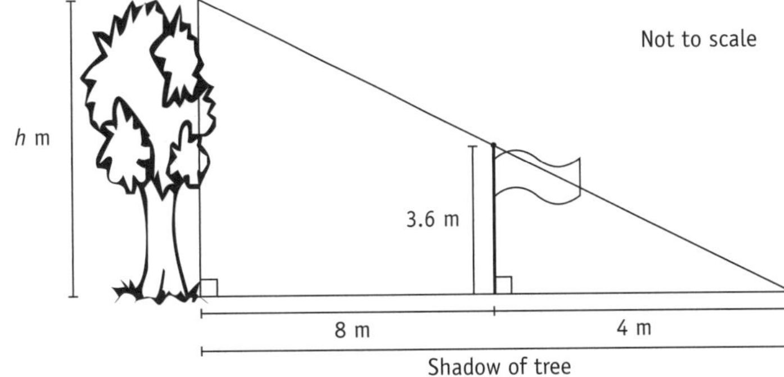

QUESTION **18** (2 marks)

A new caravan is purchased for $76 500.
Each year the caravan depreciates in value by 12%.
Using the declining-balance formula, calculate the salvage value of the caravan after 8 years. **2 marks**

QUESTION **19** (4 marks)

Dexter has a credit card with the conditions listed below.

- There is no interest-free period.
- Interest is charged at the end of each month at 14.5% p.a., compounded daily, from the date of purchase (included) to the last day of the month (included).
- The minimum payment is calculated as 2% of the closing balance on the last day of the month.

Dexter's credit card statement for August is shown, with some figures missing.

Statement period: **1 August to 31 August**		
Date	**Details**	**Amount ($)**
1 August	Opening balance	**0**
6 August	Television	**1890**
31 August	Interest charged	________
31 August	Closing balance	________
Minimum payment:	________	

After finding the interest charged and the closing balance, calculate the minimum payment. **4 marks**

QUESTION **20** (7 marks)

The diagram shows the top view of a swimming pool which is in the shape of a rectangle and a semicircle.

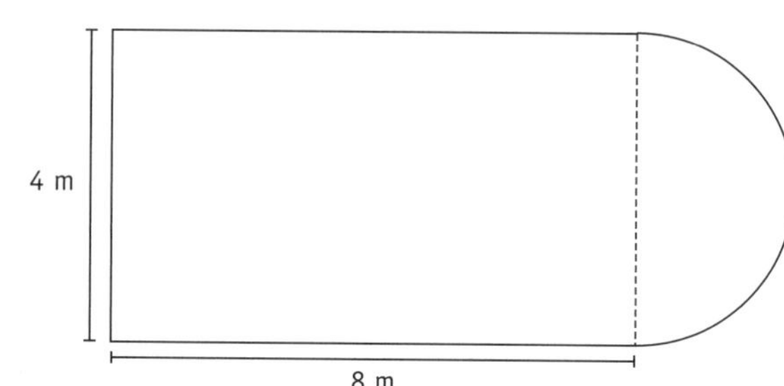

a Show that the area of the surface of the water is approximately 38.28 m^2. **3 marks**

b The pool is to be filled to a depth of 1.6 m. Find the volume of water in the pool, to the nearest kilolitre. (Use 1 m^3 = 1000 L.) **2 marks**

c A hose is used to fill the pool. How long will it take to fill the pool to a depth of 1.6 m if the hose used discharges at the rate of 20 L per minute? Give your answer to the nearest hour. **2 marks**

QUESTION **21** (2 marks)

If $P = \frac{a(16 - b)}{c}$, what is the value of a if $P = 24$, $b = 6$ and $c = 1.5$. **2 marks**

Sample HSC Examination 2

QUESTION **22** (5 marks)

A landscaping project to transform the backyard of a house is planned. The costs, including GST, are detailed below:

- pergola costing \$18 460
- concrete slab:
  - » concrete 9.5 m^3 at \$280/m^3
  - » concreters: 2 workers for 6 hours each at \$80/h
- flooring:
  - » tiles: 36 m^2 at \$45/m^2 (includes 10% for breakage)
  - » tiler (includes glue, etc): 32 m^2 at \$105/m^2
- glass pool fence and gate installation: 8 m at \$420/metre
- garden shed:
  - » demolition and removal of existing shed: \$220
  - » installation of new shed: \$2140
- outdoor furniture: \$3200.

What is the total cost of the project? **5 marks**

QUESTION **23** (7 marks)

The graph shows the line of best fit in an experiment where the burning time of a candle was measured.

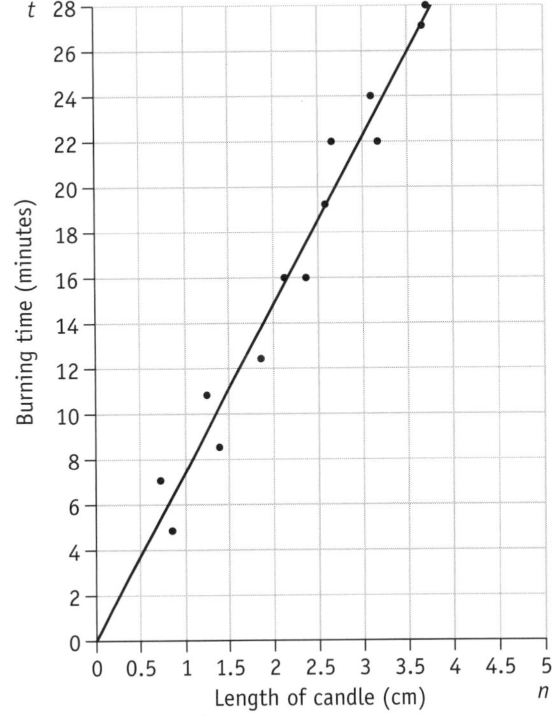

a Name the dependent variable. **1 mark**

b What is the vertical intercept of this line? **1 mark**

c What is the significance of this intercept? **1 mark**

d What is the gradient of the line, correct to one decimal place? **2 marks**

e Write the equation of the line in terms of the given variables. **2 marks**

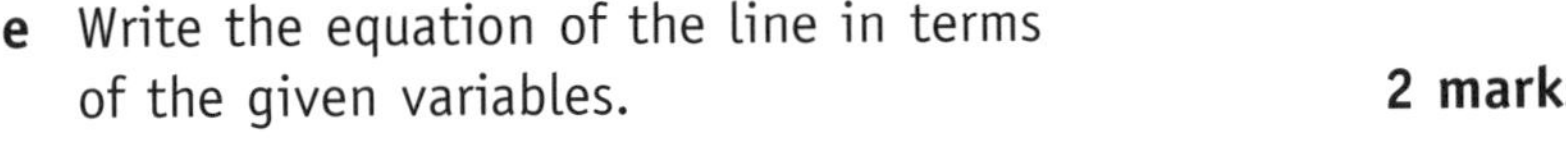

Sample HSC Examination 2

Question 24 (4 marks)

A class of Year 12 students recorded their bodyweight and displayed the data in the back-to-back stem-and-leaf plot shown:

Boys		Girls
	4	9
6 2	5	3 4 6 7 8 9
8 7 4 2 1 0	6	0 0 2 3 8
5 5 4 0	7	2 5
1	8	
7	9	

a Determine the upper and lower quartiles for the boys' data. **2 marks**

b Determine whether there is an outlier in the boys' data. **2 marks**

Question 25 (3 marks)

Calculate the difference in the surface area of a tennis ball (radius 3.5 cm) and a soccer ball (radius 10 cm). Give your answer correct to two significant figures. **3 marks**

Question 26 (3 marks)

Students in two classes recorded how many sit-ups they could complete in one minute. The results are shown on the box-and-whisker plots below.

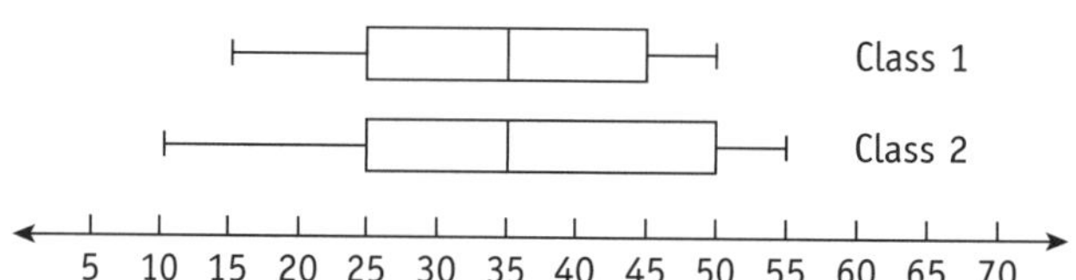

Compare the data for both classes by commenting on the difference in range, median and interquartile range. **3 marks**

Question 27 (3 marks)

The angle of depression of B from A is 42°. AC is 20 m, and $AM = MC$. Find the size of angle MBC, to the nearest degree. **3 marks**

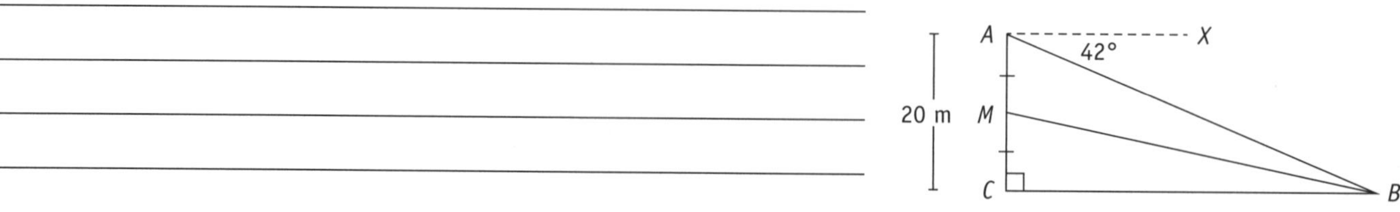

Sample HSC Examination 2

Question **28** (7 marks)

The population of people living on an island is increasing exponentially. The population of the island is modelled using the formula $P = 30\,000(1.035)^t$, where P is the population and t is the number of years after 2018.

a What was the initial population (in 2018)? **1 mark**

b Complete the table of values below using $P = 30\,000(1.035)^t$. **1 mark**

t	0	5	10	15	20	25
P		35 631	42 318		59 694	

c Draw the population graph for the table of values. **2 marks**

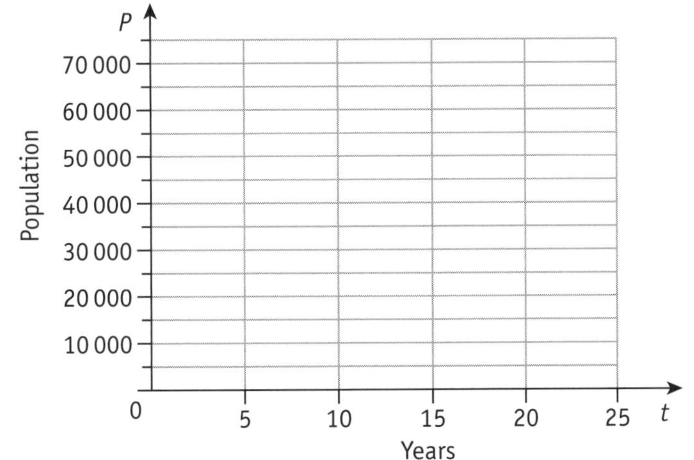

d Use the graph to estimate the population in 2030, to the nearest 1000. **1 mark**

e Estimate the time taken for the population to reach 55 000. **1 mark**

f Use your calculator to estimate the population in 2050, to the nearest 100. **1 mark**

Sample HSC Examination 2

QUESTION **29** (5 marks)

A school has 10 classrooms that require access to water. The classrooms are represented by vertices *A* to *J* on the network diagram below. Water is presently piped to vertex *X*. The dotted edges represent possible water pipe routes between adjacent classrooms. The weights on edges represent the length of pipe required, in metres.

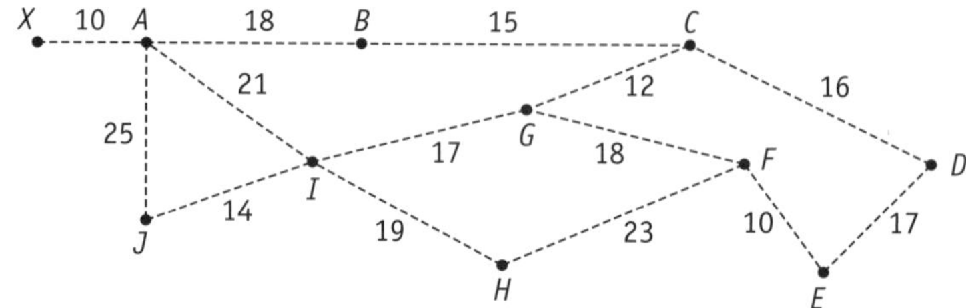

All classrooms are to be connected using the smallest length of pipe possible. The cost of the water pipe plus laying is $123.40/m.

The principal used a minimum spanning tree to determine the lowest cost of water pipe required. However, the price was too expensive. She decided to reduce costs by not connecting classroom *B* to the water supply. What will be the savings for the school under the new plan of not connecting classroom *B*? **5 marks**

Sample HSC Examination 3

Total time: 2 hours **Total marks: 80**

SECTION I

Marks: 10

Instructions

- Attempt Questions 1 to 10.
- Allow about 15 minutes for this section.
- Each question is worth 1 mark.
- Fill in only ONE CIRCLE for each question.

1 What is the range of the following scores?
16, 8, 2, 19, 31, 18, 8, 25

Ⓐ 9 Ⓑ 11 Ⓒ 29 Ⓓ 33

2 What is the bearing of P from Q?

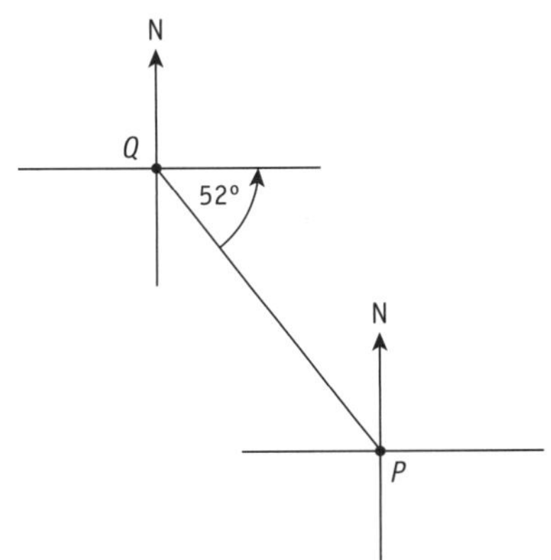

Ⓐ S52° E Ⓑ N52° W Ⓒ S38° E Ⓓ N38° W

3 What is the gradient of the line $y = 3 - 2x$?

Ⓐ 3 Ⓑ −2 Ⓒ $\frac{3}{2}$ Ⓓ $-\frac{2}{3}$

4 A set of eight scores has a lower quartile and 14 and an upper quartile of 20. Which of these is an outlier for the data set?

Ⓐ 5 Ⓑ 6 Ⓒ 29 Ⓓ 30

5 The point (5, 200) lies on the parabola with equation $y = kx^2$. What is the value of k?

Ⓐ $k = \sqrt{40}$ Ⓑ $k = 8$ Ⓒ $k = 16$ Ⓓ $k = 40$

6 The scale on an aerial photo is given as 1 : 250 000. Two mountain peaks are 5 cm apart on the photo. What is the actual distance between the two peaks in kilometres?

Ⓐ 1.25 km Ⓑ 5 km Ⓒ 12.5 km Ⓓ 50 km

7 A car is purchased for $32 890. It will depreciate at the rate of 20% p.a. Using the declining-balance method, which of these is closest to the salvage value of the car after 5 years?

(A) $10 777 (B) $12 471 (C) $22 113 (D) $29 601

8

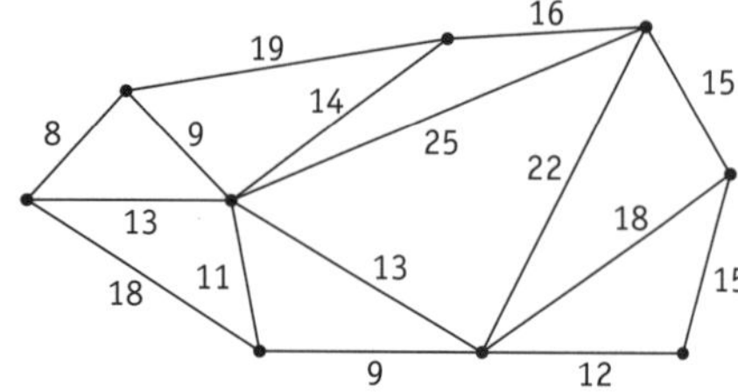

Daniel used the network diagram above to draw this spanning tree.

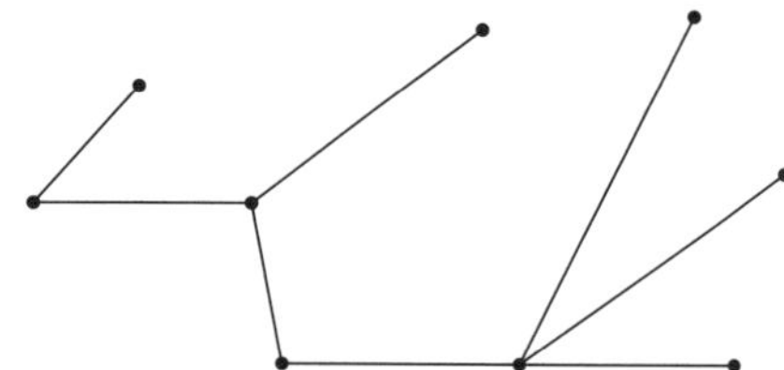

Which of these is the total weight of Daniel's spanning tree?

(A) 8 (B) 84 (C) 97 (D) 107

9 Tony made this spinner. The arrow is spun twice.
What is the probability that an even number is spun twice?

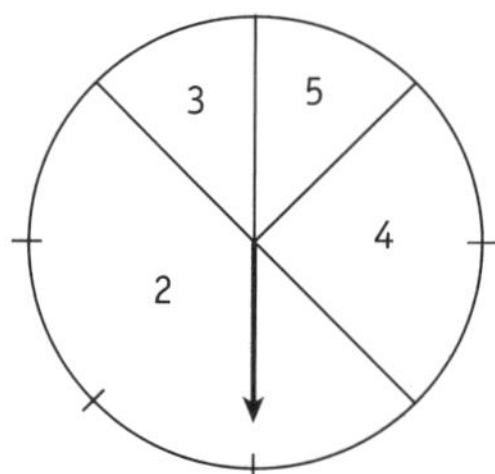

(A) $\frac{1}{4}$ (B) $\frac{1}{2}$

(C) $\frac{9}{16}$ (D) $\frac{6}{8}$

10 The results of an analysis of reasons given by workers who arrive late to the office over a 6-month period are displayed below. A Pareto chart is to be drawn for the data.

Reasons for lateness to work			
Reasons	**Frequency**	**Cumulative frequency**	**Cumulative percentage**
Traffic	40		
Slept in	28		p
Low on petrol	8		
Family issues	4		

What is the value of p?

(A) 28 (B) 68 (C) 80 (D) 85

Sample HSC Examination 3

SECTION II

Marks: 70

Instructions

- Attempt Questions 11 to 29.
- Allow about 1 hour and 45 minutes for this section.
- Answer the questions in the spaces provided. These spaces provide guidance for the expected length of response.
- Your responses should include relevant mathematical reasoning and/or calculations.

QUESTION **11** (2 marks)

Ava invested $8500 at 2.75% p.a. for 18 months.

How much simple interest did Ava earn? **2 marks**

QUESTION **12** (4 marks)

For his casual job James earns an hourly rate of $27.25.

James is paid:

- normal rate for any hours between 8 am and 4 pm
- time and a half for hours between 4 pm and 10 pm
- double time for hours between 10 pm and 8 am.

The following table outlines James's work in one week:

James's work diary		
Day	**Start time**	**Finish time**
Monday	9:00 am	1:00 pm
Tuesday	–	–
Wednesday	8:00 pm	12:00 am
Thursday	1:00 pm	6:00 pm
Friday	–	–
Saturday	9:00 am	12:00 pm

Find James's pay for:

a Monday **1 mark**

b the whole week **3 marks**

QUESTION **13** (2 marks)

A fertiliser mixture contains 2 parts nitrogen, 2 parts potash and 3 parts phosphate by mass. How many kilograms of phosphate are in a bag of fertiliser which has a mass of 42 kg? **2 marks**

QUESTION **14** (4 marks)

The height of a ball which is thrown vertically upwards with a speed of 20 m per second is given by the equation $h = 20t - 5t^2$, where t is the time in seconds after the ball has been thrown and h is the height in metres.

a Complete the table. **1 mark**

t	0	1	2	3	4
h					

b Use the table to draw the graph of the equation $h = 20t - 5t^2$. **2 marks**

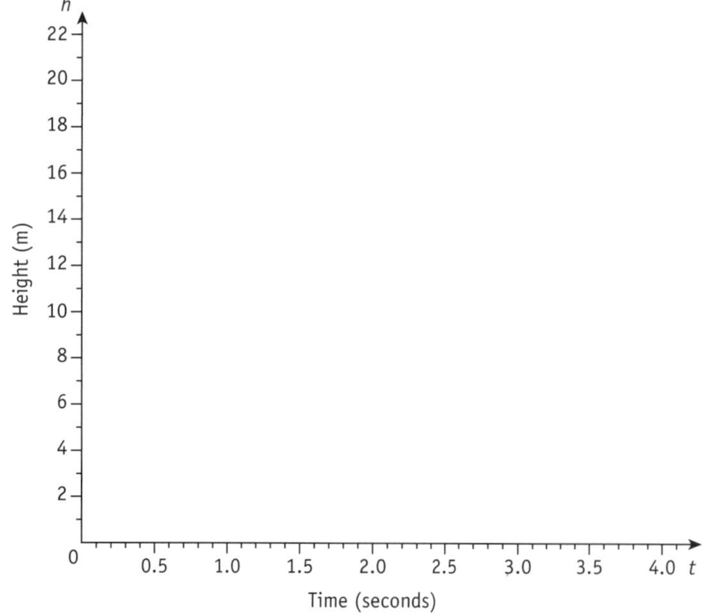

c What was the maximum height of the ball? **1 mark**

Sample HSC Examination 3

QUESTION **15** (3 marks)

Marty wants to invest \$25 000 for 4 years.

He has three investment options:

- option 1: 4% p.a. compounded annually
- option 2: 3.8% p.a. compounded quarterly
- option 3: 3.6% p.a. compounded monthly.

Determine Marty's best investment strategy. Support your answer with calculations. **3 marks**

QUESTION **16** (6 marks)

The equation $P = 1250(1.06)^t$ represents the population P of a town t years after 1980.

a Substitute $t = 0$ into the equation to find the population in 1980. **1 mark**

b Complete the table for $P = 1250(1.06)^t$, writing the values of P as whole numbers. **2 marks**

t	0	10	20	30	40
P		2239			12 857

c Use the table to graph the equation on the diagram below. **2 marks**

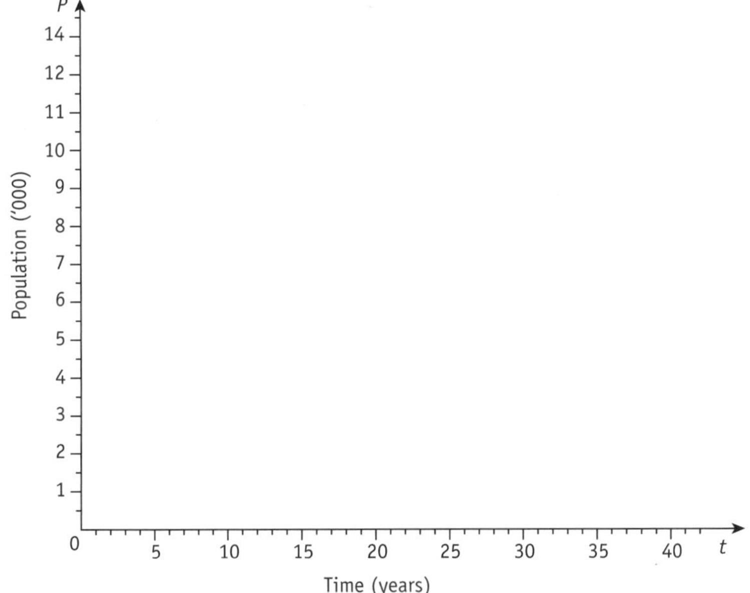

d Hence, estimate the population of the town in 2015, to the nearest hundred. **1 mark**

Sample HSC Examination 3

QUESTION **17** (2 marks)

Water is poured into a container at a constant rate.

A graph has been drawn to show the increase in the depth of water in the container over time.

In the space below, draw a possible container that matches the graph. **2 marks**

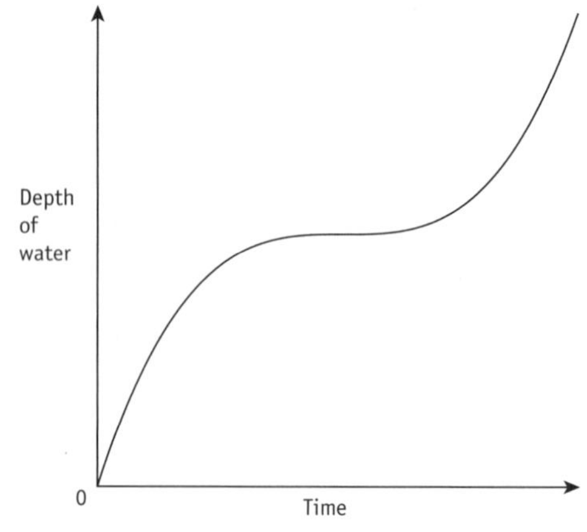

QUESTION **18** (2 marks)

The length of Sailor Sam's anchor rope is 25 m.

It is attached to the bottom of the lake but after the boat has drifted in the current, the angle the rope makes with the surface of the water is 52°.

At what depth (d) is Sam's anchor, correct to two decimal places? **2 marks**

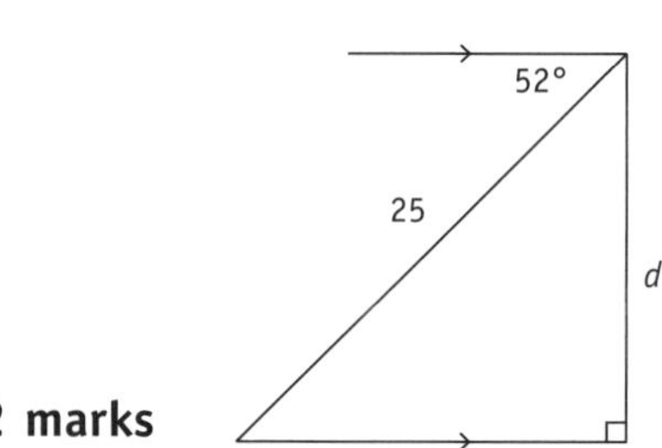

QUESTION **19** (2 marks)

A tree 12 m tall casts a shadow 16 m long.

What is the angle of elevation of the sun, to the nearest minute? **2 marks**

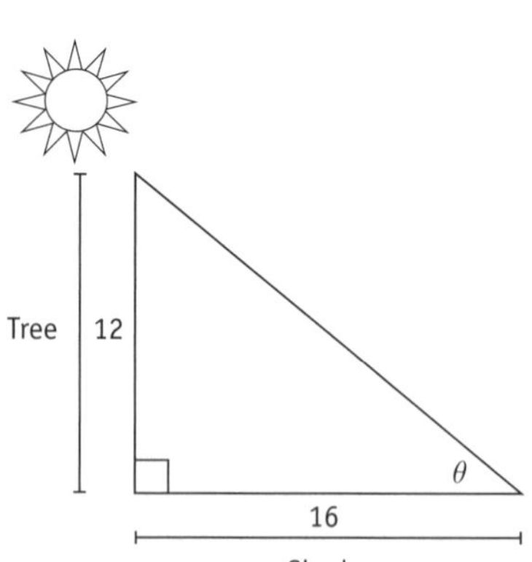

QUESTION **20** (4 marks)

Two women, Adalee and Keily, plan to cycle a 40-km course.

Both girls commence riding the course at 10 am.
Adalee is averaging 25 km/h.

Keily is riding at a slower pace, averaging 20 km/h.

a At what time will Adalee finish the course? **2 marks**

b When Adalee finishes the course, how far does Keily still need to ride? **2 marks**

QUESTION **21** (4 marks)

Calculate the lengths of *TQ* and *TP* and hence find the area of the triangle *PQR*, to the nearest square metre. **4 marks**

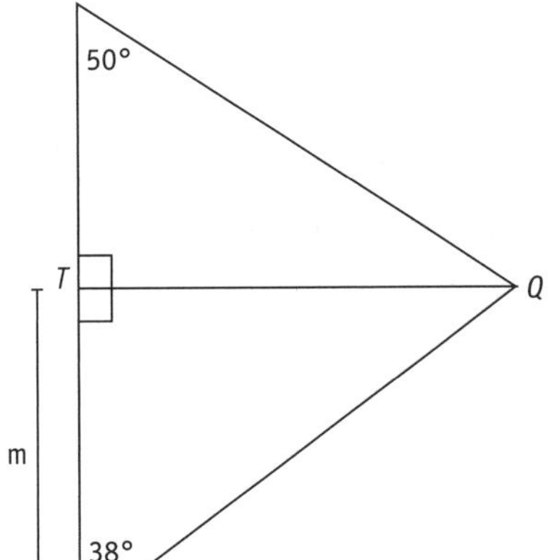

QUESTION **22** (5 marks)

Noah and Cordelia both live in Marrangaroo. On a certain day, Noah leaves Marrangaroo and drives to Gilgandra. Later in the day he leaves Gilgandra and drives the same route back to Marrangaroo.

Noah's travel graph is shown below.

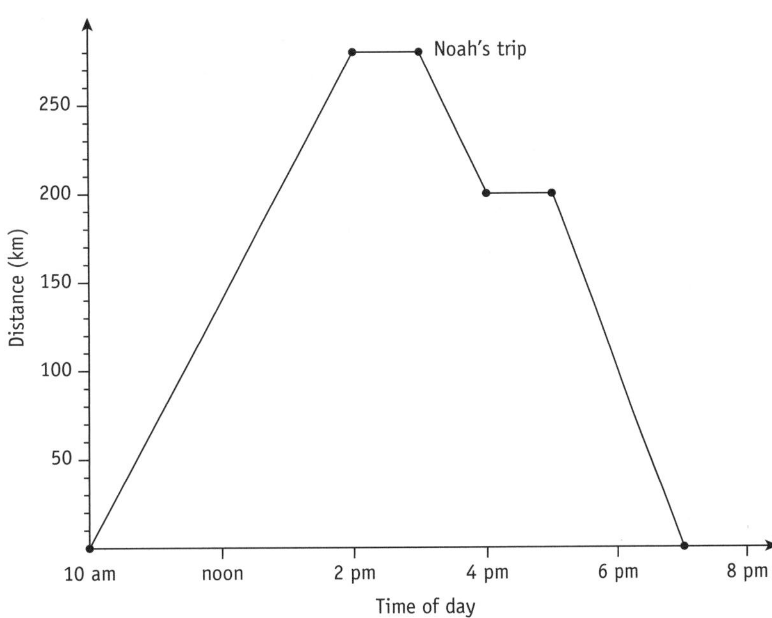

a What was his average speed between 10 am and 2 pm? **1 mark**

b What was the total distance travelled in the day? **1 mark**

c Cordelia leaves Marrangaroo at 2 pm and drives at an average speed of 100 km/h to meet Noah. At 5 pm Cordelia drives back to Marrangaroo, averaging 80 km/h. Determine the time she arrives back in Marrangaroo and draw a travel graph representing her journey. **3 marks**

QUESTION **23** (2 marks)

Triangles *ABC* and *ADE* are similar. *AC* = 8 m, *CE* = 4 m and *BC* = 6 m

Find the length of *DE*. **2 marks**

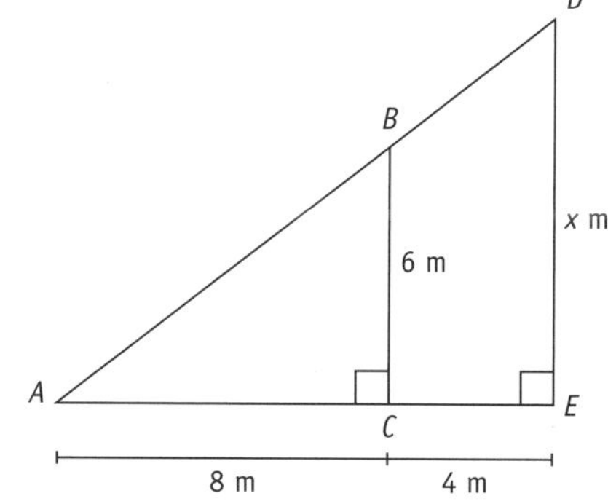

QUESTION **24** (4 marks)

Nick borrowed $28 000 for a new car.

He was charged 6% p.a. compound interest, calculated monthly on the balance owing.
He made monthly repayments of $500.

Nick used the spreadsheet below to track the amount owing at the end of each month.

Nick's reducible balance personal loan				
Amount borrowed: $28 000				
Interest rate: 6% p.a., monthly reducible				
Monthly repayments: $500				
Month	**Principal (*P*)**	**Interest (*I*)**	**Principal + Interest (*P* + *I*)**	**Amount owing (*P* + *I* – *R*)**
1	28 000.00	140.00	28 140.00	27 640.00
2	27 640.00	138.20	27 778.20	27 278.20
3	27 278.20	136.39	27 414.59	26 914.59
4	26 914.59	134.57	27 049.16	26 549.16
5				
6				*X*

Use calculations to complete the table and find the value of X, which is the amount Nick owes after six months. **4 marks**

Question **25** (6 marks)

Charlotte collected data on the ages (a) and heights (h) of a group of 10- to 16-year-old females.

She recorded the data on a scatterplot and drew a line of best fit to model the relationship between age and height of females.

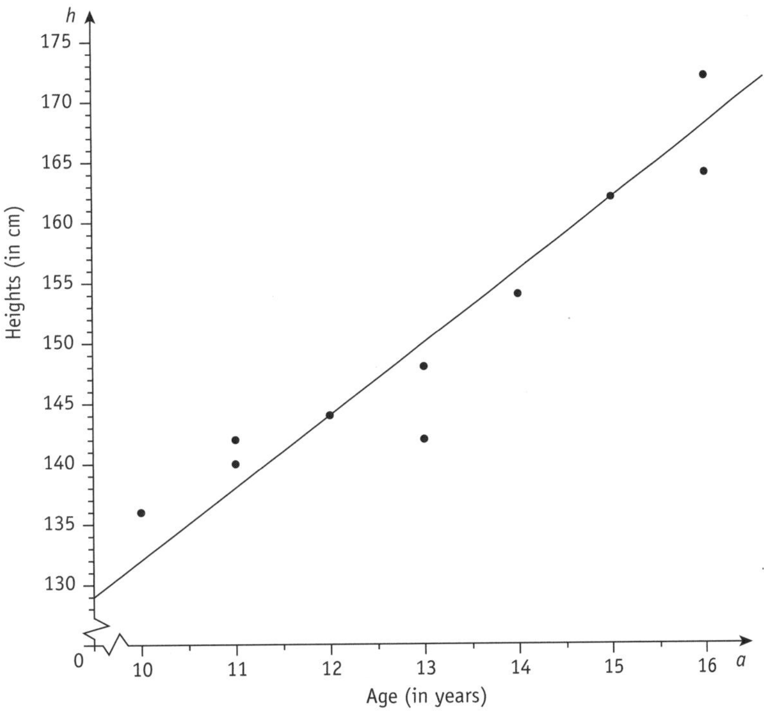

a Determine the gradient of Charlotte's line. **2 marks**

---

b Explain the meaning of the gradient of her line in this context. **1 mark**

---

c Determine the equation of Charlotte's line. **2 marks**

---

d Why would this model not be useful to estimate the height of a 25-year-old female? **1 mark**

___

___

QUESTION **26** (3 marks)

The results of a mathematics test are shown in the back-to-back stem-and-leaf plot.

Compare the results for females and males, commenting on measures of central tendency and spread. **3 marks**

Scores in Maths test		
Females		**Males**
	4	7
8 6	5	6 8
8 5 5	6	3 5 9
9 6 4	7	2
8 2	8	3
4	9	0

___

___

___

___

QUESTION **27** (4 marks)

The following table shows the distances in kilometres between towns *P*, *Q*, *R*, *S* and *T*.

The dashes shown in the table means these towns are not connected directly.

		To:				
		P	***Q***	***R***	***S***	***T***
From:	***P***	–	20	–	–	–
	Q	45	–	40	–	–
	R	15	40	–	25	40
	S	–	60	20	–	–
	T	–	–	70	30	–

a Use the table to complete the network diagram by recording weights on the directed vertices. **2 marks**

Q

P • *S* • *T*

R

b Find the shortest distance between P and T. **2 marks**

QUESTION **28** (6 marks)

The table shows an estimated number of kilojoules burned per kilogram of body mass per 30 minutes in different activities.

Activity	Energy used in 30 minutes
Walking at 6 km/h	9.22 kJ/kg
Cycling at 20 km/h	18.43 kJ/kg
Swimming at 50 m/min	20.73 kJ/kg
Jogging at 10 km/h	21.19 kJ/kg
Running at 16 km/h	32.25 kJ/kg

a Dayle weighs 68 kg and cycles at an average speed of 20 km/h for 45 minutes. How much energy has she used, to the nearest kilojoule? **2 marks**

b George, who weighs 83 kg, eats a hamburger from a fast-food restaurant which contains 870 kilocalories. For what distance must George walk at 6 km/h to burn off the energy contained in the hamburger, to the nearest kilometre? (1 kilocalorie = 4.184 kJ) **4 marks**

Sample HSC Examination 3

QUESTION **29** (5 marks)

A solid is made up of a cone and a hemisphere.

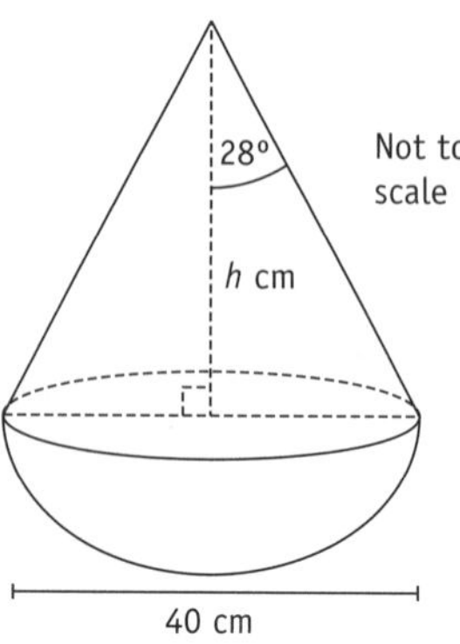

a Show that the height of the cone, correct to one decimal place, is 37.6 cm. **2 marks**

---

b Hence find the volume of the solid, correct to three significant figures. **3 marks**

Answers

CHAPTER 1—Financial mathematics: Investments

PAGE 1 **1 a** 0.5% **b** 1.5% **c** 3% **d** 2% **2 a** 0.75% **b** 0.625% **3 a** 2% **b** 1.25% **4 a** 60 months **b** 12 quarters **c** 16 six-months **d** 6 four-months **5 a** 16 quarters **b** 2.25% **6 a** 10% **b** 10.8% **c** 13% **d** 16.79%

PAGE 2 **1 a** $720 **b** $3360 **c** $14 400 **d** $354 **e** $384.38 **f** $14 760 **g** $21 125 **h** $13 530 **2 a** $600 **b** $3600 **3 a** 4 years **b** 5 years **4 a** 4.2% **b** 3.5%

PAGE 3 **1 a** $4500 **b** $3400 **2 a**

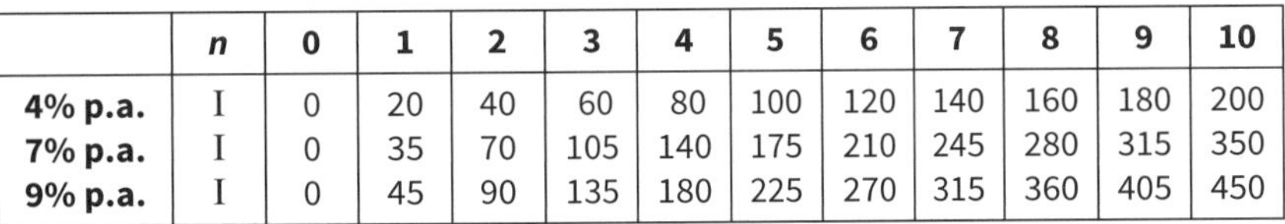

	n	0	1	2	3	4	5	6	7	8	9	10
4% p.a.	I	0	20	40	60	80	100	120	140	160	180	200
7% p.a.	I	0	35	70	105	140	175	210	245	280	315	350
9% p.a.	I	0	45	90	135	180	225	270	315	360	405	450

b

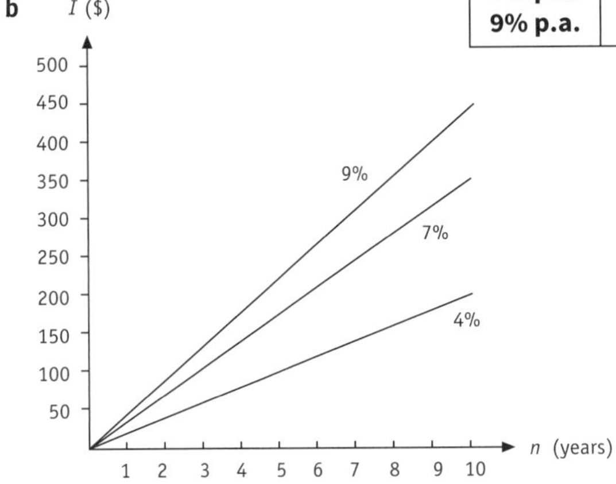

PAGE 4 **1 a** $10 960.69 **b** $2960.69 **2 a** $2469.49 **b** $5594.33 **c** $30 653.59 **d** $7689.65 **e** $174 754.99 **f** $1388.53

PAGE 5 **1 a** $4764.06 **b** $15 280.15 **2 a** $5832 **b** $13 381.03 **c** $19 965 **d** $15 109.02 **3 a** $5723.08 **b** $30 825.95

PAGE 6 **1 a** $8236.71 **b** $3350.24 **c** $29 065.89 **d** $59 665.44 **2 a** $35 143.71 **b** $67 884.05

PAGE 7 **1 a** $2500 **b** $6400 **c** $12 000 **2 a** $3325.29 **b** $2871.87 **3 a** $13 021.77 **b** $12 996.78

PAGE 8 **1 a** $2814.20 **b** $6701.20 **c** $20 426.40 **d** $10 247.25 **2 a** $7306.74 **b** $13 319.13

PAGE 9 **1 a** $1700 **b** 9 years **c** $1200 **d** The future value will increase at a faster rate. The future value doubles approximately every 4 years, so after 14 years it will be close to $12 000. **2 a** 18% p.a. becomes 9% per six months. Using the formula $A = P(1 + r)^n$ then $A = 1000(1 + 0.09)^n = 1000(1.09)^n$ where n is the number of six-month periods.

b

n	2	4	6	8	10	12	14	16	18	20
A($)	1188	1412	1677	1993	2367	2813	3342	3970	4717	5604

c

Future value ($)
6000
5000
4000
3000
2000
1000
Compounded monthly
Compounded six-monthly
1 2 3 4 5 6 7 8 9 10
n (years)

d The difference will increase, the future value is greater for the investment which has the interest compounded monthly.

PAGE 10 **1 a** $15.15 **b i** $12 701.11 **ii** $32 312.55 **c i** $5402.22 **ii** $32 176.19 **2 a** $34 676.14 **b i** $31 370.62 **ii** $30 478.23 **iii** $34 115.38

PAGE 11 **1 a** $768.75 **b** $2.54 **2** $399 408 **3** $40 119 **4** $581.56 **5** 13.23% **6** $545 701 **7** $2295

PAGES 12–14 **1** A $(3200 \times 0.0425 \times 1.25)$ **2** D $(16\,000(1.003)^{24})$ **3** C $\left(20\,000 = PV(1.01^{12}),\ PV = \dfrac{20\,000}{1.01^{12}}\right)$ **4** C $(230(1.026)^5 - 230)$ **5** C (Difference between $10\,000(1.004)^{60}$ and $10\,000(1.0045)^{60}$) **6 a i** $\$5000 \times 1.407 = \$7035$ **ii** $\$3400 \times 1.062 = \$3610.80$ **b i** $\$12\,000 \times 2.144 - \$12\,000 = \$13\,728$ **ii** $\$9600 \times 1.051 - \$9600 = \$489.60$ **7 a** $\$16\,000(1.04)^7 = \$21\,054.91$ **b** $\$24\,000(1.005)^{60} = \$32\,372.40$ **c** $\$8000(1.02)^{16} = \$10\,982.29$ **d** $\$11\,400(1.0075)^{36} = \$14\,918.56$ **8 a** $\dfrac{\$6000}{1.05^{12}} = \$3341.02$ **b** $\dfrac{\$15\,000}{1.01^{80}} = \$6766.77$ **c** $\dfrac{\$9500}{1.005^{120}} = \$5221.51$ **d** $\dfrac{\$32\,000}{1.004^{300}} = \$9661.31$ **9 a** $\$5000(1.04)^{18} = \$10\,129$ **b** $\$5000(1.04)^{65} = \$63\,994$ **10 a** $\dfrac{\$120\,000}{1.005^{180}} = \$48\,898$ **b** $\dfrac{\$500\,000}{1.005^{360}} = \$83\,021$

11 Option A: Simple interest = $28\,000 \times 0.05 \times 3 = 4200$, Future Value = $28\,000 + 4200 = 32\,200$; Option B: Future Value = $28\,000(1.048)^3 = 32\,228.63$; Option C: Quarterly interest rate = $4.4\% \div 4 = 1.1\%$, No. of quarters = $3 \times 4 = 12$, Future Value = $28\,000(1.011)^{12} = 31\,928.01$. The best investment strategy is Option B.

Answers

Chapter 2—Financial mathematics: Depreciation and loans

Page 16 1 a 0.22 b 0.165 c 0.1875 2 a 15% b 23.5% c 9.8% 3 a i $S = 28\,500(1 - 0.2)^2 = \$18\,240$ ii $S = 28\,500(1 - 0.2)^5 = \$9339$ b $12\,600 = V_0(1 - 0.12)^6$. $V_0 = 12\,600 \div 0.88^6 = \$27\,132$ c $19\,500 = 37\,800(1 - r)^4$. So $(1 - r)^4 = \dfrac{19\,500}{37\,800} = 0.5159$. Taking the fourth root of both sides, $1 - r = 0.8475$. This gives $r = 1 - 0.8475 = 0.1525$, or $15\frac{1}{4}\%$.
d $14\,400 = 45\,000(1 - 0.15)^n$. So $0.85^n = \dfrac{14\,400}{45\,000} = 0.32$. Now you need to try various values of n to get as close as possible to answering $0.85^n = 0.32$. This gives $n = 7$ years.

4

Year	Net book value ($)	Depreciation ($)	Final value ($)
1	30 000	0.12 × 30 000 = 3600	30 000 − 3600 = 26 400
2	26 400	0.12 × 26 400 = 3168	26 400 − 3168 = 23 232
3	23 232	0.12 × 23 232 = 2788	23 232 − 2788 = 20 444
4	20 444	0.12 × 20 444 = 2453	20 444 − 2453 = 17 991
5	17 991	0.12 × 17 991 = 2159	17 991 − 2159 = 15 832
6	15 832	0.12 × 15 832 = 1900	15 832 − 1900 = 13 932
7	13 932	0.12 × 13 932 = 1672	13 932 − 1672 = 12 260
8	12 260	0.12 × 12 260 = 1471	12 260 − 1471 = 10 789

a The amount becomes less in subsequent years. This is because the same percentage of depreciation applies to a lesser book value.
b With the straight-line method the amount of depreciation remains the same from year to year.

Page 17 1 a $20 000 b $20 000 − $12 000 = $8000 c With $S = 12\,000$, $V_0 = 20\,000$, $n = 1$, then $12\,000 = 20\,000(1 - r)^1$. This gives $1 - r = \dfrac{12\,000}{20\,000} = 0.6$, so $r = 1 - 0.6 = 0.4$. So $r = 40\%$. d $S = 20\,000(1 - 0.4)^6 = \$933.12$ e Extending the graph to the 6th year should give a value just under $1000. 2 a $18 000 b $14 000 c 4 years (where the curves intersect on the graph)
d Not necessarily. Ryan's car might be worth more than Bella's according to the depreciation schedule, but this doesn't take into account potential accidents, car maintenance level or the type of car, which all affect its value.

Page 18 1 a $8000 b $466.67 2 a $600 b $2400 c $1080 d $3480 e $96.67 3 a $14 780 b $246.33

Page 19 1 a $9675 b $22 575 c $376.25 2 a $78 750 b $4775 3 a $12 600 b 6.5% p.a.

Page 20 1 a $298 570.77 b i $299 052.33 ii $1619.87 iii $300 672.20 iv $298 812.20 c i $1860 × 6 = $11 160
ii $300 000 − $298 570.77 = $1429.23 2 a $18 553.33 b i $17 102.81 ii $9791.96 iii $5372.78 c $148.42 d Mia has paid the loan out. e $891.83

Page 21 1 a i 260 × $7.4581 = $1939.11 ii 312 × $8.3669 = $2610.47 iii 435 × $5.6754 = $2468.80
b i 327 × $6.9921 = $2286.42 ii $2286.42 × 360 = $823 111.20 iii $823 111.20 − $327 000 = $496 111.20
c i 385 × $7.1643 × 240 − $385 000 = $276 981 ii 482.5 × $6.9921 × 360 − $482 500 = $732 028
d 245 × ($8.3669 − $7.4581) = $222.66 2 a 460 × $6.3233 = $2908.72 b $2908.72 × 360 − $460 000 = $587 139
3 $3802.52 × 360 − $670 000 = $698 906

Page 22 1 a $280 000 b 21 years c i $2398 × 12 × 30 = $863 280 ii $863 280 − $400 000 = $463 280
2 a 6 years b i $1199 × 26 × 24 = $750 000 ii $750 000 − $400 000 = $350 000 3 a $280 000 − $80 000 = $200 000
b i $2398 × 12 × 18 + $200 000 = $717 968 ii $463 280 − ($717 968 − $400 000) = $145 312

Page 23 1 a $10 000 b i $10 000 − $4627.18 = $5372.82 ii 11.50% ÷ 365 = 0.031 51% c $\dfrac{139}{4627.18} \times 100\% = 3.0\%$
2 a 18% ÷ 365 = 0.049 315% b 27 days c $\$960\,(1 + 0.000\,493\,15)^{27} = \$972.86$ d $12.86
3 a 13.2% ÷ 365 = 0.036 16% b i $\$216.80 \times 1.000\,3616^{14} - \$216.80 = \$1.10$ ii $\$104.90 \times 1.000\,3616^{5} - \$104.90 = \$0.19$
c $\$186.70 \times 1.000\,3616^{22} + \$217.90 + \$187.20 \times 1.000\,3616^{8} + \$105.09 = \$698.92$

Page 24 1 a $416.40 b $\$216.4 \times 1.000\,7575^{19} - \$216.4 = \$3.14$ 2 a $143.93 b $120.00

Pages 25–28 1 B ($\$26\,990 \times 0.84^4$) 2 D 3 C ($\$1250 \times 1.000\,443^{16}$) 4 D ($910 × 12 × 4 − 0.8 × $41 250)
5 B ([$18 000 + $18 000 × 0.08 × 10] ÷ 120) 6 a i $2300 ii 0.058 877%
b $\$125 \times (1.000\,547\,67)^{15} - \$125 + \$340 \times (1.000\,547\,67)^{12} - \$340 = \$3.27$ c $7694.61 d 0.01 × $2305.39 + $5.39 = $28.44
7 a i $32 956.25 ii $56.58 b $351 c The loan is completely paid out in the 17th repayment d $1621.28 8 a 310 × $7.068 = $2191.08
b $2476.44 ÷ 360 = $6.879. From the table, this is 5.5% p.a. c 2278.78 ÷ $5.558 × 1000 = $410 000
d 340 × $6.321 × 12 × 30 = $773 690.40 e 400 × $8.056 × 12 × 20 − $400 000 = $373 376

f 460 × \$5.678 × 12 × 30 − \$460 000 = \$480 276.80 and 460 × \$6.879 × 12 × 20 − \$460 000 = \$299 441.60 gives a saving of \$180 835.20 **g** Original: 280 × \$5.368 = \$1503.04, New: 200 × \$7.164 = \$1432.80. Decrease of \$70.24
9 a \$42 500 × 0.82^3 = \$23 433.14 **b** \$42 500 × 0.82^5 × 0.88^5 = \$8315.18

Chapter 3—Algebra: Simultaneous linear equations

Page 29 **1 a**

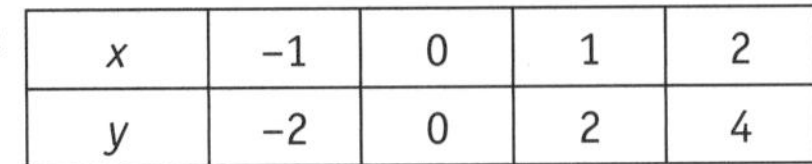

x	−1	0	1	2
y	−2	0	2	4

b

x	−1	0	1	2
y	0	1	2	3

c

x	−1	0	1	2
y	−3	−1	1	3

d

x	−1	0	1	2
y	6	5	4	3

e

x	−1	0	1	2
y	4	2	0	−2

f

x	−2	−1	0	1
y	−3	0	3	6

2

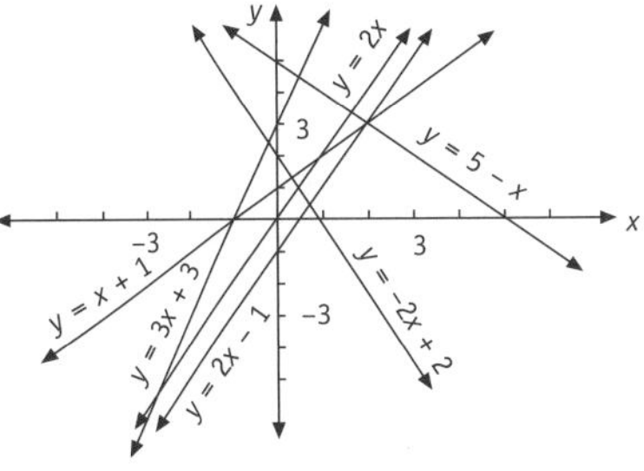

Page 30 **1 a** 5 **b** The fixed amount of the pocket money per week, \$5 **c** 2 **d** Liam's mother pays him \$2 per hour when he helps her. **2 a** 90 **b** Barton is 90 km from Aden. **c** −15 **d** Dorian rides at a constant 15 km/h. **e** $d = -15t + 90$ or $d = 90 - 15t$

Page 31 **1 a i** $y = 20x$

x	0	10	20	30	40
y	0	200	400	600	800

ii $y = 600 - 10x$

x	0	10	20	30	40
y	600	500	400	300	200

b

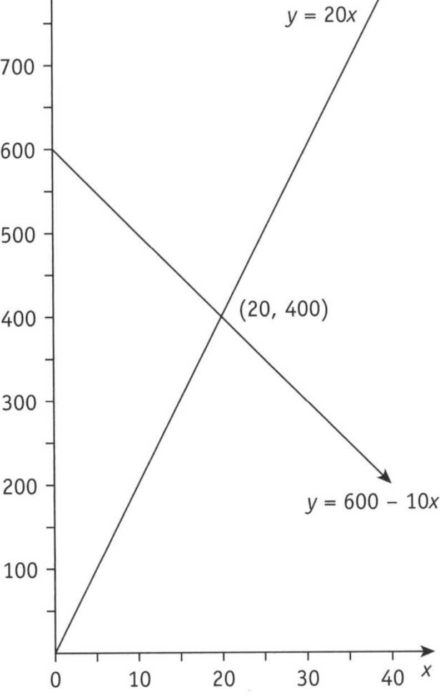

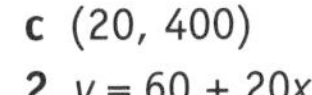
c (20, 400)

2 $y = 60 + 20x$

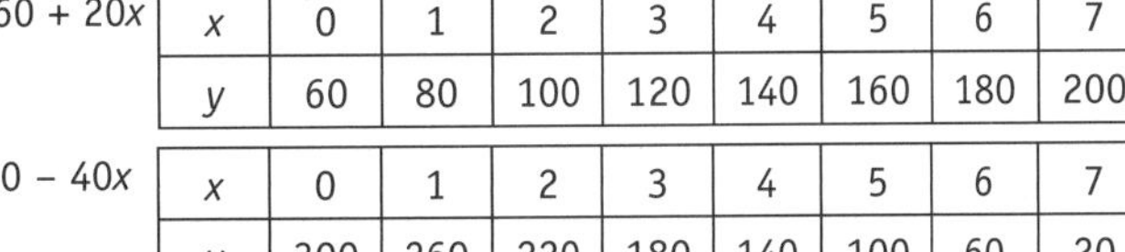

x	0	1	2	3	4	5	6	7
y	60	80	100	120	140	160	180	200

$y = 300 - 40x$

x	0	1	2	3	4	5	6	7
y	300	260	220	180	140	100	60	20

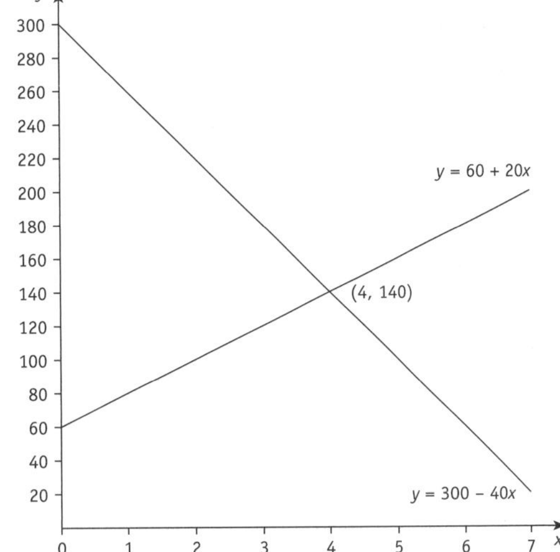

The point of intersection is (4, 140).

Answers

PAGE 32 1 **a** $C = 50 + 6m$ where C = cost in dollars and m = metres of turf. **b**

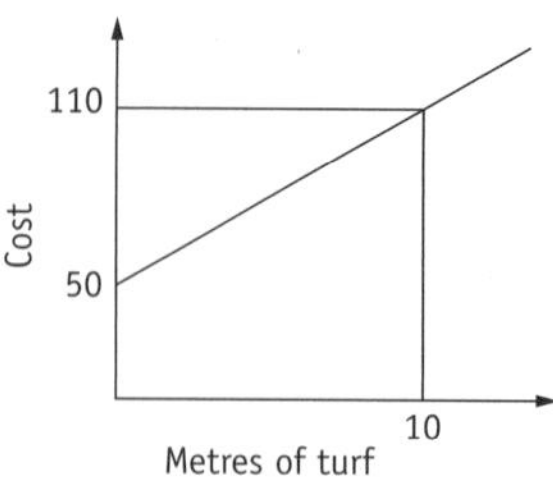

c Cost of 100 m = 50 + 6 × 100 = \$650 **d** The model only calculates cost of the turf and delivery, not other costs such as labour.

2 **a** $C = 200 + 5n$ **b**

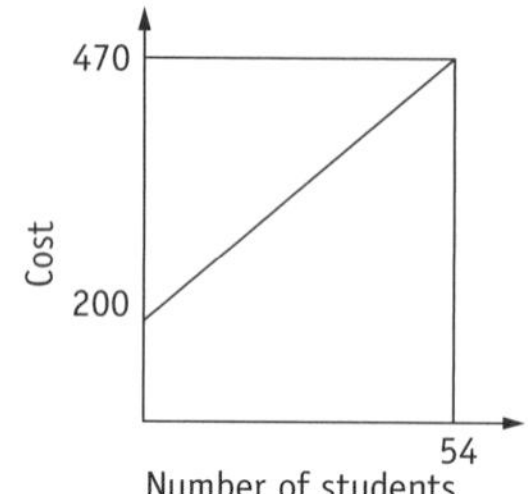

c Cost = 200 + 50 × 5 = \$450 **d** Only 54 people can travel on the bus. Numbers greater than 54 require the hire of another bus, which may involve a higher booking fee.

PAGE 33 1 **a** $C = 60n$ **b** $C = 7x$ 2 **a** $C = 4 + 2n$ **b** \$28 3 $P = 70 + 50n$ 4 **a** $d = 70t$ **b** 420 km 5 **a** $80n$ is the distance travelled in n hours and this distance is being subtracted from 280 km. **b** 80 km 6 **a** $C = 120 + 8n$ **b** $S = 16n$ 7 **a** $C = 1000 + 30n$ **b** $S = 200n$ **c** $P = 170n - 1000$ 8 **a** $d = 60t$ **b** $d = 240 - 80t$

PAGE 34 1 **a** See graph **b** 40 **c** the cost of making each cake **d** (\$40 × 20 + \$250) ÷ 20 = \$52.50 **e** See graph **f** See graph

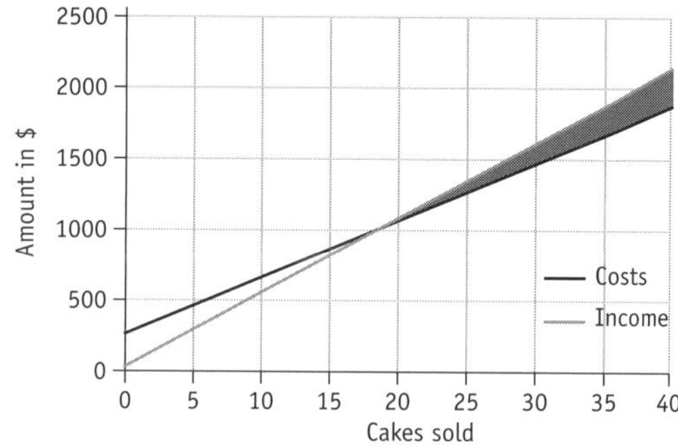

2 **a** 10 **b** Zone 2 **c** Zone 1 **d i** the cost of producing each T-shirt **ii** the fixed costs of production **e i** the income from each shirt sold **ii** There is no income if no T-shirts are sold. **f** $I = 10x$ **g** $C = 5x + 50$

PAGE 35 1 **a** There is a one-off cost of \$2400 and a cost of \$40 for each person who attends.
b Each person who attends pays \$120 per ticket. **c**

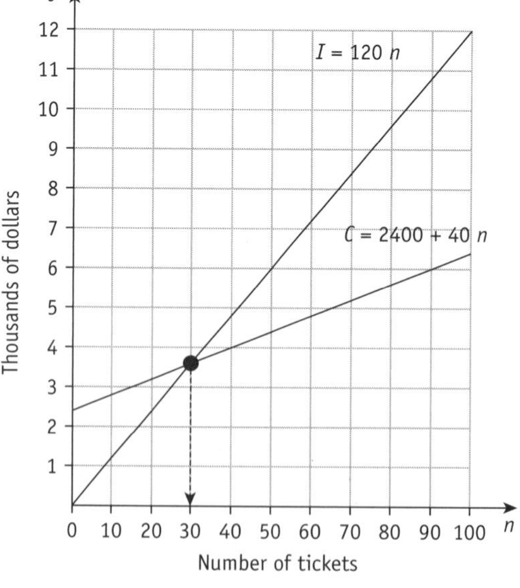

d 30 tickets **e** \$3200
f No, the maximum profit is \$5600.
2 **a** \$13 **b** 1.5 cents/h
c

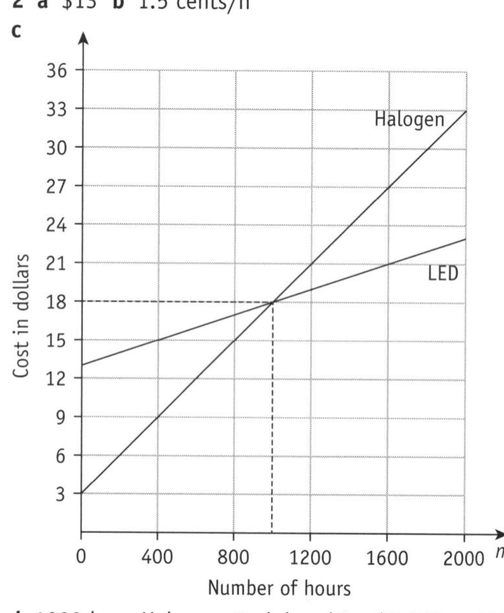

d 1000 h **e** Halogen: 2 globes \$6 + \$0.015 × 4000 = \$66. LED: \$13 + \$0.005 × 4000 = \$33. Cheaper by \$33.

Answers

Pages 36–39 1 B 2 C 3 A 4 C 5 D **6 a i**

x	−1	0	1	2
y	−4	−1	2	5

ii

x	−1	0	1	2
y	4	5	6	7

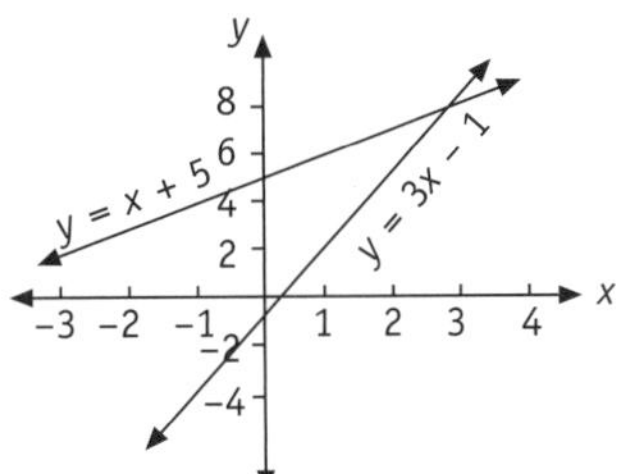

b 3 **7 a** ii **b** iv **c** i **d** iii

8 a

Number of jugs	0	4	8	12	16	20
Cost ($)	8	20	32	44	56	68

b and **g**

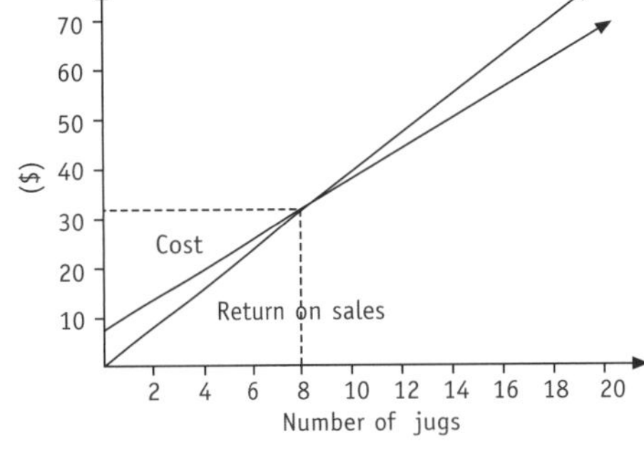

c 8, $8 is the fixed cost of making a jug of lemonade **d** 3, $3 is the additional cost per jug of lemonade **e** $50 **f** 16 jugs **h** The lines intersect at (8, 32); the break-even point is where 8 jugs of lemonade are produced and sold. **9 a** 4 cakes **b i** loss of $20 **ii** profit of $30 **c i** $I = 20n$ **ii** $C = 40 + 10n$ **10 a i** $C = 100 + 15n$ **ii** $S = 40n$

b

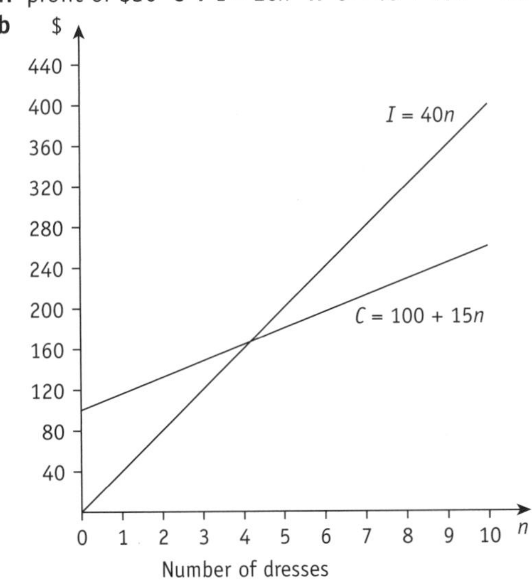

c 4 dresses **d** $50

Chapter 4—Algebra: Graphs of practical situations

Page 40 **1 a**

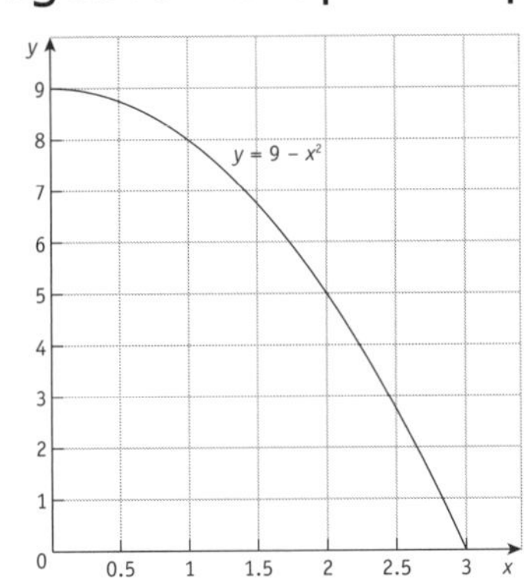

b

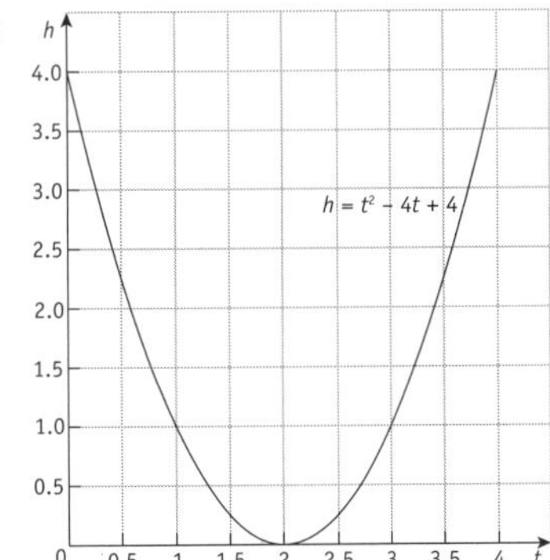

2 a $y = 16 - x^2$

x	0	1	2	3	4
y	16	15	12	7	0

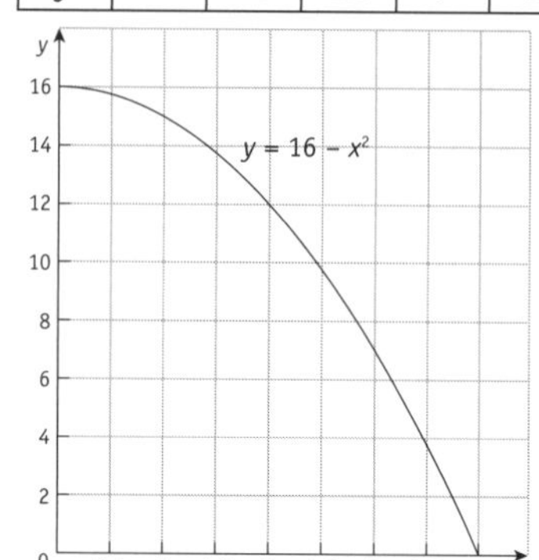

b $h = 8 + 2t - t^2$

t	0	1	2	3	4
h	8	9	8	5	0

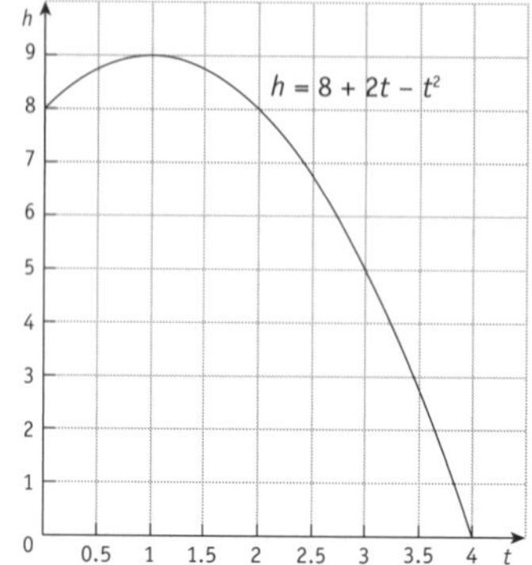

PAGE 41 **1 a**

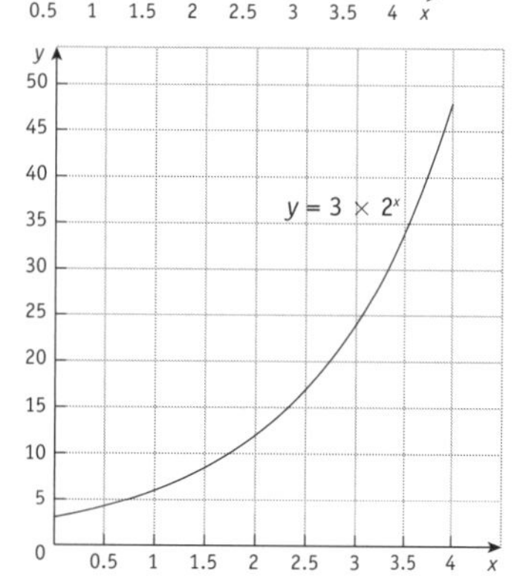

b

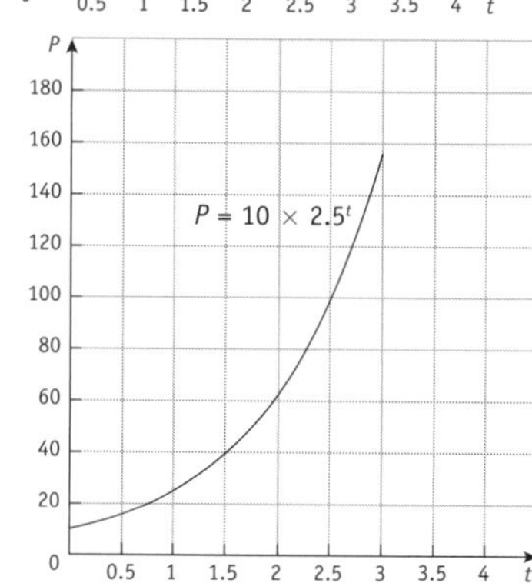

2 a $y = 120 \times 2.5^x$

x	0	1	2	3	4
y	120	300	750	1875	4687.5

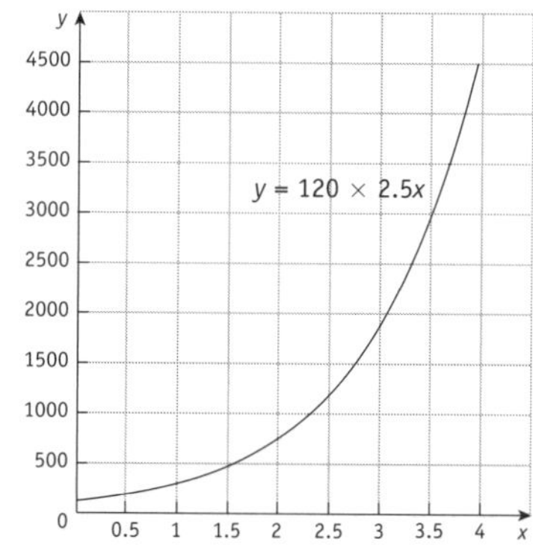

b $P = 20\,000(1.5)^t$

t	0	1	2	3	4
P	20 000	30 000	45 000	67 500	101 250

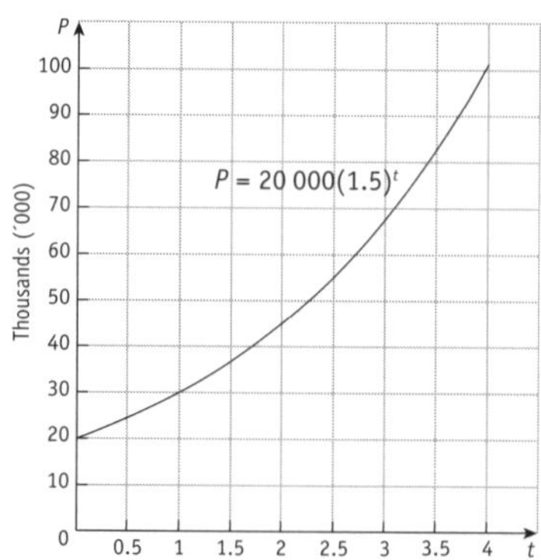

Answers

Pages 42

1 a

t	0	1	2	3	4	5
d	0	60	120	180	240	300

b

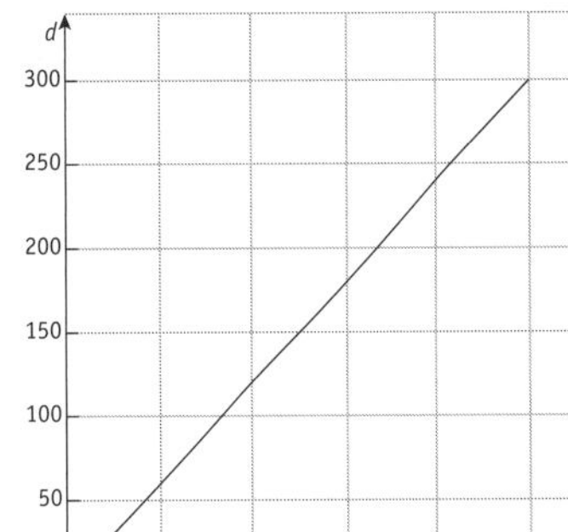

2 a

s	0	50	60	80	100
d	0	250	300	400	500

b

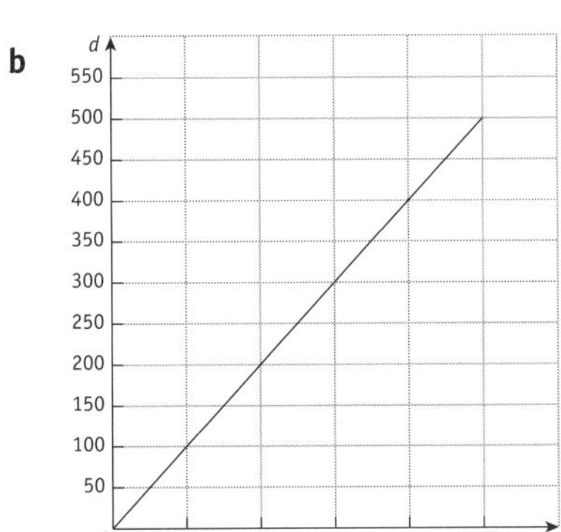

3 a

s	30	40	60	80	90	100
t	12	9	6	4.5	4	3.6

b

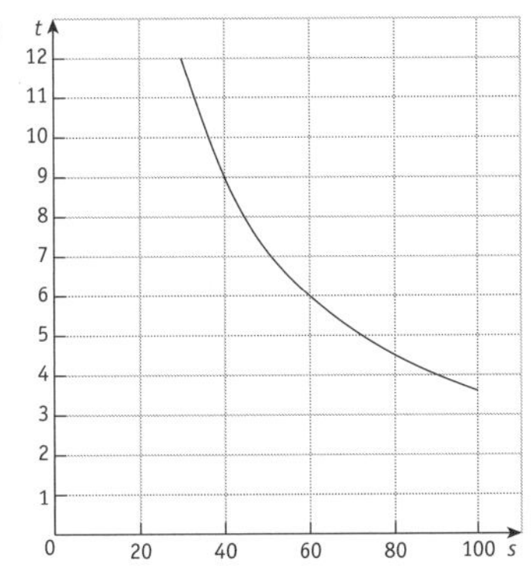

Page 43

1 a i 80 km/h **ii** 100 km/h **iii** 90 km/h **b** 360 km **c** 5 h **d** 360 ÷ 5 = 72 km/h **2 a** 1 km **b** 2 km/h

c

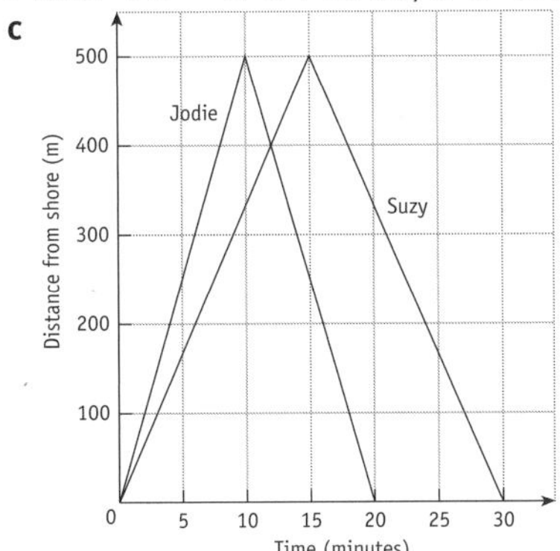

d 330 m

Page 44

1 a

Depth of water

Time

b

Depth of water

Time

c

Depth of water

Time

d

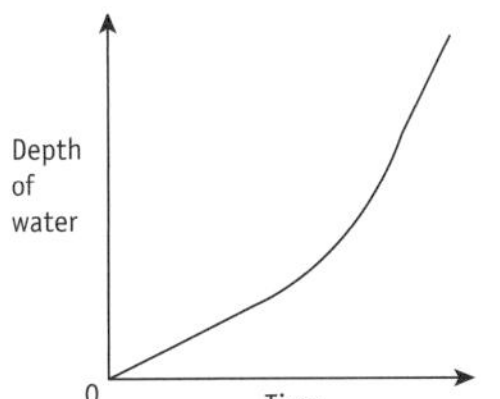

2 a

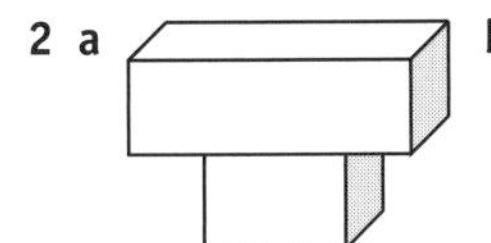

b

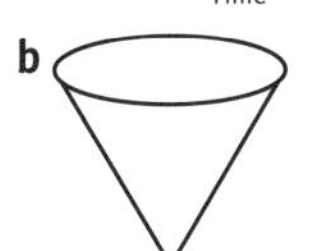

c

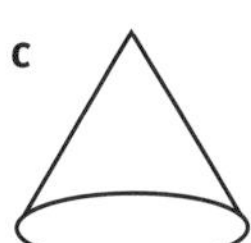

d

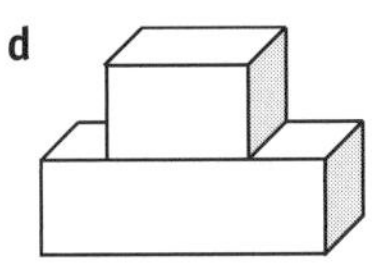

Answers

Pages 45–49 **1** D **2** B ($P = 200(1.06)^1 = 212$) **3** C **4** D **5** B ($280 \div 4 = 70$ km/h) **6 a** $280 + 280 = 560$ km **b** 2 h **c** $9 - 2 = 7$ hours **d i** 100 km/h **ii** 40 km/h **e** Section C **7 a**

x	0	1	2	3	4	5
y	25	24	21	16	9	0

b

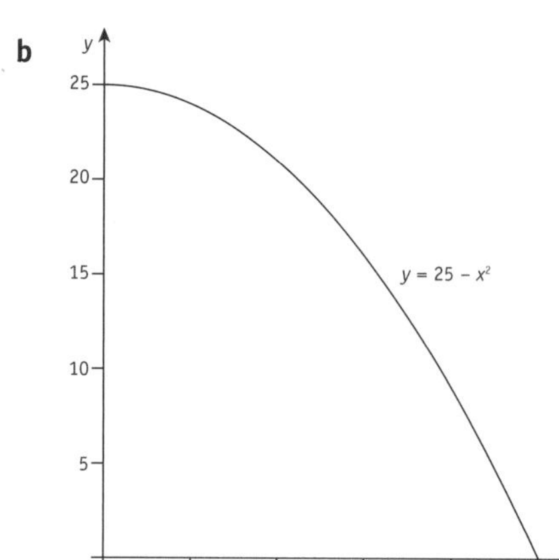

8 a 36 m **b** 20 m **c** 5 seconds

9 a

t	0	1	2	3	4	5	6
m	10	20	40	80	160	320	640

b

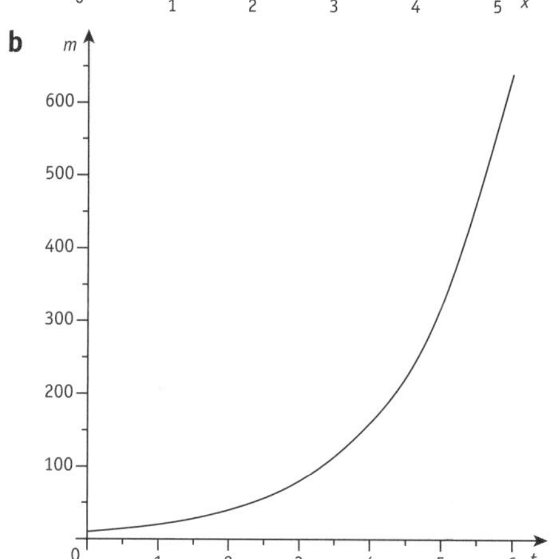

c Substitute $t = 12$: $m = 10 \times 2^{12} = 40\,960$ mice

10

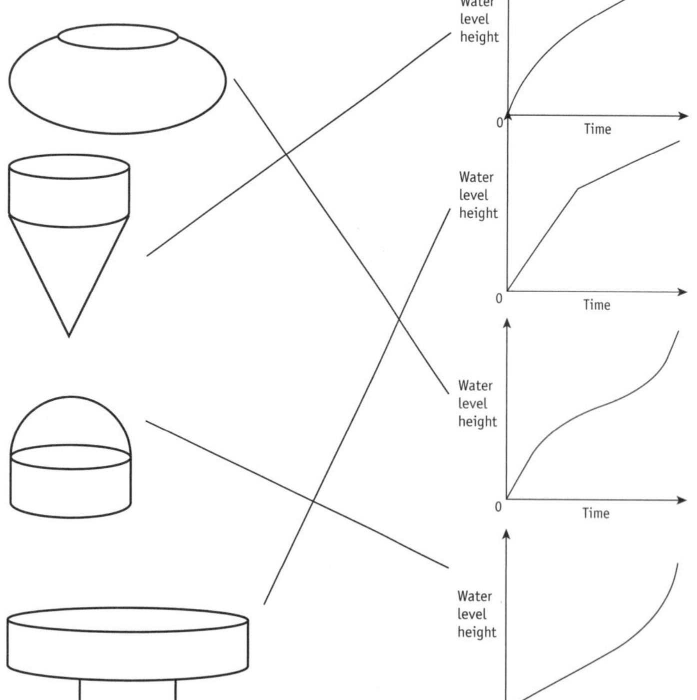

11 a

b 6 m **c** 4 seconds **d** 6.75 m

12 a

t	0	1	2	3	4	5
P	24 000	25 920	27 994	30 233	32 652	35 264

b

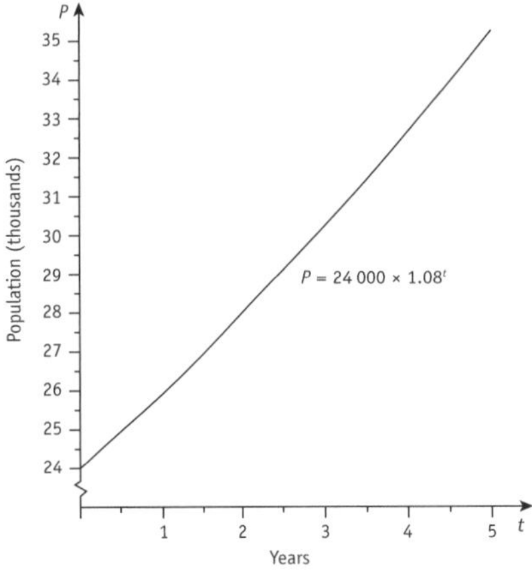

13 a

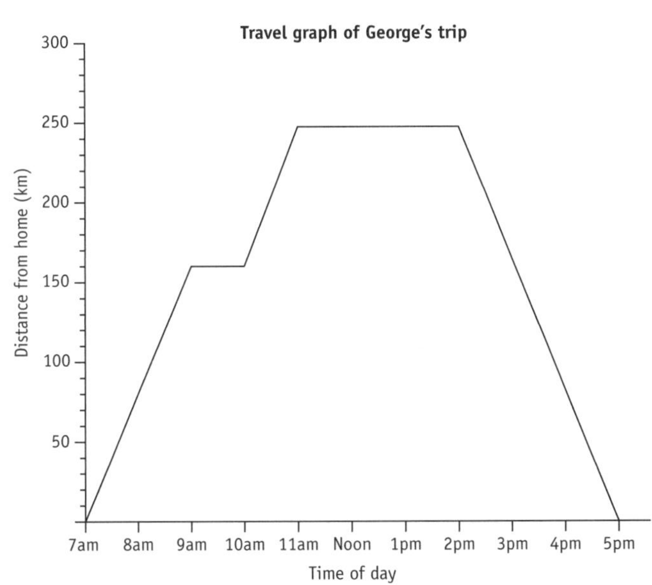

b 250 + 250 = 500 km **c** 250 ÷ 3 = 83 km/h

Chapter 5—Measurement: Right-angled triangles

Page 50 **1 a** $x^2 = 7^2 + 10^2$, $x = \sqrt{149} = 12.21$ (2 dec. pl.) **b** $y^2 = 8.7^2 + 11^2$, $y = \sqrt{196.69} = 14.02$ (2 dec. pl.) **c** $a^2 = 6^2 + 9^2$, $a = \sqrt{117} = 10.82$ (2 dec. pl.) **2 a** $p^2 = 16^2 - 12^2$, $p = \sqrt{112} = 10.58$ (2 dec. pl.) **b** $y^2 = 10^2 - 7^2$, $y = \sqrt{51} = 7.14$ (2 dec. pl.) **c** $m^2 = 17.6^2 - 12^2$, $m = \sqrt{165.76} = 12.87$ (2 dec. pl.) **3** $w^2 = 56^2 - 40^2$, $w = \sqrt{1536} = 39.2$ (1 dec. pl.) **4** $d^2 = 20^2 + 31^2$, $d = \sqrt{1361} = 37$ km (nearest whole) **5** $x^2 = 26^2 - 24^2$, $x = \sqrt{100} = 10$, Area $= \frac{1}{2} \times 24 \times 10 = 120$ cm^2

Page 51 **1 a** $\sin\theta = \frac{3}{5}$, $\cos\theta = \frac{4}{5}$, $\tan\theta = \frac{3}{4}$ **b** $\sin\theta = \frac{12}{13}$, $\cos\theta = \frac{5}{13}$, $\tan\theta = \frac{12}{5}$ **c** $\sin\theta = \frac{15}{17}$, $\cos\theta = \frac{8}{17}$, $\tan\theta = \frac{15}{8}$ **2 a** 0.7071 **b** 0.3443 **c** 0.9272 **3 a** 0.7310 **b** 0.9332 **c** 0.9053 **4 a** 10.18 **b** 19.36 **c** 17.26 **d** 5.42 **e** 49.16 **f** 35.39 **5 a** 7.98 **b** 39.74 **c** 33.15 **d** 17.39 **e** 18.39 **f** 39.36 **6 a** 27° **b** 54° **c** 31° **7 a** 47°52' **b** 20°15' **c** 63°50'

Page 52 **1 a** $\frac{x}{12} = \cos 42°$, $x = 8.92$ **b** $\frac{y}{15} = \sin 53°$, $y = 11.98$ **c** $\frac{m}{42} = \tan 48°$, $m = 46.65$ **d** $\frac{p}{15} = \sin 67°$, $p = 13.81$ **e** $\frac{k}{13} = \tan 27°$, $k = 6.62$ **f** $\frac{q}{16} = \tan 41°$, $q = 13.91$ **g** $\frac{t}{35} = \tan 51°16'$, $t = 43.64$ **h** $\frac{x}{45} = \sin 73°4'$, $x = 43.05$ **i** $\frac{x}{13.4} = \cos 43°18'$, $x = 9.75$ **2 a** $\frac{28}{a} = \sin 49°$, $a = 37.10$ **b** $\frac{18}{b} = \tan 56°$, $b = 12.14$ **c** $\frac{16.9}{c} = \cos 26°$, $c = 18.80$

Page 53 **1 a** $\sin\theta = \frac{12}{19}$, $\theta = 39°$ **b** $\tan\theta = \frac{7}{8}$, $\theta = 41°$ **c** $\cos\theta = \frac{10}{15}$, $\theta = 48°$ **d** $\tan\theta = \frac{16}{11}$, $\theta = 55°$ **e** $\sin\theta = \frac{17.9}{21.2}$, $\theta = 58°$ **f** $\cos\theta = \frac{32.8}{41.3}$, $\theta = 37°$ **2 a** $\tan\theta = \frac{12}{7}$, $\theta = 59°45'$ **b** $\sin\theta = \frac{17}{23}$, $\theta = 47°39'$ **c** $\cos\theta = \frac{15}{41}$, $\theta = 68°32'$

Answers

PAGE 54 1 a

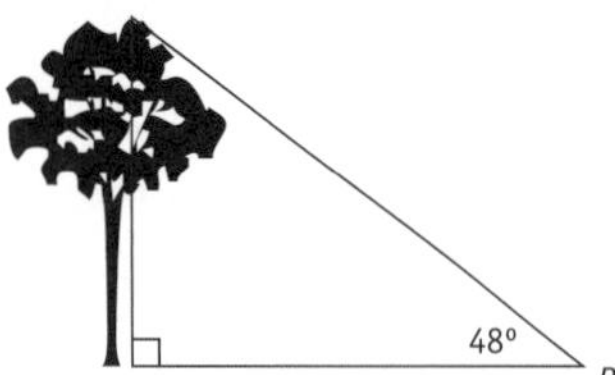

b

2 a $\frac{h}{40} = \tan 24°$, $h = 18$ m (nearest m) b $\frac{8}{d} = \tan 54°$, $d = 5.81$ m (2 dec. pl.)

3 a $\frac{20}{x} = \tan 54°$, $x = 14.53$ m (2 dec. pl.) b $\frac{4}{x} = \tan 12°$, $x = 18.82$, which is 19 m.

PAGE 55 1 a 5.85 m b 16.72 m 2 a 64 m b 52° c 45.86 m 3 a 4 cm b i 15 cm ii 17 cm

PAGE 56 1 a

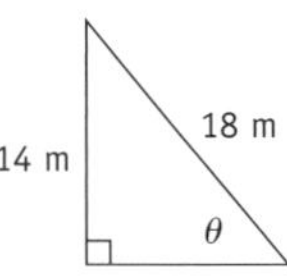

b 51° 2 a

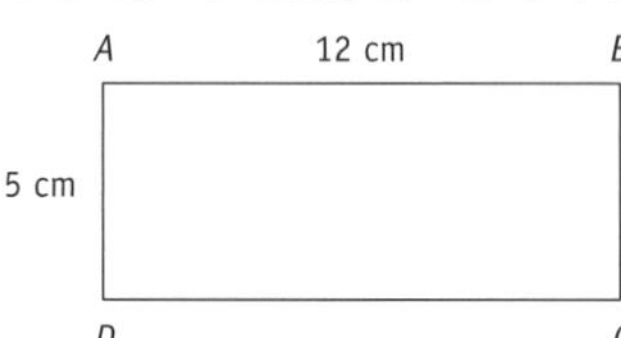

b 13 cm c 22°37'

3 a

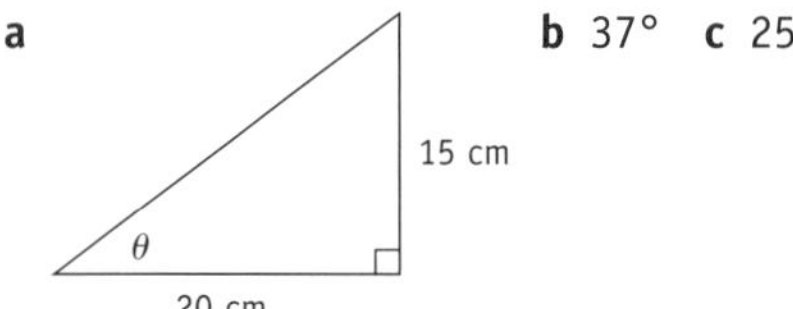

b 37° c 25 m

PAGE 57 1 a 126° b 230° c 051° d 312° e 204° f 148° 2 a S40°E b N35°W c N70°E d S40°W e S62°E f N35°W
3 a 100° b 300° c 255°

PAGE 58 1 a 166.59 km b 224.17 km 2 N 49° E 3 a 83.36 km b 86.32 km 4 301° T or N 59° W
5 1928 m 6 7.07 km

PAGES 59–62 1 D 2 B (180 − 30 = 150) 3 B (180 − 150 = 30, 70 + 30 = 100) 4 A $\left(\frac{1}{2} \times 20 \times 20 = 200\right)$

5 C (180 − 130 = 50, 360 − 50 = 310) 6 $\frac{12}{x} = \sin 54°$, $x = 14.83$ 7 $\frac{x}{1.4} = \sin 55°$, $x = 1.147$,

Distance = 1147 + 1400 = 2547 m 8 a $AD^2 = 5^2 - 3^2 = 16$, $AD = 4$, $CD^2 = 6^2 - 4^2 = 20$, $CD = \sqrt{20}$,

Perimeter = $\sqrt{20} + 3 + 5 + 6 = 18.47$ cm b $\sin\theta = \frac{4}{6}$, $\theta = 41°49'$ 9 $\frac{36}{d} = \tan 48°$, $d = 32$ m 10 $x^2 = 4^2 + 8^2$, $x = 8.94$ m

11 a $\frac{h}{16} = \tan 42°$, $h = 14.41$ m (2 dec. pl.) b $\tan\theta = \frac{14.41}{12}$, $\theta = 50°$ (nearest whole) 12 $\frac{10}{x} = \tan 38°$, $x = 12.80$ m (2 dec. pl.)

13 Smallest angle opposite smallest side, $\cos\theta = \frac{24}{26}$, $\theta = 23°$ (nearest whole) 14 Let $\angle RPQ = \theta$. $\tan\theta = \frac{28}{32}$, $\theta = 41°$ (nearest whole),

bearing is 180° − 41° = 139° 15 a

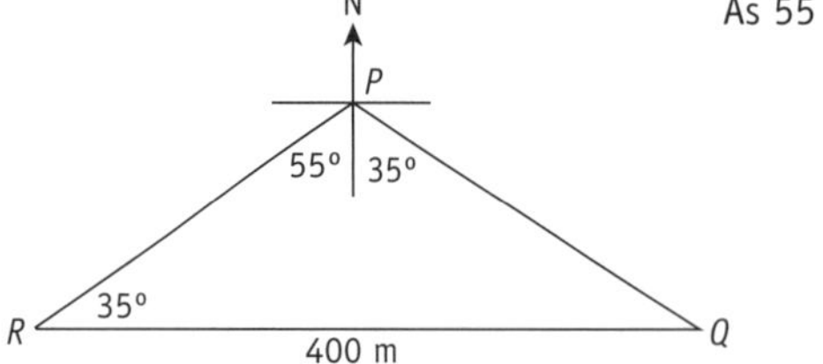

As 55 + 35 = 90, then $\angle RPQ$ is 90°

b $\frac{x}{400} = \sin 35°$, $x = 229$ km (nearest km) c Time = $\frac{229}{480}$ = 29 minutes (nearest min) 16 a $\frac{12}{d} = \tan 24°$, $d = 26.95$ m (2 dec. pl.)

b 26.95 − 4 = 22.95, $\tan\theta = \frac{12}{22.95}$, $\theta = 28°$ (nearest whole). Angle of depression is 28°.

CHAPTER 6—Measurement: Rates

PAGE 63 1 a 72 km/h = 72 000 m/h = 1200 m/minute = 20 m/s b 60 km/h = 60 000 m/h = 1000 m/minute = 16.67 m/s
c 110 km/h = 110 000 m/h = 1833.33 m/minute = 30.56 m/s 2 a 10 m/s = 600 m/min = 36 000 m/h = 36 km/h
b 8 m/s = 480 m/min = 28 800 m/h = 28.8 km/h c 25 m/s = 1500 m/min = 90 000 m/h = 90 km/h
3 a 6 km/h = 6000 m/h = 100 m/minute = 1.67 m/s b 68 km/h = 68 000 m/h = 1133.33 m/minute = 18.89 m/s
c 46 km/h = 46 000 m/h = 766.67 m/minute = 12.78 m/s d 6 m/s = 360 m/min = 21 600 m/h = 21.6 km/h

Answers

e 0.6 cm/s = 0.006 m/s = 0.36 m/min = 21.6 m/h **f** 100 m/minute = 6000 m/h = 6 km/h **4 a** 6 mL/minute = 360 mL/h = 0.36 L/h **b** 15 mL/minute = 900 mL/h = 0.9 L/h **c** 600 mL/minute = 36 000 mL/h = 36 L/h **5 a** 0.45 L/h = 450 mL/h = 7.5 mL/min **b** 0.8 L/h = 800 mL/h = 13.33 mL/min **c** 23 L/h = 23 000 mL/h = 383.33 mL/min **6 a** $0.25 \div 5 \times 60 \times 60 = 180$ mL/h **b** $180 \times 24 = 4320$, 4.32 L/day **c** 1.58 kL/year **7 a** 12 L/minute = 720 L/h = 0.72 kL/h **b** 26 L/minute = 1560 L/h = 1.56 kL/h **c** 165 L/minute = 9900 L/h = 9.9 kL/h **8 a** 1.8 kL/h = 1800 L/h = 30 L/min **b** 3.2 kL/h = 3200 L/h = 53.33 L/min **c** 0.08 kL/h = 80 L/h = 1.33 L/min

PAGE 64 **1 a** $2.25 ÷ 4.4 = $0.5341 **b** $3.45 ÷ 1.75 = $1.9714 **c** $5.40 ÷ 6.4 = $0.8438 **d** $1.60 ÷ 4.2 = $0.3810 **2 a** $3.40 ÷ 6 = $0.5667 **b** $2.80 ÷ 5.1 = $0.5490 **c** $6.80 ÷ 3.9 = $1.7436 **d** $2.18 ÷ 12.5 = $0.1744

3

Store	Size	Price	Price/100 g
A	560 g	$7.90	$1.41
B	453 g	$6.45	$1.42
C	380 g	$5.80	$1.53
D	280 g	$4.90	$1.75
E	220 g	$3.75	$1.70

4 8-rolls: $0.73/roll, 12-rolls: $0.70/roll, 6-rolls: $0.68/roll, 24-rolls: $0.61/roll. **5** cashews: $3.00/100 g, hazelnuts: $3.10/100 g, pecans: $4.70/100 g, brazil nuts: $5.49/100 g **6** 4 for $2.85 as a dozen costs $8.55

PAGE 65 **1 a** $154 **b** $154 + $25 × 2 = $204 **c** $154 + $25 × 3 = $229 **2** $125 − $65 = $60, 60 ÷ 12 = 5. As 10 × 5 = 50, the job took up to 50 min. The latest time was 11:10 am. **3 a** $3.60 + $2.19 × 12.6 = $31.19 **b** $3.60 + $2.19 × 18.8 + $0.944 × 5 = $49.49 **4 a** 365 − 80 = 285, $285 ÷ 3 = $95/h **b** $80 + $95 × 5 = $555 **c** 792.5 − 80 = 712.5, $712.5 ÷ $95 = 7.5 h **5 a** $2.75 + $1.20 × 8 + $0.35 × 12 = $16.55 **b** Time = 30 ÷ 50 = 0.6 h = 36 min, $2.75 + $1.20 × 30 + $0.35 × 36 = $51.35 **6** 512.36 − 392 = 120.36, 120.36 ÷ 3.54 = 34 km

PAGE 66 **1 a** 92 km/h **b** 108 km/h **c** 760 km/h **2 a** 19 km **b** 116 km **c** 212.5 km **3 a** 3 h 20 min **b** 30 min **c** 10 h 45 min **4 a** tortoise: $60 \div 60 = 1$ m, hare: $36 \times 1000 \div 60 = 600$ m **b** The hare wins as the tortoise takes 30 seconds to complete the race, while the hare only takes 10 seconds. **5 a** $1\frac{5}{6} \times 90 = 165$ km **b** $165 \div 75 = 2.2$ h = 2 h 12 min. Hence, 6:37 pm **6 a** Total speed is 190 km/h, Distance = $190 \times 0.5 = 95$ km **b** Time = $\frac{114}{190} = 0.6$ h = 36 min. Hence, 3:06 pm **7 a** $40 \times 1000 \div 60 \div 60 \times 0.95 = 11$ m **b** $90 \times 1000 \div 60 \div 60 \times 0.95 = 24$ m **c** $110 \times 1000 \div 60 \div 60 \times 0.95 = 29$ m

PAGE 67 **1 a** 180 km in 2 h = 90 km/h **b** $90 \times 3.5 = 315$ km **2 a i** 60 km/h **ii** 100 km/h **b i**

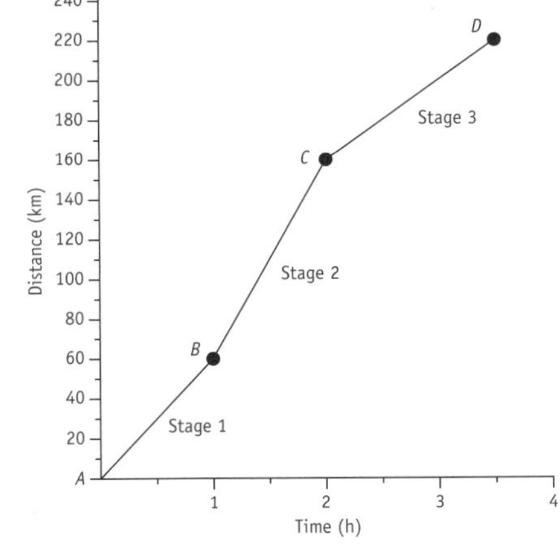

ii 220 km **3** 60 m in 4 s = 15 m/s = 54 km/h **b** B: 20 m in 4 s = 5 m/s = 18 km/h. Difference is 36 km/h or 72 km in 2 hours **4 a** 12.6 km **b**

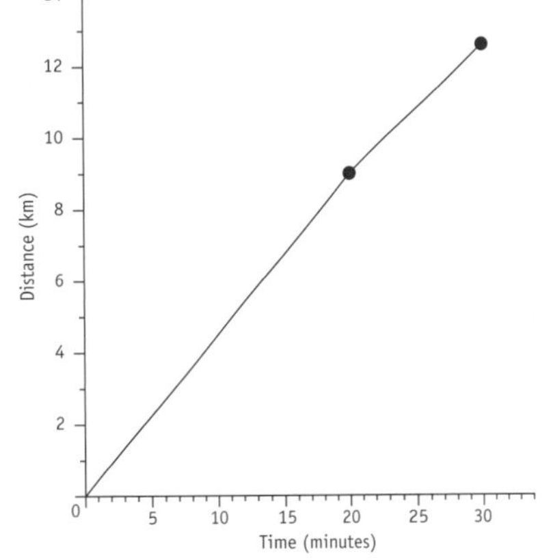

PAGE 68 **1 a** $45 \div 6.5 = 6.9$ L/100 km **b** $38 \div 5.86 = 6.5$ L/100 km **2 a** $7.6 \times 3.8 = 28.88$ L **b** $7.6 \times 5.15 = 39.14$ L **3 a** $45 \div 8.3 \times 100 = 542$ km **b** $60 \div 1.729 \div 8.3 \times 100 = 418$ km **4 a** 8.1 × 1.8 × $1.669 = $24.33 **b** 8.1 × 4.35 × $1.669 = $58.81 **5** (7.3 − 5.9) × 6.05 × $1.799 = $15.24 **6 a** 45 ÷ 1.549 ÷ 6.8 × 100 = 427 km **b i** 7.9 × 2.9 × $1.439 = $32.97 **ii** 7.9 × 2.9 × $1.619 = $37.09 **c** ($1.549 − $1.439) × 6.9 × 4.8 = $3.64 **d** Lincoln: 60 ÷ 1.439 ÷ 8.3 × 100 = 502 km, Sarah: 60 ÷ 1.619 ÷ 7.7 × 100 = 481 km. Lincoln drove 21 km further.

Answers

PAGE 69 **1 a** 64 bpm **b** 76 bpm **2 a** Average **b** 54–59 bpm **c** 62–65 bpm **3 a** 195 bpm **b** 175 bpm **c** 145 bpm **4 a** 208 – 0.7(30) = 187 bpm **b** 208 – 0.7(60) = 166 bpm **c** 208 – 0.7(86) = 148 bpm
5

Heart rates after activity based on age			
	Target heart rate zone (bpm)		
	Moderate activity	Intense activity	Maximum heart rate
Age	50-70% max.	70-85% max.	(bpm)
20	100–140	140–170	200
30	95–133	133–162	190
40	90–126	126–153	180
50	85–119	119–145	170
60	80–112	112–136	160
70	75–105	105–127.5	150

PAGE 70 **1 a** 119 mm/Hg **b** 73 mm/Hg **2 a** normal **b** mild hypertension **c** high to normal
3

Time of day	Systolic (mm/Hg)	Diastolic (mm/Hg)
03:00	135	105
06:00	130	100
09:00	120	95
12:00	125	90
15:00	125	90
18:00	120	85
21:00	115	90
00:00	110	95

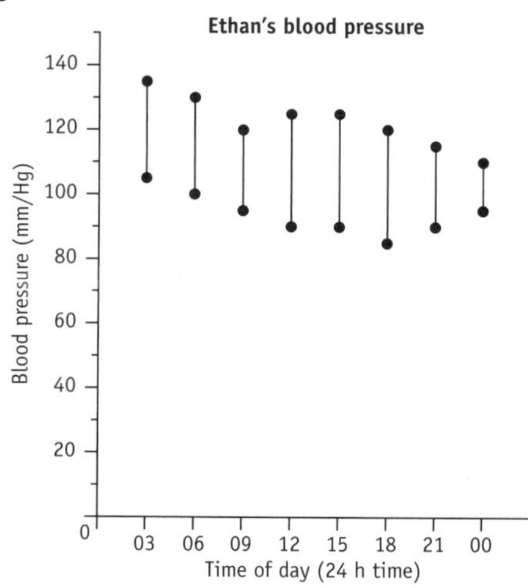

PAGES 71–75 **1** D (3.5 ÷ 150 = 0.023, 5.3 ÷ 220 = 0.024, 7.8 ÷ 380 = 0.021, 11.3 ÷ 560 = 0.020) **2** B (40 ÷ 5.4 = 7.4) **3** C (94 × 1.75 = 164.5) **4** A (128 – 116 = 12, $\frac{12}{128} \times 100\% = 9.375\%$) **5** B (40 m in 4 seconds = 10 m/s = 36 km/h) **6** 480 – 110 = 370 km long-distance driving, 6.8 × 3.7 + 8.4 × 1.1 = 34 (nearest whole) **7 a** From 9:30 am to 12:30 pm is 3 h. Distance = 80 × 3 = 240 km **b** From 1 pm to 5 pm is 4 h. Speed = 240 ÷ 4 = 60 km/h **8** Scooter: as 30 ÷ 20 = 1.5, the fuel/trip = 1.5 L. This is 3 L per day or 15 L per week. Weekly cost = \$1.60 × 15 = \$24. Bus: \$3.20 × 10 = \$32. The scooter is cheaper by \$32 – \$24 = \$8 per week. **9 a** \$36 × 3 = \$108 **b** \$26 × 12 + 0.5 × \$26 × 12 = \$468 **c** 12-monthly: \$26 × 12 = \$312, 10 months: \$30 ×10 = \$300. Better to buy the 10 months of monthly memberships. **10 a** 540 ÷ 100 = 5.4, then 46.44 ÷ 5.4 = 8.6. The car uses 8.6 L/100 km **b** 46.44 × \$1.599 = \$74.26 **11** \$33.80 × 36 + \$33.80 × 2 × 8 = \$1757.60 **12** 10:45 am to 1:45 pm is 3 h + half an hour to 2:15 pm. Speed = 260 ÷ 3.5 = 74.3 km/h. **13 a** *A*: 1 pm to 4 pm = 3 h.

Answers

Speed = 180 ÷ 3 = 60 km/h. *B*: 1:30 pm to 3:30 pm = 2 h. Speed = 180 ÷ 2 = 90 km/h. Motorist *B* is faster by 30 km/h **b** *B* passes *A* at 2:30 pm, 90 km from the destination **c** Time = 180 ÷ 72 = 2.5. Motorist *C* arrives at 5 pm.

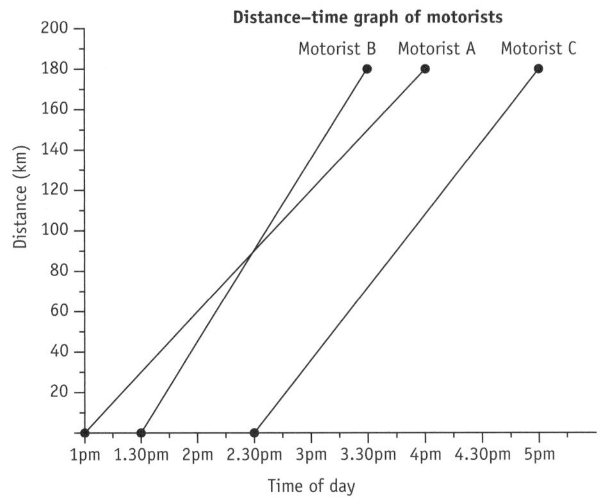

14 Distance = 28 × 2 = 56 km per day, 56 × 5 = 280 km per week, 280 × 2 = 560 km per fortnight. Petrol = 560 ÷ 100 = 5.6 lots of 100 km, 5.6 × 9.6 = 53.96 L. Cost = 53.96 × \$1.729 = \$92.95 (2 dec. pl.)

15

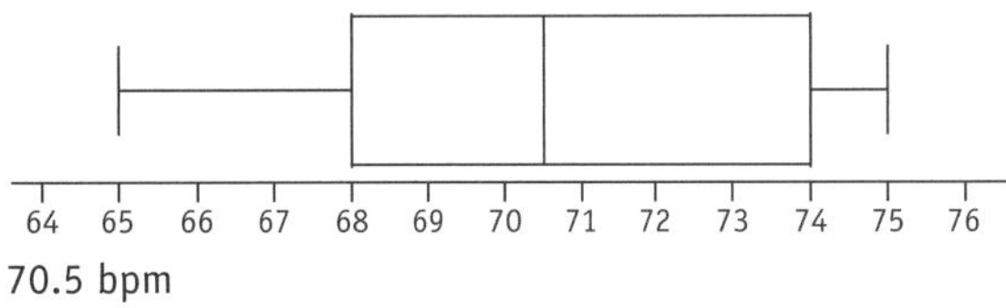

70.5 bpm

16 a 120 km in 1.5 h. Average speed = 120 ÷ 1.5 = 80 km/h **b** Time = 240 ÷ 100 = 2.4 h = 2 h 24 min. 3 pm + 2 h 24 min = 5:24 pm.
17 \$3.80 + \$2.14 × 24.3 + \$0.85 × 3 = \$58.35 (2 dec. pl.)

Chapter 7—Measurement: Scale drawings

Page 76 **1 a** 4:5 **b** 1:3 **c** 1:4 **d** 4:1 **e** 1:3:5 **f** 1:10:20 **2 a** 1:7 **b** 1:4 **c** 3:10 **d** 3:5 **e** 1:100 **f** 2:1 **3 a** 2:3 **b** 6:5 **c** 4:5:6 **4 a** 13:7 **b** 7:20 **c** 13:20 **5 a** 6:5 **b** 1:1 **6 a** 4:11 **b** 11:15 **7 a** 20:1 **b** 1:19

Page 77 **1 a** 12 cm **b** 120 cm **2** 90 **3 a** \$45 **b** \$84 **4 a** \$144 000 **b** \$45 000 **5** 10.8 kg **6** 9:8 **7 a** 190 mL **b** 4.94 L **c** 6.65 L

Page 78 **1 a** \$7.50, \$12.50 **b** 0.4 kg, 0.6 kg, 0.2 kg **c** 144 min, 240 min, 96 min **2** 200 mL, 800 mL, 1000 mL **3** \$8000, \$7000, \$5000 **4** 12 kg cement, 36 kg sand, 72 kg gravel **5** \$48 000 **6** 35 cm **7 a** 2:4:5 **b** 5000 **8** 1.4 ha

Page 79 **1 a** 1:1000 **b** 1:100 **c** 1:10 000 **d** 1:10 000 **e** 1:250 **f** 1:20 **g** 1:20 000 **h** 1:5 **i** 1:6000 **2 a** 1 m **b** 3 m **c** 5 m **d** 0.8 m **e** 0.6 m **f** 1200 m **3 a** 8 m **b** 50 m **c** 6 km **d** 95 m **e** 8.3 km **f** 63.25 km **4 a** 5 cm **b** 4 cm **c** 12.6 cm **d** 8 mm **e** 30 cm **f** 28.35 cm **5 a** 1 km **b** 8.4 km **c** 26 cm

Page 80 **1 a** Scale factor = $\frac{3}{2}$; $\frac{x}{6} = \frac{3}{2}$, $x = 6 \times \frac{3}{2} = 9$ **b** Scale factor = $\frac{15}{6} = \frac{5}{2}$; $\frac{x}{4} = 4 \times \frac{5}{2}$, $x = 10$ **2 a** Scale factor = $\frac{12}{8} = \frac{3}{2}$ **b** $\frac{BC}{6} = \frac{3}{2}$, $BC = 6 \times \frac{3}{2} = 9$ cm **c** $CQ = 3$ cm **3 a** $\frac{x}{12} = \frac{8}{10}$, $x = 12 \times \frac{8}{10} = 9.6$ **b** $\frac{x}{8} = \frac{6}{10}$, $x = 8 \times \frac{6}{10} = 4.8$ **4** $\frac{d}{9} = \frac{2.8}{6}$, $d = 9 \times \frac{2.8}{6} = 4.2$

Page 81 **1 a** 30 m **b** 13 m **c** 390 m^2 **d** 3.5 m **e** 104 m^2 **2 a** 1:45 **b** 1.26 m **3 a**

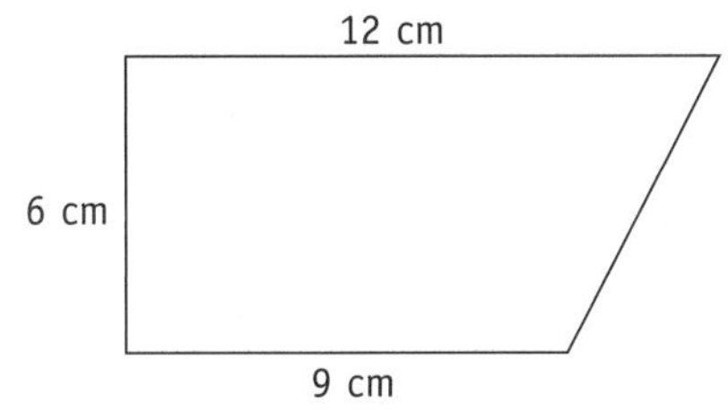

b 135 m

Page 82 **1** length = 20 m, width = 10 m (Remember the question asks for the dimensions of the swimming pool, not the brick surrounds.) **2** Approx. 6 m **3** $P = 937 + 350 + 282 + 514 + 379 + 842 + 978 = 4282$ m **4 a** 2700 m **b** 1200 m (This number is approximate.) **c** $A = 50 \times 50 = 2500$ m^2 **d** $A = 167 \times 2500 = 417\,500$ m^2 or 41.75 ha **e** $A = 29 \times 2500 = 72\,500$ m^2 (This number is approximate.) **f** $\frac{72\,500}{417\,500} \times \frac{100}{1} = 17.4\%$

Page 83 **1 a** 8 m **b** sliding door **c** walk-in robe **d** 3.865 m by 2.93 m **e** 7 cm **f** 10 cm **g** south **h** 4835 **i** \$176 900 **j** west

Page 84 **1 a** 260 km **b** 260 km **c** Around 85 km **d** C (The actual area is 48 670 km^2.) **2 a** 34 m **b** $A = 20 \times 34 = 680$ m^2 (Your area could be slightly different.) **c** Around 150 m **3 a** 15 km^2

Answers

b Now 1 km^2 = 1 km × 1 km = 1000 m × 1000 m = 1 000 000 m^2. So this catchment area covers 15 000 000 m^2. Volume of water can be found using $V = Ah$ = 15 000 000 × 0.075 = 1 125 000 m^3. Given 1 m^3 = 1000 L, then 1 125 000 m^3 = 1 125 000 000 L = 1125 ML.

Pages 85–89 **1** D $\left(\frac{x}{y} = \frac{17}{9}, x = y \times \frac{17}{9}\right)$ **2** A (12 : 8 = 3 : 2) **3** C (30 000 ÷ 1500 = 20) **4** B (6 cm = 90 km, 1 cm = 15 km, 3.8 × 15 = 57) **5** D (140 : 70 = 2 : 1) **6** Scale factor = $\frac{15}{6} = \frac{5}{2}$. $w = 5 \times \frac{5}{2} = 12\frac{1}{2}$ Width is $12\frac{1}{2}$ m

7 **a** 1 unit = 8 m **b**

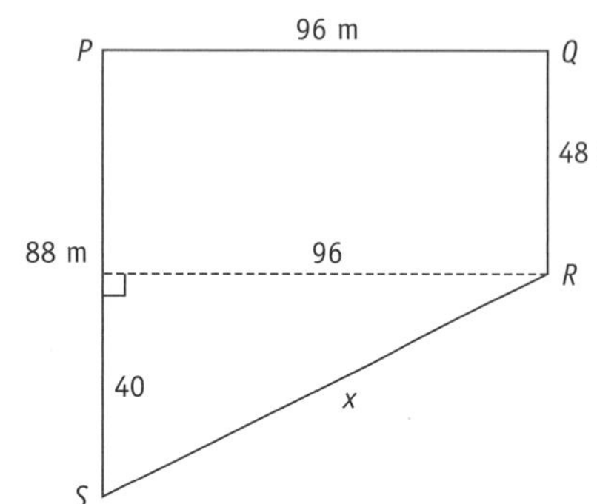

Area = $\frac{1}{2} \times 96(48 + 88)$ = 6528 m^2 **c** $x^2 = 96^2 + 40^2$. x = 104
Perimeter = 104 + 88 + 96 + 48 = 336 m
Cost = \$11.45 × 336 = \$3847.20 **8** 3 + 4 + 5 = 12. 12 parts = 60, so the girls are aged 15, 20 and 25. In 10 years time, the girls' ages will be 25, 30 and 35. This means the ratio will be 5 : 6 : 7.
9 Yard area = 16 × 10 = 160 m^2. As 3 + 2 = 5, garden area = $\frac{2}{5} \times 160$ = 64 m^2. Side of garden = 8 m, perimeter = 32 m.
10 Scale factor = $\frac{14.6}{1.7}$. $s = 4 \times \frac{14.6}{1.7}$ = 34.35 m **11** **a** 16.8 km
b 28 cm **12** **a** 70 mm
b 3.865 × 2.93 + 3.96 × 3.865 + 3.865 × 3.4 = 40 m^2
c 19.3 × 12.8 – 14.5 × 8 = 131.04 m^2 **13** **a** Using Pythagoras, $3^2 + 4^2 = 5^2$ means the distance is 5 m.

b Scale factor is $\frac{24}{4}$ = 6, hence x = 6 × 5 = 30 m **14** **a** Scale factor is $\frac{15}{25} = \frac{3}{5}$ **b** $PQ = \frac{3}{5} \times 7 = \frac{21}{5}$ = 4.2 cm **c** $QR = \frac{3}{5} \times 24 = \frac{72}{5}$ = 14.4 cm; Area = $\frac{1}{2} \times 4.2 \times 14.4$ = 30.24 cm^2 **15** **a** AS = 40, AB = 200. Scale factor = $\frac{200}{40}$ = 5
b BC = 5 × 24 = 120 cm = 1.2 m

Chapter 8—Statistical analysis: Statistical investigation process for a survey

Page 90 **1** The subject of the question is not clear: does it refer to the board or the workers? **2** This is a leading question; the question implies the answer the interviewer wants. **3** **a** It is easy to process and a definite answer is obtained. **b** This style of question allows for a range of opinions. **4** Questions should be relevant, precise, clearly worded and unambiguous. **5** Avoid expressing an opinion in the questions, being too vague and avoid giving too many choices in the one question.

Page 91 **1** **a** bar graph **b** histogram **c** line graph **2** **a** radar chart **b** dot plot **c** sector graph **3** A bar graph. If the length was 125 mm then 1 mm would represent 1 student. The section for walking would be 32 mm, the section for cycling 15 mm, etc. **4** **a** It is easy to see the order, from the most common to least common hair colour. **b** The number of people with each hair colour cannot be determined.

Page 92 **1** **a** The vertical axis is not clearly labelled. The increments on the vertical axis are not consistent. Do sales represent the number of the products sold, the income received from the sales or the mass of the products sold? Products A, B, C and D may not be directly comparable; for example, product C may be sales of 150-g jars of coffee and product A may be sales of 1-kg containers. **b** This misrepresents the data because the volume of sales and the income received may be greatest for product A. **c** On the graph, label the vertical axis as the number of products sold (in 1000s) and include a description of each product. **2** With the section of the vertical axis shown, it appears sales of product B are at least 50% greater than each of the other products. If the full axis was drawn, a smaller variation would be seen; the maximum difference in sales is 20%. **3** There is no label on the vertical axis so this could be the percentage of the population liking or disliking each product. The products are not described and with the graph style is it the height or the surface area which is meant to represent the product variable?

Page 93 **1** **a** 224 **b** 223.5 **c** 222.2 **2** **a** 8 **b** 8.5 **c** 9.6 **3** **a** 7 **b** 7 **c** 7.2 **4** **a** 5 **b** 5 **c** 5
5 **a**

Heights (cm)	Class centre (cm)	Number of students
143–149	146	2
150–156	153	5
157–163	160	7
164–170	167	13
171–177	174	10
178–184	181	3

b 164 – 170 **c** 6631 ÷ 40 = 165.775 **6** 5, 5, 6, 8 **7** Total age of 4 people = 4 × 12 = 48. Total age of 5 people = 5 × 14 = 70. Age of new person = 70 – 48 = 22 **8** Mean = (21 + 22 × 4 + 23 × 3 + 24 × 5 + 25 × 2) ÷ 15 = 348 ÷ 15 = 23.2. The mean guess is 23.2°.

Answers

PAGE 94 **1 a** 31 − 4 = 27 **b** 20.1 − 0.3 = 19.8 **c** 4.9 − (−2.5) = 7.4 **2 a** 5, 7, 9, 10, 12, 19, 32. IQR = 19 − 7 = 12 **b** 16, 19, 26, 30, 54, 76. IQR = 54 − 19 = 35 **c** 6, 8, 10, 11, 17, 18, 24, 40. IQR = 21 − 9 = 12 **3 a** 17 − 13 = 4 **b** 93 − 64 = 29 **c** 38 − 32 = 6 **4 a** 16 − 13 = 3 **b** 80 − 67 = 13 **c** 37 − 34 = 3 **5 a** 3 + 12 = 15 **b** 21 − 12 = 9 **c** −4 + 12 = 8 **6 a** Q_1 = 18, Q_3 = 28. IQR = 28 − 18 = 10. Check 18 − 1.5 × 10 = 3 which is larger than 2. Yes, 2 is an outlier. **b** Q_1 = 10, Q_3 = 34. IQR = 34 − 10 = 24. Check 34 + 1.5 × 24 = 70 which is not larger than 63. No, 63 is not an outlier.

PAGE 95 **1 a** 10, higher, 9 **b** 8, higher, 3 **c** 8, higher, 7 **d** 7, equal to, 7 **2 a** 26, higher, 24 **b** 160, lower, 174 **c** 160, lower, 168 **3 a** 17, higher, 16 **b** 5, higher, 4 **c** 8, higher, 7 **4 a** 6, lower, 7 **b** 3, equal to, 3 **c** 14, lower, 19

PAGES 96–100 **1** A **2** C (13 − 2 = 11) **3** D **4** B **5** B (37 − 19 = 18) **6 a** A person's favourite team may not be on the list. **b** The person may not drink cola-flavoured drinks. **c** There are two aspects to comment on: size and location. Better to have two separate questions. **d** The question is biased suggesting that parents who do not have their children vaccinated do not care.
7 a It is not random because each student on the bus does not have an equal chance of being selected. **b** The opinions of students who sit at the back of buses may differ from those who sit at the front **8** No, a major housing development could be planned for Kurraglen. **9 a** 7.9375 **b** 8 **c** 17 **d** 7.5 **10 a** 1.9 **b** 0, 2 and 3 **c** 6 **d** 2 **e** By comparing the means, Australia has shown a significant increase in the number of gold medals won in the more recent Olympics. The earlier results are more consistent (by comparing the standard deviations) but they are consistently low. **11 a** 19 **b** 18.5 **c** IQR = 23 − 16 = 7
12 a The highest number is 8 more than 1. 1 + 8 = 9 **b** IQR = 7 − 3 = 4 **c**

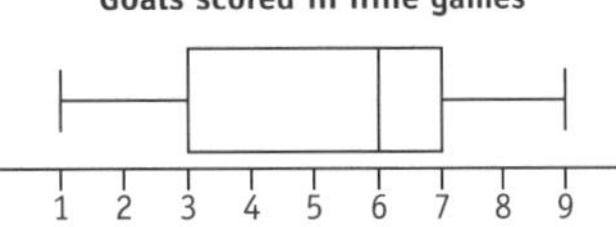

13 a 11, equal to, 11 **b** 32, less, 33 **c** 35, less, 37 **d** 33, more, 29 **14** The range of pre-test marks (6) was less than the range of post-test marks (7). The median of pre-test marks (4) was less than the median of post-test marks (6). The IQR of pre-test marks (5 − 3 = 2) was less than the IQR of post-test marks (9 − 5 = 4). **15** The range of boys' average times (100) was more than the range of girls' average times (90). The median of boys' average times (60) was more than the median of girls' average times (30). The IQR of boys' average times (40) was the same as the IQR of girls' average times (40).

CHAPTER 9—Statistical analysis: Exploring and describing data from two variables

PAGE 101 **1 a i** 3 **ii** 4 **b**

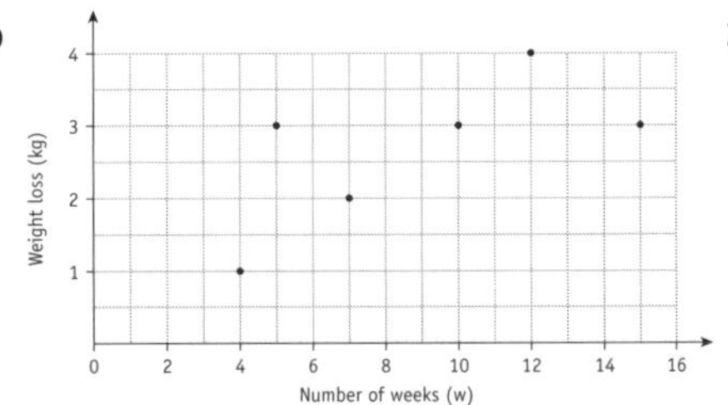

2 a i 5 **ii** 7 **b**

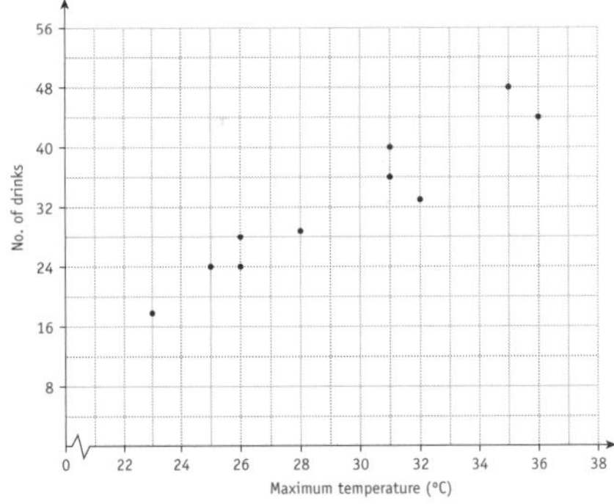

3

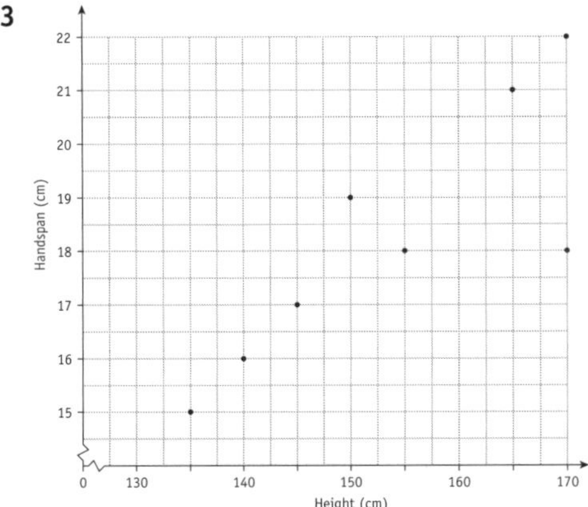

Answers

PAGE 102 **1** (iii), (iv), (i), (v), (ii) **2** **a** (v) **b** (iii) **c** (i) **d** (ii) **e** (iv) **3** **a** linear **b** non-linear **c** non-linear **d** linear **e** non-linear **4** **a** **b**

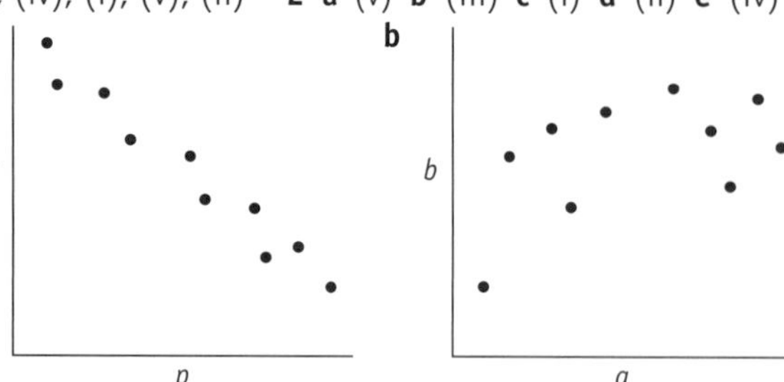

PAGE 103 **1** **a** **b** **c**

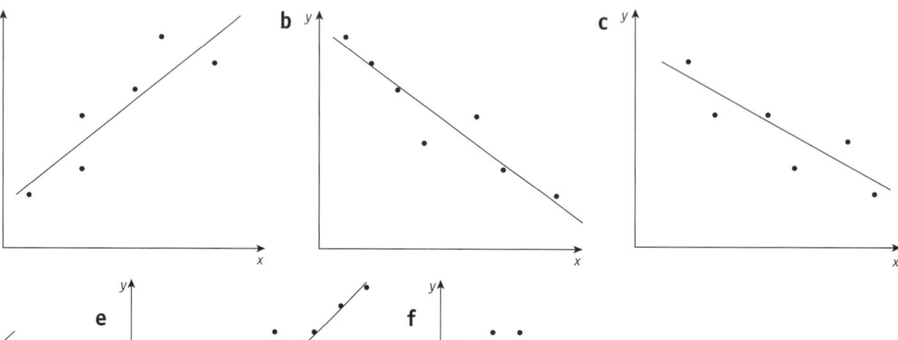

d **e** **f**

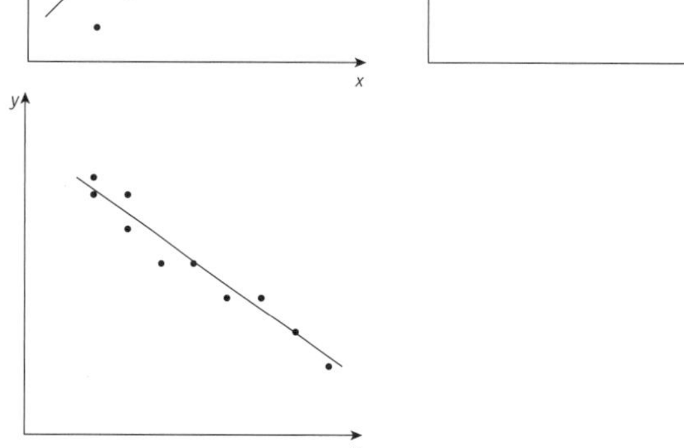

g **h**

PAGE 104 **1** **a** $y = 6x + 12$ **b** $y = 0.45x - 23$ **c** $y = \frac{4}{5}x + 8$ **2** **a** $63 = 5(8) + c$. This means $c = 63 - 40 = 23$. **b** $54 = 5(21) + c$. This means $c = 54 - 105 = -51$. **c** $3.6 = 5(12.8) + c$. This means $c = 3.6 - 64 = -60.4$. **3** **a** gradient $= \frac{68 - 18}{22 - 12} = 5$ **b** gradient $= \frac{25 - 23}{16 - 8} = 0.25$ **c** gradient $= \frac{11 - 63}{39 - 26} = -4$ **4** **a** gradient $= \frac{72 - 37}{51 - 26} = 1.4$. The equation is in the form $y = 1.4x + c$. Substituting (26, 37) gives $37 = 1.4 \times 26 + c$. This means $c = 0.6$, so the equation is $y = 1.4x + 0.6$. **b** gradient $= \frac{16 - 48}{20 - 12} = -4$. The equation is in the form $y = -4x + c$. Substituting (12, 48) gives $48 = -4 \times 12 + c$. This means $c = 96$, so the equation is $y = -4x + 96$. **c** gradient $= \frac{265 - 101}{51 - 19} = 5.125$. The equation is in the form $y = 5.125x + c$. Substituting (19, 101) gives $101 = 5.125 \times 19 + c$. This means $c = 3.625$, so the equation is $y = 5.125x + 3.625$. **5** **a** gradient $= \frac{15 - 7}{192 - 187} = 1.6$. The gradient is 1.6. **b** $7 = 1.6(187) + c$ means $c = -292.2$ **6** **a** gradient $= \frac{6400 - 4300}{2.4 - 1.6} = 2625$. The gradient is 2625. The equation is $s = 2625d + c$. Substituting (1.6, 4300) gives $4300 = 2625(1.6) + c$. This means $c = 100$. **b** As $c = 100$, then 100 skiers will come to the resort even though there is no snow.

Answers

PAGE 105 **1 a** 65 bpm **b** 111 bpm **c** 14 km/h **d** The model is only relevant for speeds up to 18 km/h. We cannot estimate Andrew's heart rate for any speed higher as we are not certain of his ability to ride faster, nor the health of his heart.

2 a b

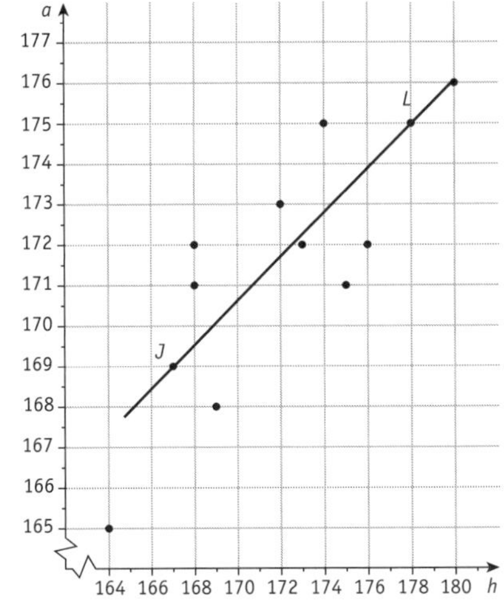

c i 174 cm **ii** 169 cm

PAGES 106–111 **1** C **2** C ($2.5 \times 6 + 46 = 86$) **3** D **4** D **5** B **6 a** Strong, negative linear association

b gradient $= \dfrac{27-33}{15-3} = \dfrac{27-33}{15-3} = -0.5$ **c** Substitute (3, 33): $33 = -0.5(3) + c$. Hence, $c = 34.5$.

d Substitute $w = 10$: $b = -0.5(10) + 34.5$, $b = 29.5$ **e** No, the model is only useful for BMIs from 25 to 36. The formula suggests $15 = -0.5w + 34.5$, which gives $w = 39$, or 39 weeks, but the program is only for 15 weeks.

7 a and **b**

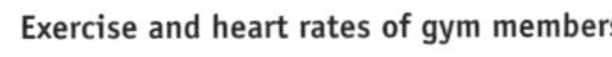

c From the graph, a resting heart rate of 46 bpm.

8 a, **b**, **c**

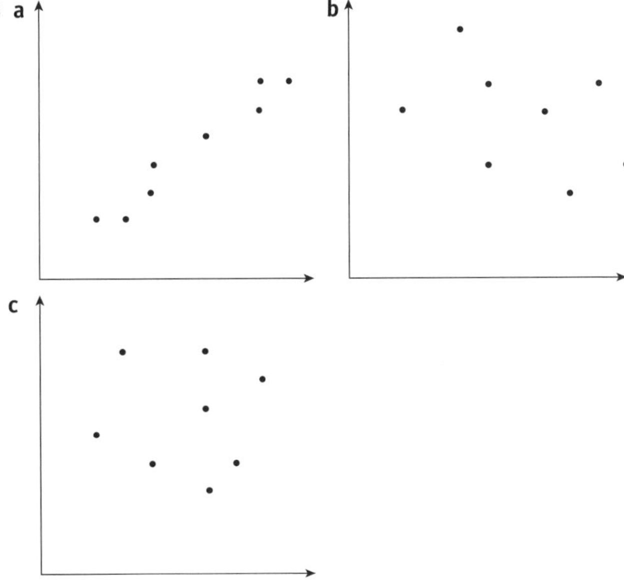

9 a and **b**

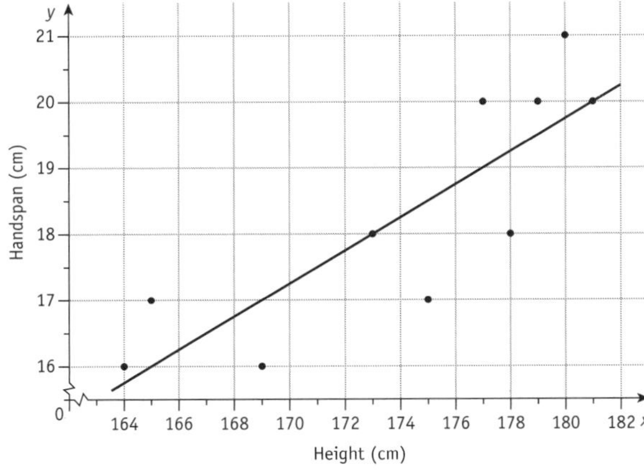

c gradient $= \dfrac{20-18}{181-173} = 0.25$ **d** Substitute $x = 172$: $y = 0.25(172) - 25.25$. $y = 17.75$. Laura's handspan is 17.75 cm

Answers

10 The association is strong, positive and linear.

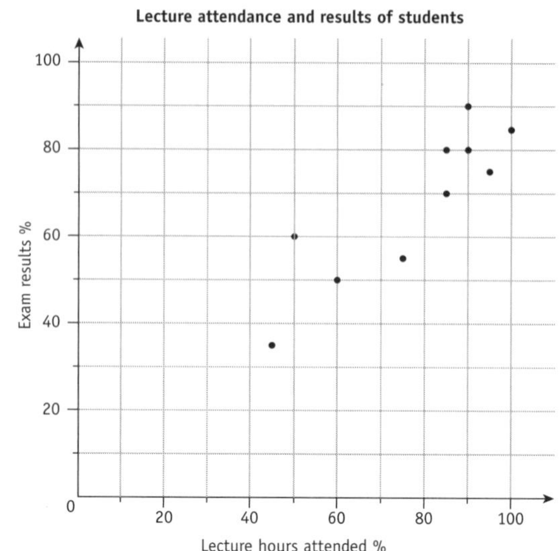

11 a

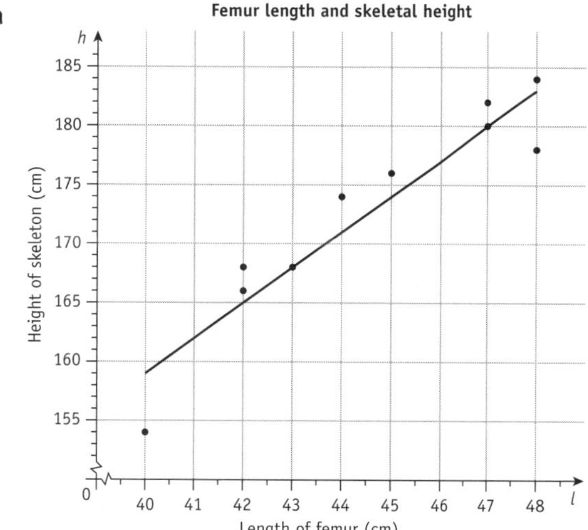

b A strong, positive linear association **c** see diagram to the left

d gradient $= \frac{180 - 168}{47 - 43} = 3$ **e** Substitute (43, 168): $168 = 3(43) + c$.

Hence, $c = 168 - 129$. $c = 39$. The equation is $h = 3l + 39$.

f Substitute $h = 177$: $177 = 3l + 39$. $3l = 138$, $l = 46$. The length is 46 cm. **g** Substitute $l = 20$: $h = 3(20) + 39$. $h = 99$. The model suggests a height of 99 cm. However, this may not be accurate as the model is only relevant for femurs which measure between 40 cm and 48 cm.

Chapter 10—Networks: Introduction to networks

Page 112 **1 a i** 4 **ii** 6 **iii** 3 **iv** *A, C, D* **b i** 5 **ii** 7 **iii** 2 **iv** *A, C, D* **c i** 6 **ii** 9 **iii** 3 **iv** *A, C* **d i** 4 **ii** 6 **iii** 3 **iv** *A, C, D*

2 a

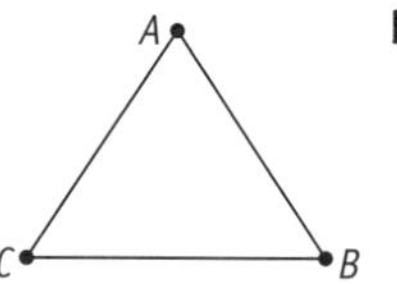

b

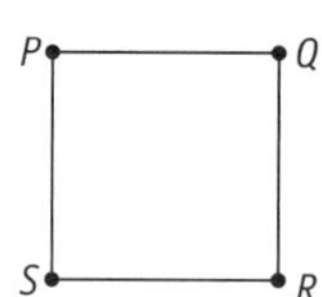

c

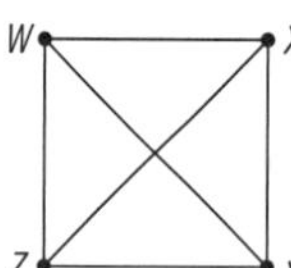

3 a 4, 6, 3 **b** 5, 8, 3 **c** 3, 6, 4

4 a *E* **b** 8 **c** 3 **d** 13 **e** 18

Page 113 **1 a**

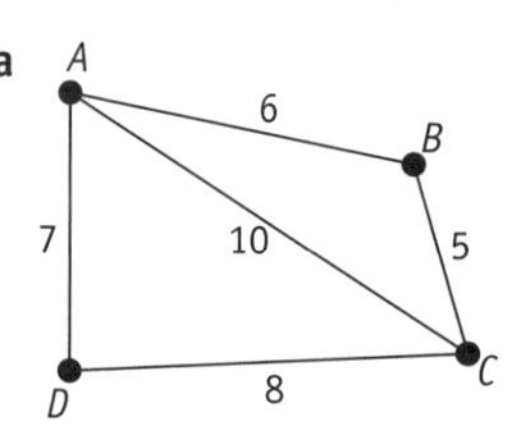

b

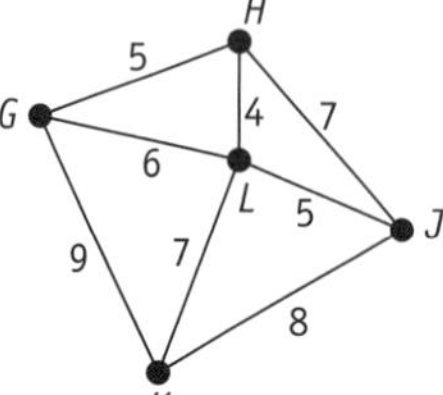

c

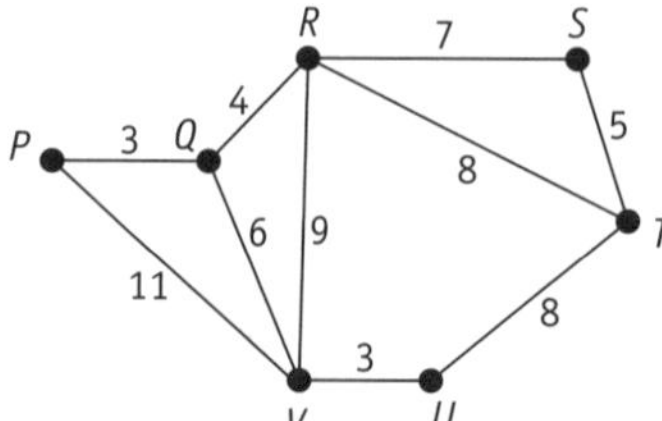

Answers

2 a

	P	Q	R	S
P	–	6	12	8
Q	6	–	9	–
R	12	9	–	11
S	8	–	11	–

b

	F	G	H	J	K
F	–	7	10	11	6
G	7	–	14	–	12
H	10	14	–	5	–
J	11	–	5	–	13
K	6	12	–	13	–

c

	A	B	C	D	E	F
A	–	8	–	–	7	5
B	8	–	3	–	–	–
C	–	3	–	2	–	6
D	–	–	2	–	9	–
E	7	–	–	9	–	–
F	5	–	6	–	–	–

3

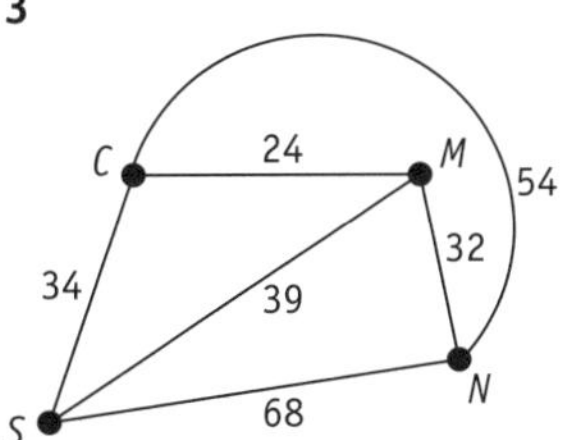

PAGE 114

1 a

		To:		
		X	Y	Z
From:	X	–	10	6
	Y	9	–	7
	Z	7	6	–

b

		To:			
		A	B	C	D
From:	A	–	8	–	–
	B	–	–	7	5
	C	–	9	–	6
	D	–	–	–	–

c

		To:					
		P	Q	R	S	T	U
From:	P	–	7	–	–	–	–
	Q	–	–	11	13	–	8
	R	–	–	–	10	–	–
	S	–	–	–	–	8	–
	T	–	–	–	7	–	6
	U	–	9	–	–	–	–

2 a

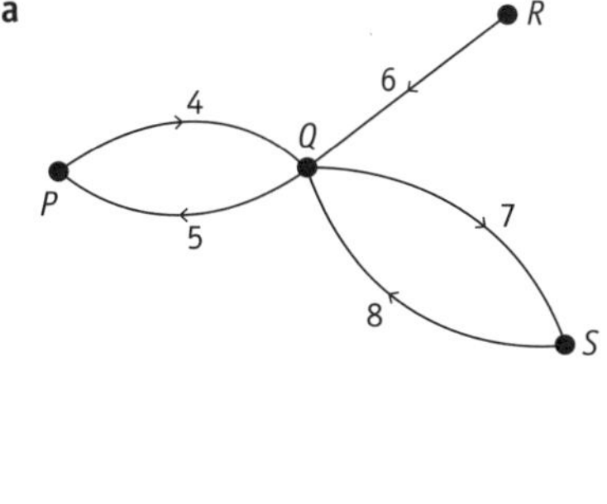

b

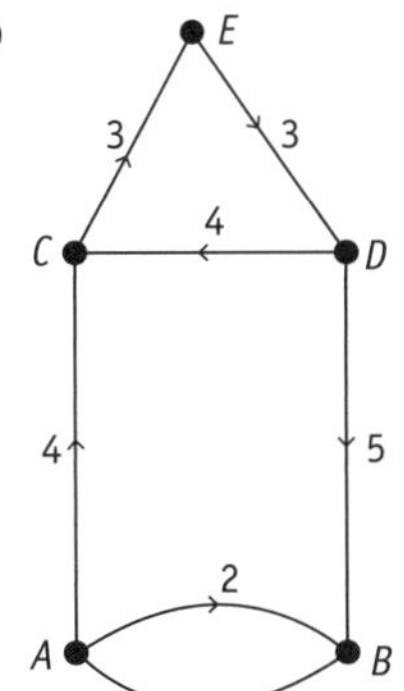

c

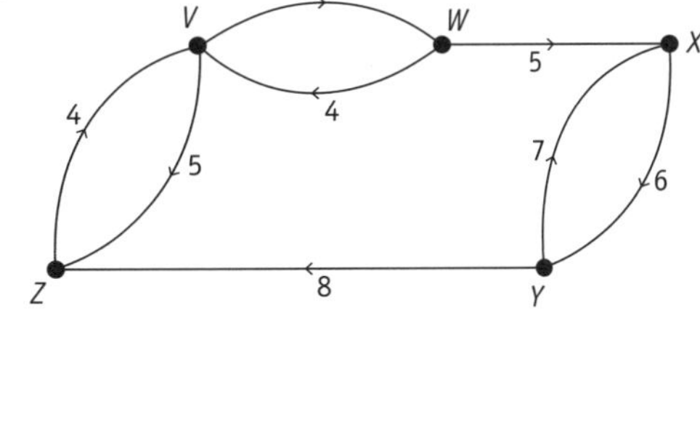

3

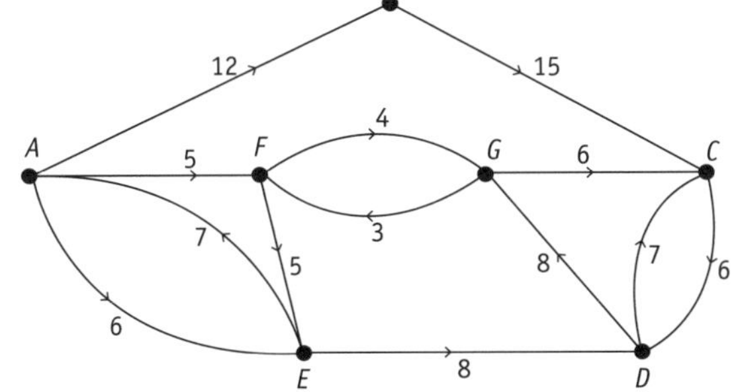

PAGE 115

1 a

b

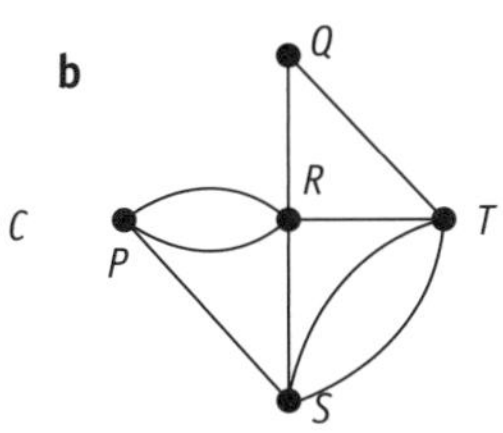

2 a

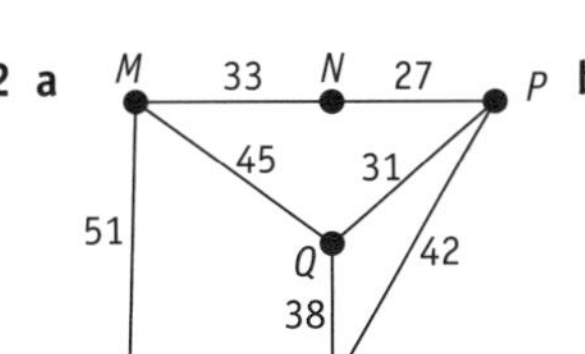

b

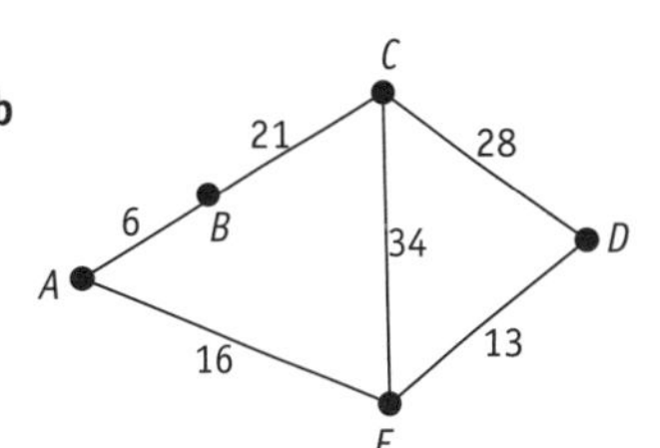

Answers

3

4 a

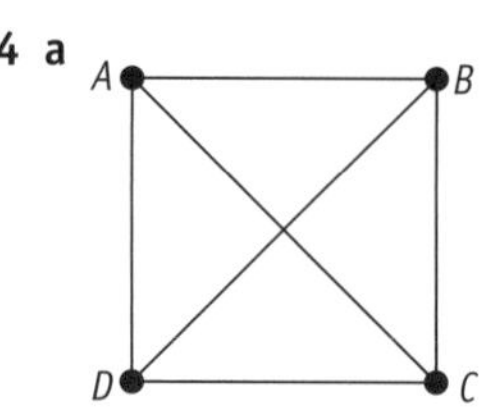

b

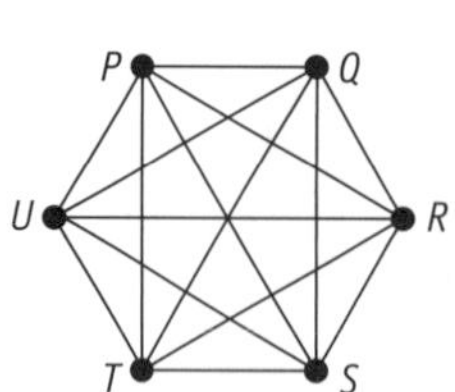

5

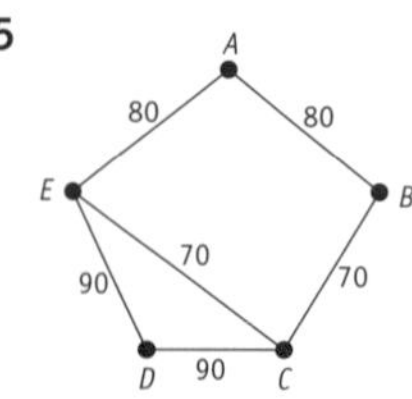

Pages 116–120 **1** B **2** D **3** C **4** A

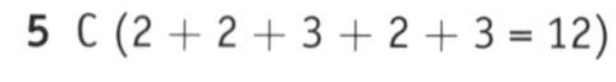

5 C (2 + 2 + 3 + 2 + 3 = 12)

6 a 8 **b** 4 **c** *A*, *C*, *D*, *F* **7**

8

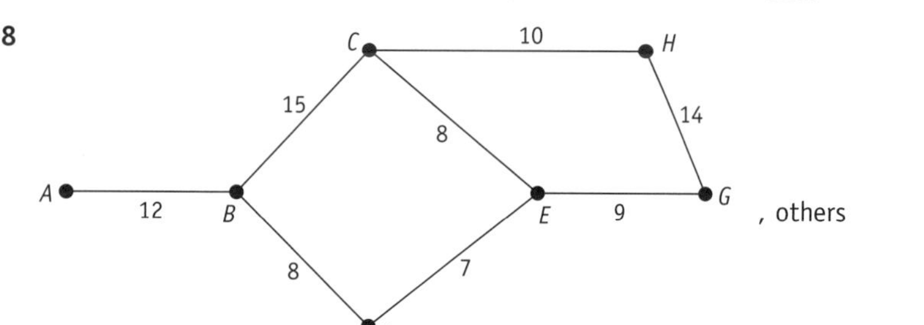

, others

9

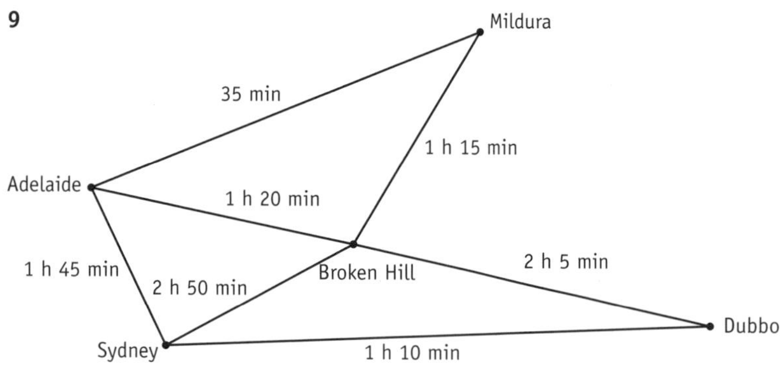

10 a 7 **b** *A*, *B*, *E* and *G*. This means 4 vertices.

11

Answers

12

	P	*Q*	*R*	*S*	*T*	*U*
P	–	12	8	10	–	–
Q	12	–	–	–	16	–
R	8	–	–	7	–	–
S	10	–	7	–	11	–
T	–	16	–	11	–	9
U	–	–	–	–	9	–

13

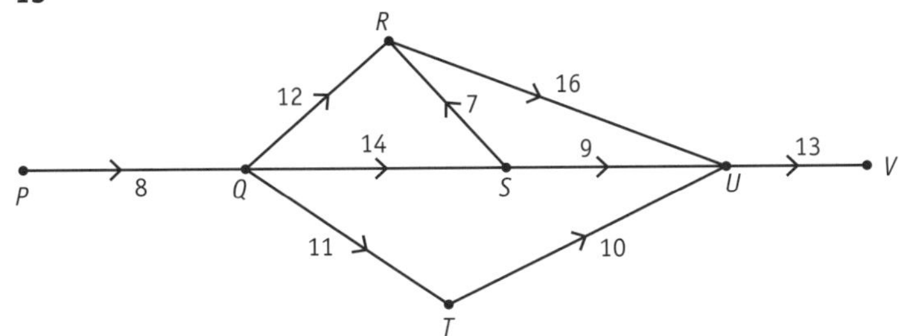

14 a *B* connects with *A, D* and *E*. It has a degree of 3. **b** *C* connects to *B* and *D*. It is degree 2. This is the only vertex with an even degree. There is only 1 vertex. **15**

		To:						
		A	***B***	***C***	***D***	***E***	***F***	***G***
From:	***A***	–	16	–	12	–	–	–
	B	11	–	–	–	–	–	–
	C	–	–	–	9	–	–	–
	D	–	–	–	–	–	7	8
	E	–	10	5	–	–	–	13
	F	–	–	–	–	6	–	–
	G	–	–	–	–	–	4	-

16

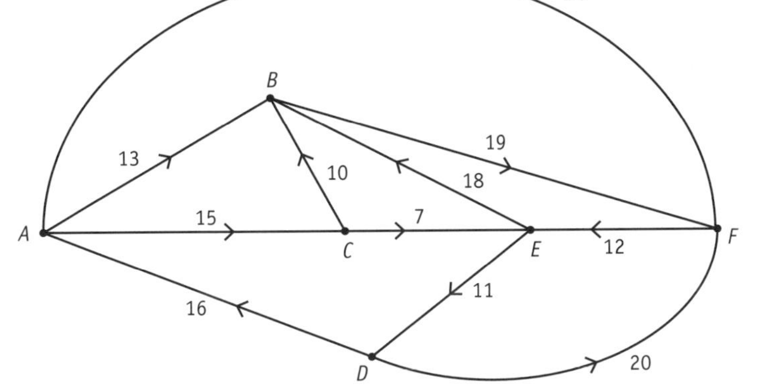

17 a

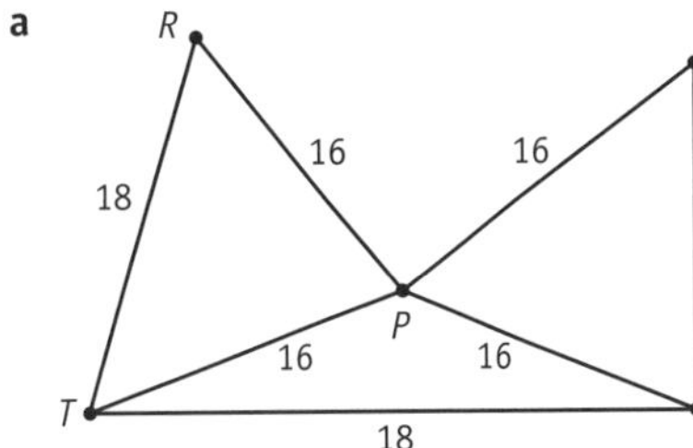

b Length = 16 × 4 + 18 × 3 = 118 m. Cost = \$16.20 × 118 = \$1911.60 **c** New cost = \$16.20 + \$1.30 = \$17.50. Length = 420 ÷ 17.5 = 24. The new pipe is 24 m long.

Chapter 11—Networks: Shortest paths

Page 121 **1 a** *R* (or *S*) **b** *D* and *C* (or *C* and *D*) **2 a** *ABE, ABCE, ADCE* **b** *AE, ABE, ABCE* **c** *ABCE, ADCE* **d** *ABJDE, AHJDE* **3 a** 2 **b** 6 **4 a** 6 **b** 4 **c** 4 **d** 4

Page 122 **1 a** *JKPMN*, 26 m **b** *ABGDE*, 55 m **c** *QVUT*, 44 m **d** *ABCD*, 31 m **e** *MNPQR*, 45 m **f** *TZWX*, 58 m **2 a** 39 min (*ABCD*) **b** 41 min (*PUTS*) **c** 37 min (*WXY*) **d** 2 h 3 min (*MVNPQR*)

Answers

Page 123 1 a

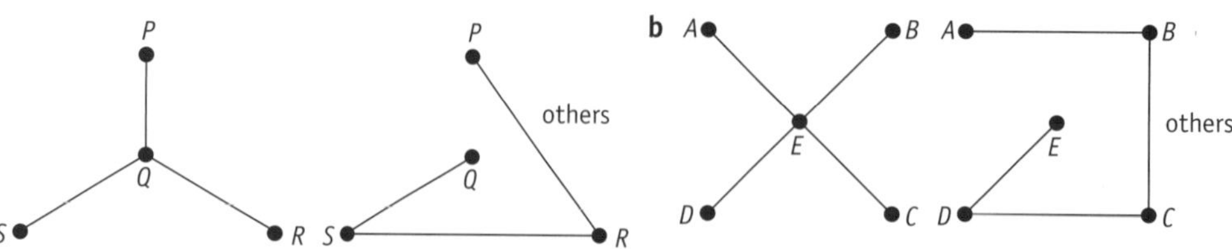

2 a 11 b 10 c 10 3 a 40 b 40 c 36 4 a 99 b 89

Page 124 1 a

Edge	Weight
PT	1
QS	2
QT	4
QR	4

b 11 c

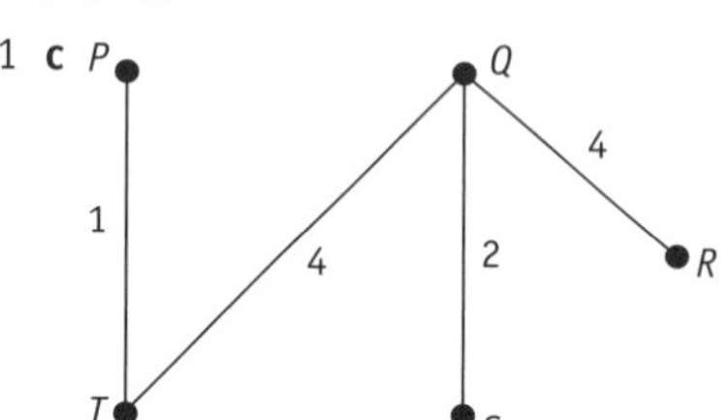

2 a

Action	Weight	Vertex visited
Start at *G*	–	*G*
Use *GA*	4	*A*
Use *AF*	2	*F*
Use *AB*	3	*B*
Use *BC*	4	*C*
Use *CD*	7	*D*
Use *DE*	10	*E*

b 30 c

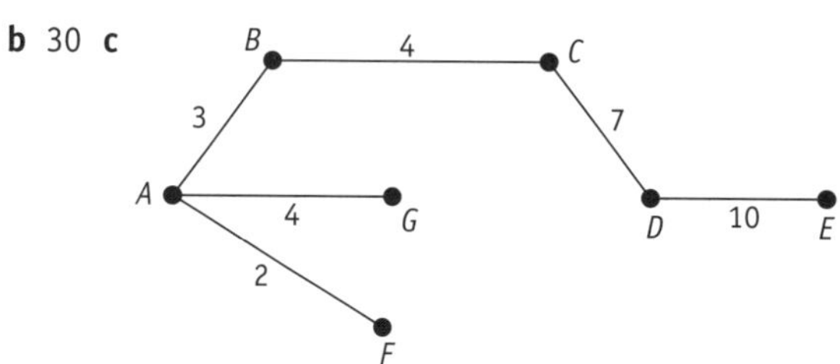

3 a 29

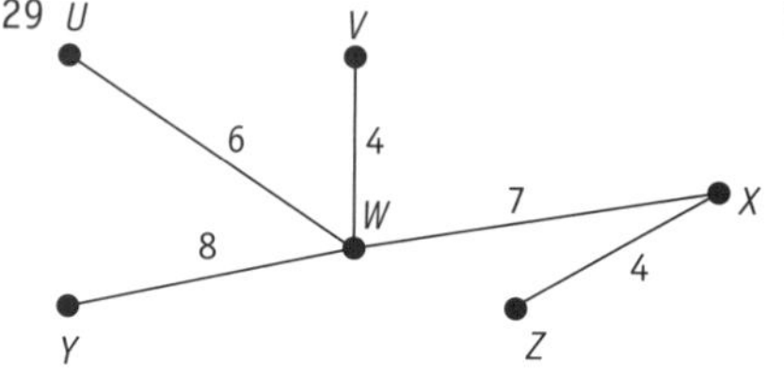

b 20

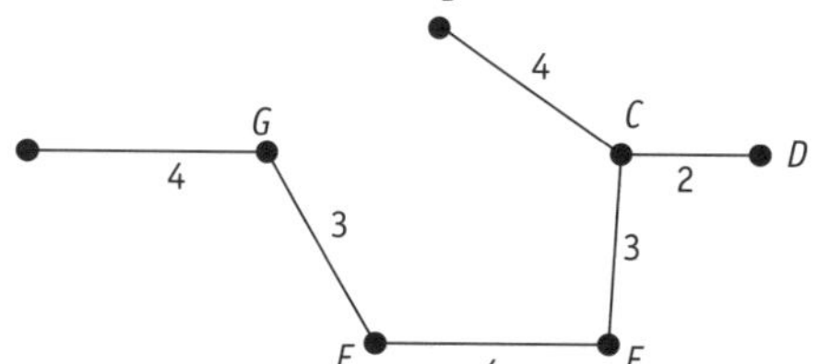

c 40

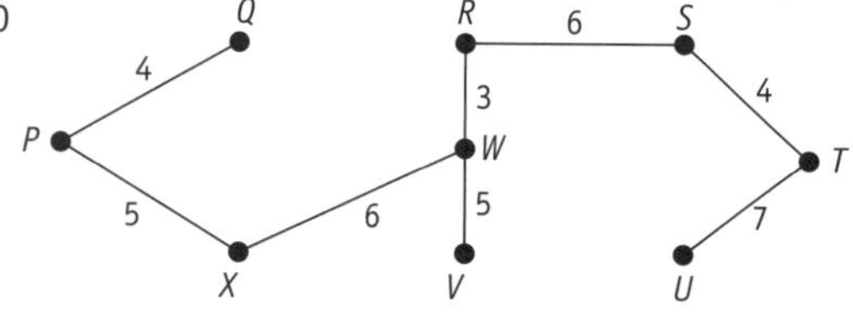

d 26

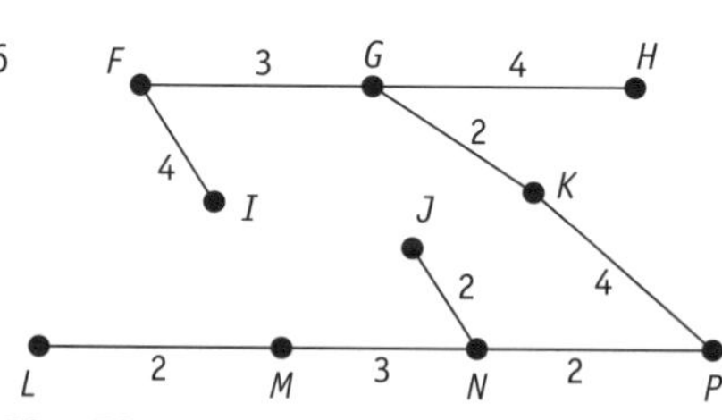

Page 125 1 51 + 39 + 45 + 35 + 40 = 210 km 2 40 + 50 + 70 + 40 + 50 + 60 + 50 + 60 = 420 m
3 a 9 + 13 + 18 + 16 + 12 + 21 + 15 + 18 + 20 = 142 m b \$85 × 142 = \$12 070 4 a 16 + 16 + 18 + 21 + 23 + 28 = 122 metres. Hence the cost is 122 × \$45 + 7 × \$580 = \$9550. b Delete 16 and 18 and include 23. Hence, 16 + 23 + 28 + 23 + 21 = 111 metres. The cost is 111 × \$45 + 6 × \$580 = \$8475, a saving of \$1075.

Pages 126–130 1 A (*ABEFGH* is 24. If *EG* = 7 then *ABEGH* = 23.) 2 B

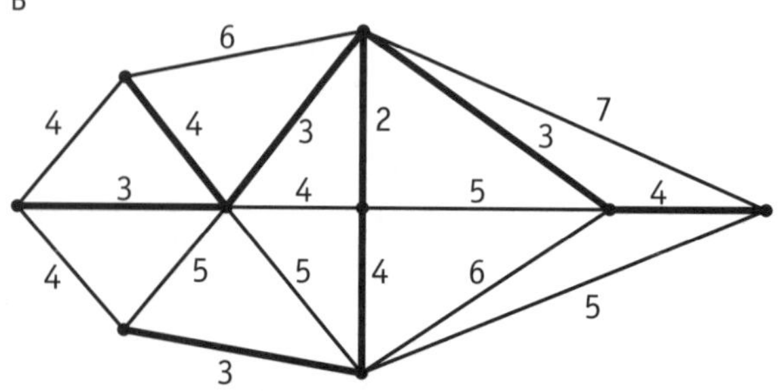

(Minimum spanning tree has a weight of 26.)

Answers

3 D **4** A **5** D

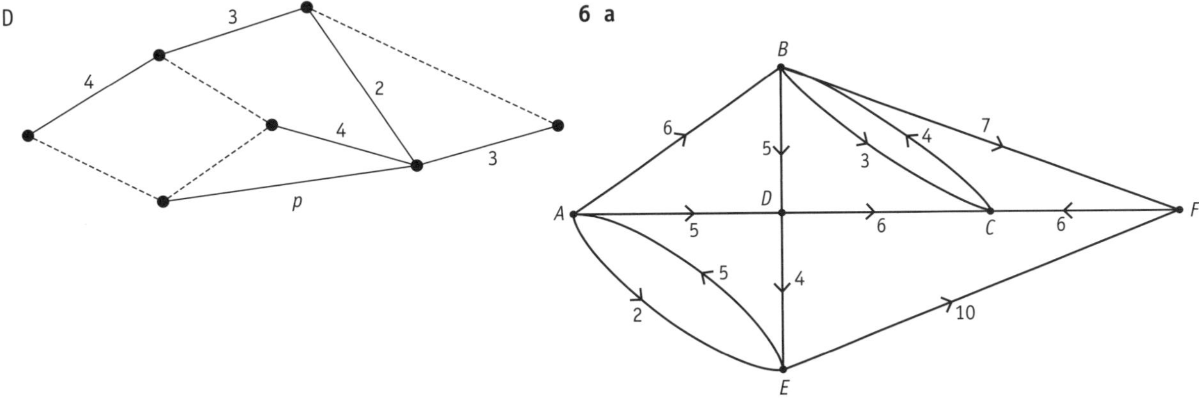

6 a

b $AEF = 2 + 10 = 12$. Shortest distance is 12 km. **c** $ABCBF = 6 + 3 + 4 + 7 = 20$. Shortest distance is 20 km.

7 a

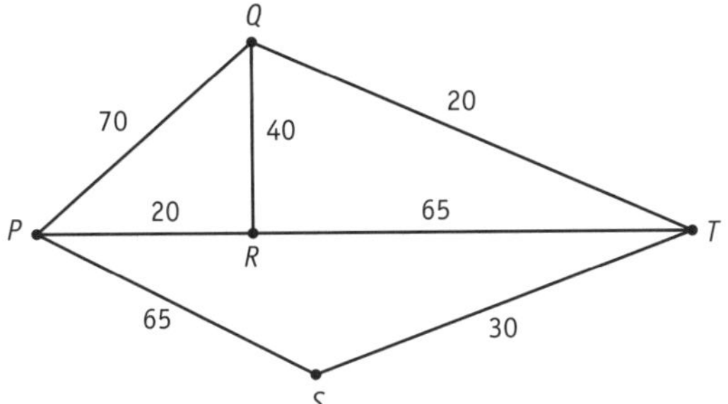

b $PRQT = 20 + 40 + 20 = 80$. The shortest time is 80 minutes.

8 a

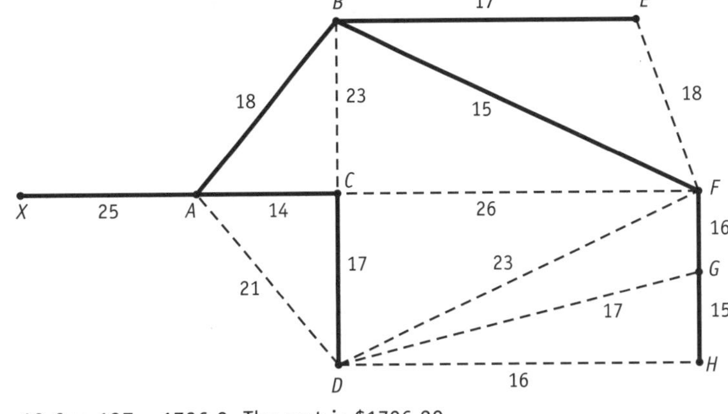

b Distance = 25 + 18 + 14 + 17 + 17 + 15 + 16 + 15 = 137. Cost = $12.6 \times 137 = 1726.2$. The cost is $1726.20.

9 a *DABCDB*, others **b** *CBAGFCDEF*, others **c** *HEDCEGFBABC*, others **10 a** 35

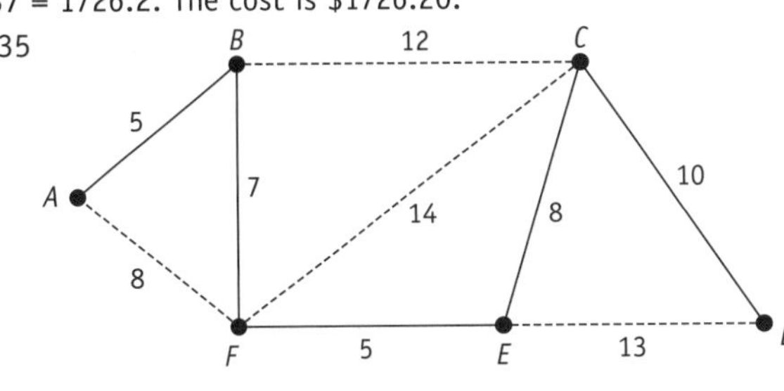

b 40

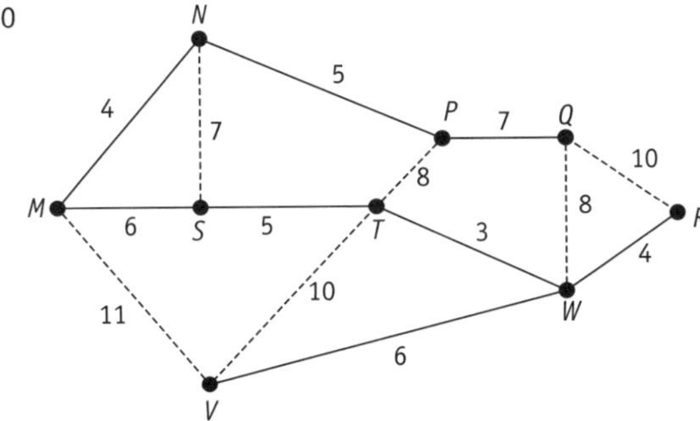

c 68

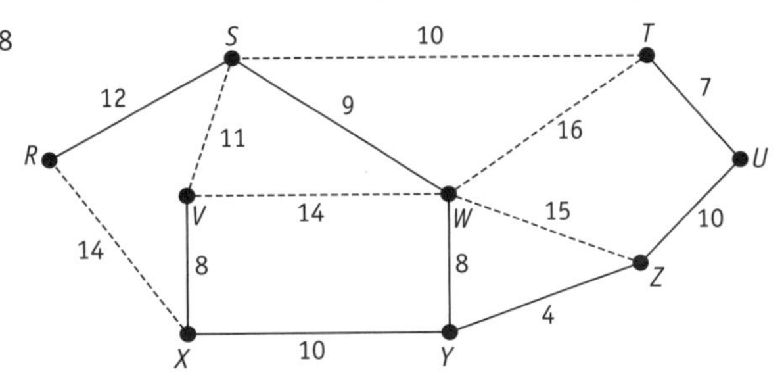

Alternatively, use *ST* and not *UZ*.

11 a \$2500 **b** Needs to start and finish at a vertex of odd degree. This means start/finish at D and F.
c \$10 000 **d** The price would now be \$10 500, an increase of \$500.

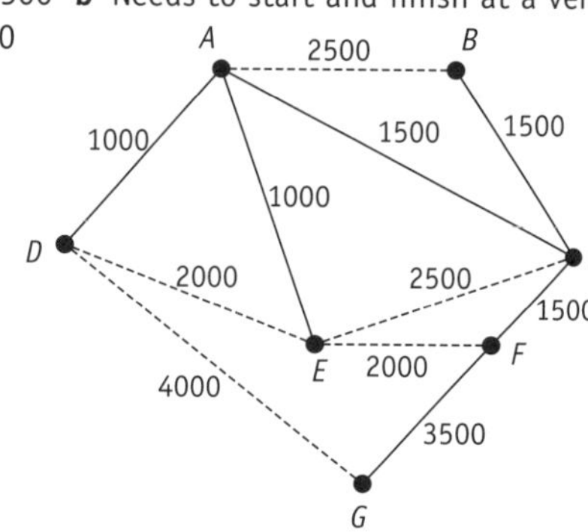

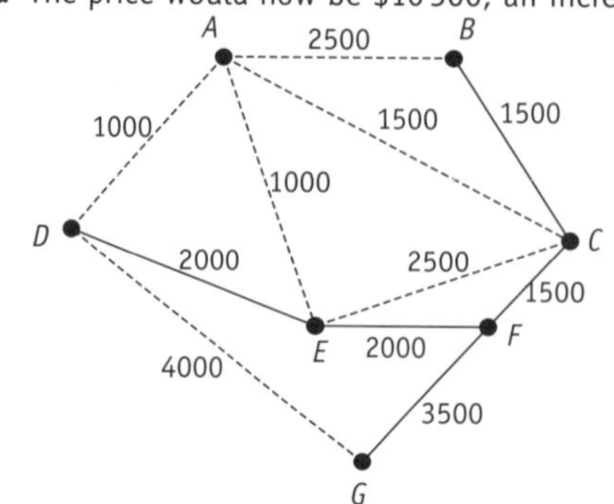

12 a 110 m **b** minimum spanning tree **c** 525 m
d \$12 600 + \$65 × 525 = \$46 725.
As \$46 725 < \$48 000, the quote is less than the budgeted cost.

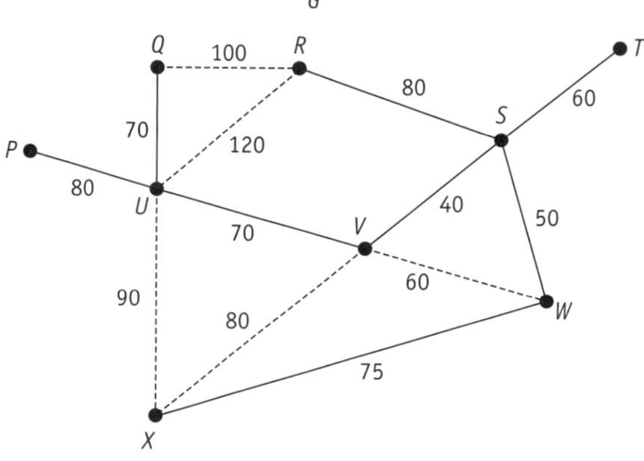

CHAPTER 12—Sample HSC Examinations

PAGES 131–140 **1** D **2** C (300 g for \$4.80 = \$1.6/100 g, 400 g for \$6.20 = \$1.55/100 g, 500 g for \$7.60 = \$1.52/100 g, 600 g for \$9.90 = \$1.65/100 g) **3** C **4** B (105 ÷ 5 x 7 = 147) **5** A **6** C (\$0.21 × 2.4 × 6 = \$3.02) **7** B $\left(\frac{1}{2}\times\pi\times5^2=\frac{25\pi}{2}\right)$ **8** A
9 B $(7000(1.0025)^{48})$ **10** D (4 + 3 + 2 + 4 + 3 + p = 20. Hence, p = 4) **11** \$28 × 3 + \$28 × 2 × 2 = \$196
12 \$480 × 0.045 × 3 = \$64.80 **13** (\$6.80 + \$2.99) ÷ 11 = \$0.89 **14** $\frac{3}{4}\times\pi\times4^2+12\times4=86$ cm^2 (nearest whole)
15 a 160 ÷ 2 = 80 km/h **b** 260 − 180 = 80 km **c**

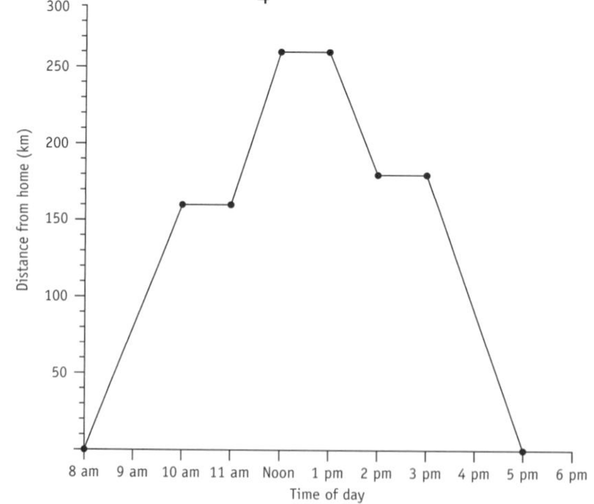

d 8 am to 5 pm = 9 hours **16** 1, 1, 20, 24, 26, 27, 27, 30, 31, 31, 32, 32, 34, 50, 52, 54, 58, 62, 86 median = 31, lower quartile = 26 and upper quartile = 52, IQR = 52 − 26 = 26, test for outlier: 52 + 1.5(26) = 91 > 86. This means that the 86-year-old is not an outlier.
17 a 40% **b** Relative freq. of even number = 10% + 20% = 30%. Number of times = 0.30 × 400 = 120
18 Total cost food/drink = 120 × 110 = 13 200; PA/music = 100 + 60 × 4.5 = 370; total cost = 1800 + 13 200 + 370 + 250 = 15 620. The cost is \$15 620. **19** $10=\frac{8(m-12)}{6}$. $8m-96=60$. $8m=156$. $m=19.5$ **20 a** Normal **b** Moderate hypertension
21 Cost for time = \$127 − \$55 = \$72. Number of quarter hours = 72 ÷ 12 = 6. Six quarter-hours = 1 hour 30 minutes.
11:50 am + 1 h 30 min = 1:20 pm

22 a

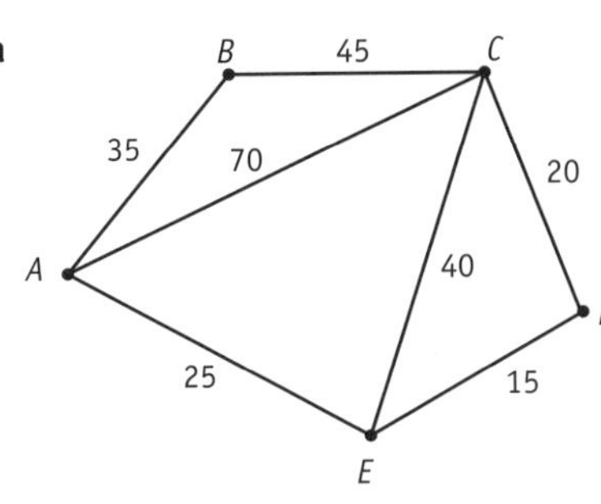

b 25 + 15 + 20 = 60 min. The path is *AEDC*. **23** $\frac{16}{d}$ = tan 54°. d = 11.62 m **24 a** $\angle ACB$ = 300° − 270° = 30° **b** $\frac{d}{250}$ = tan 30°. d = 144 km **c** By Pythag. AC = 288.68 km. Time = 0.722 h = 43 minutes. Hence 9:43 am **25 a** $x = 9$ **b** $x = 61$ **26 a** 213 **b** While they may not use much water, others do. This table is an average taken from many water users. **c** 18 × 365 = 6570 L **d** 213 × 365 = 77 745 L = 77.745 kL **e** $\frac{21.1}{100}$ × 213 = 45 L, so it is washing clothes. **f** $\frac{29}{213} \times \frac{100}{1}$ = 13.6% **g i** Saving $\frac{20}{100}$ × 35 = 7 L/day. **ii** 7 × 365 = 2555 L **27 a** \$80 000 × $(1 - 0.12)^{10}$ = \$22 280 **b** 80 000 − 10 × 6000 = 20 000. The salvage value using declining-balance formula is \$2280 higher. **28** Time = 8.5 hours plus 2 = 10.5 hours. Speed = $\frac{8160}{10.5}$ = 777 km/h **29 a** 220 − 165 = 55 **b** 208 − 0.7a = 165. Hence 0.7a = 443 and a = 61

30

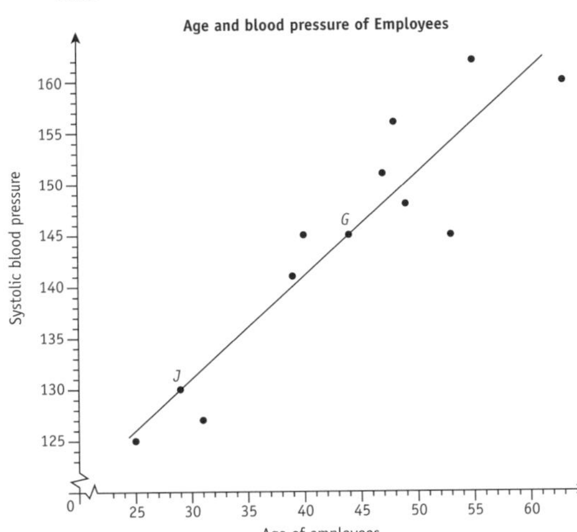

Gradient of line = $\frac{145-130}{44-29}$ = 1, equation is $p = a + c$, line passes through (29, 130): substituting a = 29 and p = 130: 130 = 29 + c, c = 101. The equation is $p = a + 101$.

PAGES 141–150 **1** D **2** C **3** C **4** D (This is an exponential graph which passes through (1, 3), thus D is the only possible equation.) **5** D (75 km = 7 500 000 cm. 7 500 000 ÷ 250 000 = 30) **6** D (Solve simultaneously, or substitute points into both equations.) **7** A **8** C (12 000 × 1.02^{12}) **9** C **10** A (15 π $(0.13^2 - 0.1^2)$ **11 a** She breaks even when 20 T-shirts are sold. **b** When 28 T-shirts sold: revenue = \$420, cost = \$340, profit = \$420 − \$340 = \$80 **12** $AB^2 = 9^2 + 11^2$, AB = 14.21. sin $\theta = \frac{14.21}{16}$, θ = 63° **13 a** 50 ÷ 6.2 = 8.06 L/100 km **b** 8.06 × 4.7 × \$1.699 = \$64 (nearest dollar) **14** Let x = length of wire. $\frac{36}{x}$ = cos 28°. x = 36 ÷ cos 28° = 40.77 m **15** $FV = 285\,000(1.034)^{30}$ = \$777 000 to nearest thousand dollars. **16** 270 − 220 = 50, $\frac{x}{84}$ = cos 50°, x = 84 × cos 50° = 54 km **17** 4 + 8 = 12. Scale factor is $\frac{12}{4}$ = 3. Height of tree = 3.6 × 3 = 10.8 m **18** Salvage value = $76\,500(1 - 0.12)^8$ = \$27 512 **19** Daily interest charge = 0.039 726%. Interest = 1890 × $0.000\,397\,26^{26}$ = \$19.62. Closing balance = \$1909.62. Minimum payment = 0.02 × 1909.62 = \$38.19 **20 a** 8 × 4 + 0.5 × π × 2^2 = 38.28 m^2 (2 dec. pl.) **b** 38.28 × 1.6 = 61.216 m^3 = 61 kL **c** Using 61 000 m^3: 61 000 ÷ 20 = 3050 minutes = 51 hours (nearest whole) **21** $24 = \frac{a(16-6)}{1.5}$, $24 = \frac{10a}{1.5}$, $10a = 36$, $a = 3.6$ **22** pergola = 18 460, slab = 280 × 9.5 + 2 × 6 × 80 = 3620, floor = 36 × 45 + 32 × 105 = 4980, fence = 8 × 420 = 3360, shed = 220 + 2140 = 2360, furniture =3200, total = 18 460 + 3620 + 4980 + 3360 + 2360 + 3200 = \$35 980 **23 a** t (Candle burning time) **b** 0 **c** With no length of candle, there will not be any burning time. **d** gradient (m) = $\frac{26-0}{3.5-0}$ = 7.4 **e** $t = 7.4\ n$. (Note b = y-intercept = 0) **24 a** Lower quartile = 61 upper quartile = 75 **b** 75 + 1.5 × 14 = 96 < 97. So 97 is an outlier **25** $4\pi \times 10^2 - 4\pi \times 3.5^2$ = 1102.6990… = 1100 cm^2 **26** The range of Class 1 (35) is less then the range of Class 2 (45). The median of Class 1 (35) is the same as the median of Class 2 (35). The IQR of Class 1 (20) is less than the IQR of Class 2 (25). **27** $BC = \frac{20}{\tan 42^\circ}$ = 22.21. If θ = < MBC, then tan $\theta = \frac{10}{22.21}$. θ = 24°.

Answers

28 a $P = 30\,000(1.035)^0 = 30\,000$

b

t	0	5	10	15	20	25
P	30 000	35 631	42 318	50 260	59 694	70 897

c

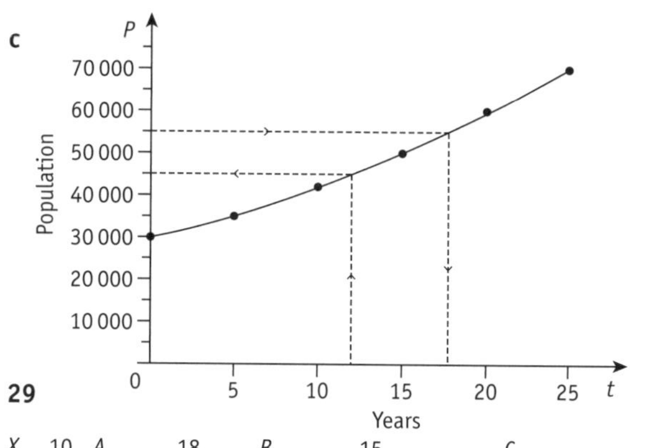

d 45 000 **e** 18 years **f** $t = 32$, $P = 30\,000(1.035)^{32} = 90\,200$

29

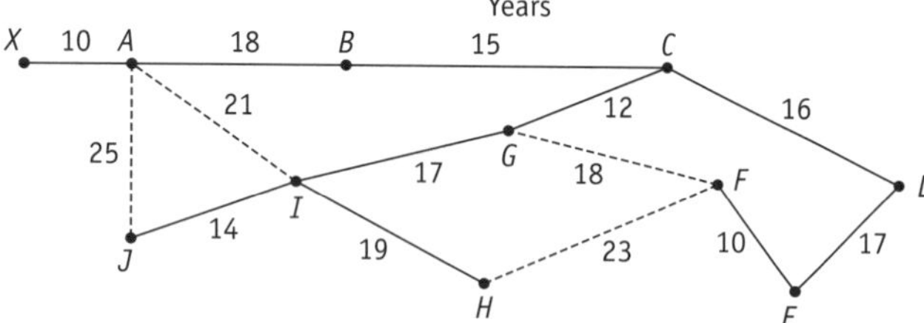

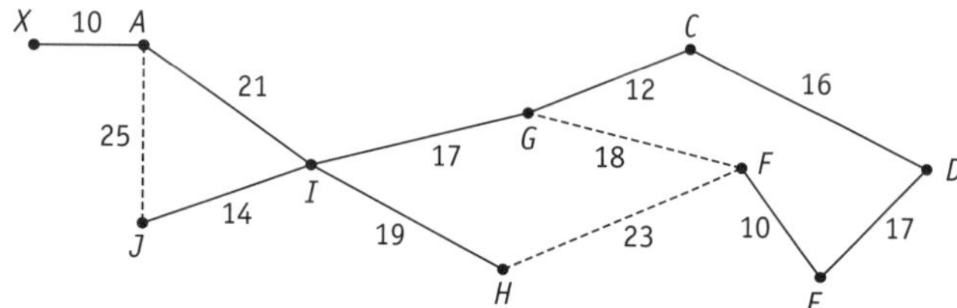

Initial length = 148 m. New length = 136 m. Savings = \$123.40 × 12 = \$1480.80

Pages 151–162 **1** C (31 − 2 = 29) **2** C **3** B **4** D (IQR = 6. 30 is more than 20 + 1.5 × 6) **5** B ($200 = k(5^2)$. $k = 8$) **6** C **7** A ($32\,890 \times 0.8^5$) **8** D (8 + 13 + 14 + 11 + 9 + 12 + 22 + 18) **9** C $\left(\frac{3}{4}\times\frac{3}{4}=\frac{9}{16}\right)$ **10** D $\left(\frac{68}{80}\times 100\% = 85\%\right)$

11 $\$8500 \times 0.0275 \times \frac{18}{12} = \$350.63$ **12 a** \$27.25 × 4 = \$109 **b** Mon.: 4 h normal, Wed.: 2 h time and a half, 2 double, Thur.: 3 h normal, 2 h time and a half, Sat.: 3 h normal. Total pay = \$27.25 × 10 + \$27.25 × 1.5 × 4 + \$27.25 × 2 × 2 =\$545

13 $\frac{3}{7} \times 42$ kg = 18 kg

14 a

t	0	1	2	3	4
h	0	15	20	15	0

b

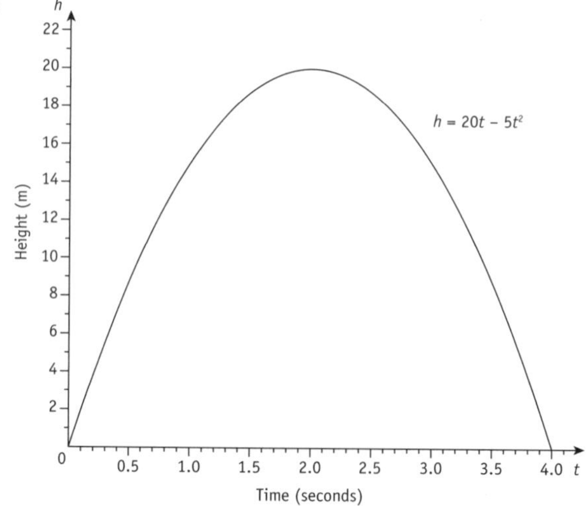

c 20 m

15 Option 1: $FV = \$25\,000 \times 1.04^4 = \$29\,246.46$; Option 2: $FV = \$25\,000 \times 1.0095^{16} = \$29\,083.31$; Option 3: $FV = \$25\,000 \times 1.003^{48} = \$28\,865.88$; Option 1 is best strategy. **16 a** 1250 **b**

t	0	10	20	30	40
P	1250	2239	4009	7179	12 857

c

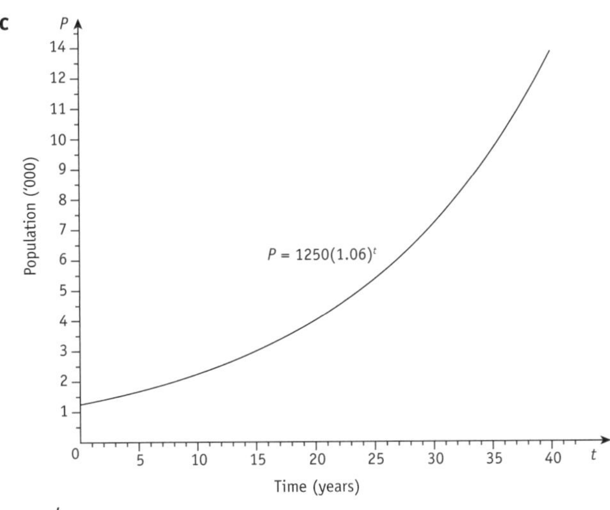

d For 2015, use $t = 35$: from the graph, population is about 9600.

17

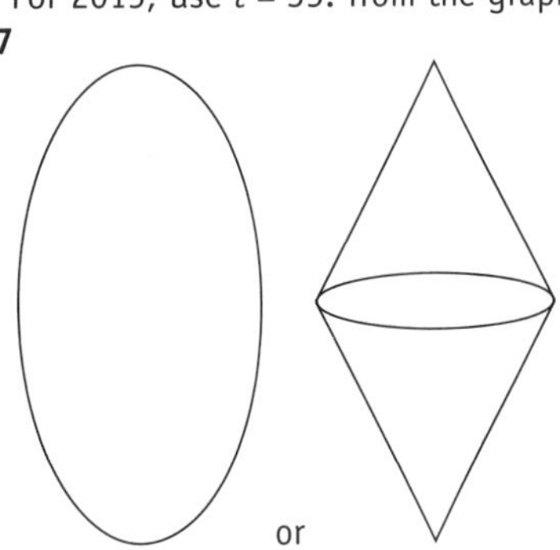

18 $\frac{d}{25} = \sin 52°$, $d = 25 \times \sin 52°$, $d = 19.70$ m **19** $\tan \theta = \frac{12}{16}$, $\theta = 36° 52'$ **20 a** Time $= \frac{40}{25} = 1.6$ h $= 1$ h 36 min; Adalee will finish at 11:36 am **b** Distance $= 20 \times 1.6 = 32$; as $40 - 32 = 8$, Keily still has 8 km to ride. **21** $\frac{TQ}{28} = \tan 38°$, $d = 28 \times \tan 38°$, $TQ = 21.88$ m. $\frac{21.87}{TP} = \tan 50°$, $TP = 21.87 \div \tan 50°$, $TP = 18.36$ m. Area $= \frac{1}{2} \times 21.87 \times (28 + 18.36) = 507$ m^2 **22 a** Speed $= 280 \div 4 = 70$; 70 km/h **b** $280 + 280 = 560$ km **c** Time $= 200 \div 80 = 2.5$ h; 7.30 pm

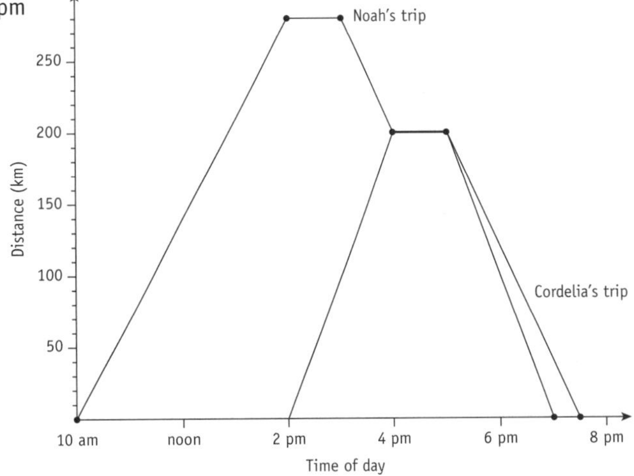

23 AE $= 8 + 4 = 12$ m, scale factor $= \frac{12}{8} = 1.5$, $x = 6 \times 1.5 = 9$. This means DE is 9 m.

24

Nick's reducible balance personal loan				
Amount borrowed: $28 000				
Interest rate: 6% p.a., monthly reducible				
Monthly repayments: $500				
Month	**Principal (*P*)**	**Interest (*I*)**	**Principal + Interest (*P* + *I*)**	**Amount owing (*P* + *I* - *R*)**
1	28 000.00	140.00	28 140.00	27 640.00
2	27 640.00	138.20	27 778.20	27 278.20
3	27 278.20	136.39	27 414.59	26 914.59
4	26 914.59	134.57	27 049.16	26 549.16
5	26 549.16	132.75	26 681.91	26 181.91
6	26 181.91	130.91	26 312.82	25 812.82

$X = 25\,812.82$

25 **a** Line passes through (12, 144) and (15, 162). Gradient $= \frac{162-144}{15-12} = 6$ **b** The height increases by 6 cm/year. **c** Equation is $h = 6a + c$. Substitute $a = 12$, $h = 144$: $144 = 6(12) + c$. Hence, $c = 72$. The line has equation $h = 6a + 72$. **d** The data and hence the model is only relevant for 10- to 16-year-old females. **26** The range of scores for the females (38) is lower than the range of scores for the males (43). The median of scores for the females (74) is higher than the median of scores for the males (65). The IQR of scores for the females (17) is lower than the IQR of scores for the males (20.5).

27 **a**

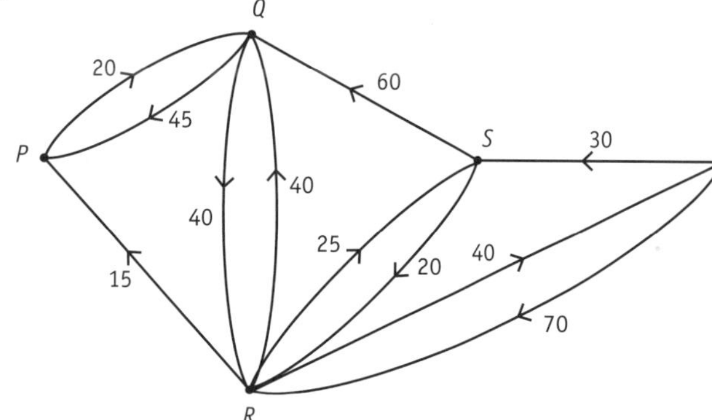

b Path is PQRT which is a distance of 100 km.

28 **a** $1.5 \times 18.43 \times 68 = 1880$ kJ **b** $870 \times 4.184 = 3640.08$ kJ. Energy from walking $= 83 \times 9.22 = 765.26$ kJ.

Number of 30-minute periods $= \frac{3640.08}{765.26} = 4.756\,66$.

Total time = 2.378 hours.

Distance $= 6 \times 2.378 = 14$ km (nearest whole)

29 **a** $\frac{20}{h} = \tan 28°$, $h = 37.6$

b $V = \frac{1}{2} \times \frac{4}{3} \times \pi \times 20^3 + \frac{1}{3} \times \pi \times 20^2 \times 37.6 = 32\,500 \text{ cm}^3$